EMERGING TECHNOLOGIES IN AGRICULTURAL ENGINEERING

Innovations in Agricultural and Biological Engineering

EMERGING TECHNOLOGIES IN AGRICULTURAL ENGINEERING

Edited by
Megh R. Goyal, PhD, PE

AAP APPLE
ACADEMIC
PRESS

Apple Academic Press Inc.
3333 Mistwell Crescent
Oakville, ON L6L 0A2 Canada

Apple Academic Press Inc.
9 Spinnaker Way
Waretown, NJ 08758 USA

© 2018 by Apple Academic Press, Inc.

First issued in paperback 2021

No claim to original U.S. Government works

ISBN-13: 978-1-77463-611-4 (pbk)
ISBN-13: 978-1-77188-340-5 (hbk)

Library and Archives Canada Cataloguing in Publication

Emerging technologies in agricultural engineering / edited by Megh R. Goyal, PhD, PE.

(Innovations in agricultural and biological engineering)
Includes bibliographical references and index.
Issued in print and electronic formats.
ISBN 978-1-77188-340-5 (hardcover).--ISBN 978-1-315-36636-4 (PDF)

1. Agricultural engineering--Technological innovations. I. Goyal, Megh Raj, editor II. Series: Innovations in agricultural and biological engineering

| S674.3.E44 2017 | 630 | C2017-901095-6 | C2017-901096-4 |

Library of Congress Cataloging-in-Publication Data

Names: Goyal, Megh Raj, editor.
Title: Emerging technologies in agricultural engineering / editor: Megh R. Goyal.
Description: Waretown, NJ : Apple Academic Press, 2017. | Includes bibliographical references and index.
Identifiers: LCCN 2017005306 (print) | LCCN 2017006953 (ebook) | ISBN 9781771883405 (hardcover : alk. paper) | ISBN 9781315366364 (ebook)
Subjects: LCSH: Agricultural engineering.
Classification: LCC S675 .E48 2017 (print) | LCC S675 (ebook) | DDC 631.3--dc23
LC record available at https://lccn.loc.gov/2017005306

Apple Academic Press also publishes its books in a variety of electronic formats. Some content that appears in print may not be available in electronic format. For information about Apple Academic Press products, visit our website at **www.appleacademicpress.com** and the CRC Press website at **www.crcpress.com**

CONTENTS

List of Contributors..vii

List of Abbreviations...ix

Preface..xi

Warning/Disclaimer...xvii

About Senior Editor-in-Chief...xix

Book Messages..xxi

Other Books on Agricultural & Biological Engineering.............xxiii

Editorial...xxv

PART I: EMERGING TECHNOLOGIES..1

1. Drainage Congestion in Hirakud Command, Odisha: Issues and Opportunities to Augment Crop Production................................3

Balram Panigrahi, Jagadish Chandra Paul, and Kajal Panigrahi

2. Irrigation Technology Options for Selected Agricultural Crops.............19

Vishal K. Chavan

PART II: ENERGY MANAGEMENT IN AGRICULTURE.............................31

3. Solar Energy: Principles and Applications....................................33

M. K. Ghosal

4. Enhanced Anaerobic Digestion by Inoculation with Compost.............107

Zhiji Ding, Nirakar Pradhan, and Willy Verstraete

5. Tractor Utilization in Vidarbha Region, India.............................129

D. V. Tathod, Y. V. Mahatale, and V. K. Chavan

6. Performance of Reversible Mold Board Plow...............................137

Yogesh V. Mahatale, Dnyaneshwar V. Tathod, and Vishal K. Chavan

7. Performance of Self-Propelled Intercultivator.............................165

Uddhao S. Kankal, M. Anantachar, and V. K. Chavan

PART III: MANAGEMENT OF NATURAL RESOURCES...................207

8. **Agro-Biodiversity Collection and Conservation from Arid and Semi-Arid Regions of India**...209

    Raju R. Meghwal and N. K. Dwivedi

9. **Watershed Development and Management**................................231

    R. K. Sivanappan

10. **Water Resources of India: Problems, Solutions, and Prospects**.............247

    R. K. Sivanappan

11. **Present Status of Water Resources in Tamil Nadu**...................275

    R.K. Sivanappan

12. **Water Conservation and Increased Crop Production Under Precision Farming**...305

    R. K. Sivanappan

PART IV: PRINCIPLES OF SOIL AND WATER ENGINEERING.............317

13. **Basics of Soil and Water Conservation Engineering**..................319

    R. K. Sivanappan

Appendices..355

Index...385

LIST OF CONTRIBUTORS

M. Anantachar
Dean College of Agricultural Engineering, University of Agricultural Sciences, Raichur, India.

Vishal K Chavan, PhD
Senior Research Fellow, AICRP for Dryland Agriculture, Dr. Panjabrao Deshmukh Krishi Vidyapeeth, Krishinagar PO, Akola 444104, Maharashtra, India. E-mail: vchavan2@gmail.com.

Zhiji Ding, PhD
Laboratory of Microbial Ecology and Technology (LabMET), Faculty of Bio-Engineering Science, Ghent University, Coupure Links 653, Gent 9000, Belgium. Tel +39 3240479407, E-mail: jimmydzj2006@gmail.com.

N. K. Dwivedi, PhD
National Bureau of Plant Genetic Resources, Regional Station, CAZRI Campus, Jodhpur 342005, Rajasthan, India. Mobile: +91 7023003868, E-mail: drnkdwivedi52@yahoo.co.in.

M. K. Ghosal, PhD
Professor, Department of Farm Machinery and Power, College of Agricultural Engineering and Technology, Orissa University of Agriculture and Technology, Bhubaneswar 751003, Odisha, India, E-mail: mkghosal1@rdiffmail.com.

Megh R. Goyal, PhD, PE
Retired Professor in Agricultural and Biomedical Engineering, University of Puerto Rico—Mayaguez Campus; Senior Technical Editor-in-Chief in Agriculture Sciences and Biomedical Engineering, Apple Academic Press Inc., PO Box 86, Rincon, PR 00677, USA. E-mail: goyalmegh@gmail.com.

Uddhao S. Kankal, MTech
Testing Engineer, Department of Farm Machinery and Power, College of Agriculture & Technology, Dr. PDKV, Akola, Maharashtra, India. Mobile: +91 9049051525, E-mail: uskankal@gmail.com.

Y. V. Mahatale, PhD
Assistant Professor, Department of Farm Power & Machinery, College of Agricultural Engineering & Technology, Warwat Road, Jalgaon, District Buldhana, Maharashtra, India. E-mail: yogeshvmahatale@gmail.com.

Raju R. Meghwal, PhD
Research Scientist, National Bureau of Plant Genetic Resources, Regional Station, CAZRI Campus, Jodhpur 342005, Rajasthan, India.

Balram Panigrahi, PhD
Professor and Head, Department of Soil & Water Conservation Engineering, College of Agricultural Engineering & Technology, Orissa University of Agriculture & Technology, Bhubaneswar, Odisha, India. E-mail: kajal_bp@yahoo.co.in.

Kajal Panigrahi, MTech
Department of Civil Engineering, National Institute of Technology, Rourkela, Odisha, India. Phone: +91 8895997206, E-mail: kajalpanigrahi@yahoo.in.

J. C. Paul, PhD

Associate Professor, Department of Soil & Water Conservation Engineering, College of Agricultural Engineering & Technology, Orissa University of Agriculture & Technology, Bhubaneswar, Odisha, India. Phone: +91 9437762584, E-mail: jcpaul66@gmail.com.

Nirakar Pradhan, PhD

Doctoral Research Fellow, Erasmus Mundus Joint Doctorate (ETECOS3) Program, University of Napoli Federico-II, Naples 80125, Italy. Tel. +39 3319068795, E-mail: nirakar.pradhan@gmail.com.

R. K. Sivanappan, PhD

Former Professor and Dean, College of Agricultural Engineering & Technology, Tamil Nadu Agricultural University (TNAU), Coimbatore; current mailing address: Consultant, 14, Bharathi Park, 4th Cross Road, Coimbatore, 641 043, Tamil Nadu, India. E-mail: sivanappanrk@hotmail.com.

D. V. Tathod, PhD

College of Agricultural Engineering and Technology, Warwat Road, Jalgaon Jamod, Dist. Buldhana, Maharashtra, India. Tel. +91 257 2200829, Mobile: +91 9604818220, E-mail: dnyanutathod@gmail.com.

Willy Verstraete, PhD

Laboratory of Microbial Ecology and Technology (LabMET), Faculty of Bio-Engineering Science, Ghent University, Coupure Links 653, Gent 9000, Belgium.

LIST OF ABBREVIATIONS

AAP	Apple Academic Press Inc.
ABE	Agricultural and biological engineering
ADS	Agricultural Development Society
AE	Agricultural engineering
ASABE	American Society of Agricultural and Biological Engineers
ASAE	American Society of Agricultural Engineers
BIS	Bureau of Indian Standards
CGWB	Central Ground Water Board
COD	Chemical oxygen demand
CWC	Central Water Commission
HP	Horsepower
HRT	Hydraulic retention time
IBT	Inter-basin transfer
IR	Infrared
ISAE	Indian Society of Agricultural Engineers
JJR	Jalgaon Jamod region
KVK	Krishi Vidya Kendra
MB	Moldboard
MHM	Million hectare-meter
MUSLE	Modified Universal Soil Loss Equation
NCAM	National Center for Agricultural Mechanization
NPP	National Perspective Plan
NWDA	National Water Development Agency
OTE	Ocean thermal energy
PAU	Punjab Agricultural University
PDs	Project directors
PF	Precision farming
PIA	Project-implementing agency
PR	Penetrometer resistance
RUSLE	Revised Universal Soil Loss Equation
SP	Seepage and percolation
SRT	Sludge loading rate
TCOD	Total chemical oxygen demand
TN	Tamil Nadu

TNAU	Tamil Nadu Agricultural University
TSS	Total suspended solids
USDA	US Department of Agriculture
USEPA	United States Environmental Protection Agency
USLE	Universal soil loss equation
VSS	Volatile suspended solids
WAS	Waste-activated sludge
WWTP	Wastewater treatment plant

PREFACE

According to the Indian Society of Agricultural Engineers (http://isae.in), "Agricultural Engineering involves application of engineering to production, processing, preservation and handling of food, feed and fiber. It also includes the transfer of engineering technology for the development and welfare of rural areas and masses. The major fields of Agricultural Engineering are: a. Farm power and machinery; b. Soil and water conservation and management engineering (watershed management, command area development, conservation, and field drainage); c. Irrigation (including surface and ground water development); d. Agricultural process engineering; e. Farm storage engineering; f. Farm buildings and structures; g. Farm use of electricity; h. Rural water supply, rural roads, rural housing, rural sewerage and other aspects of rural engineering."

Since the Stone Age, human beings without any formal education have known the importance of agricultural tools. They knew that crops had to be planted near the banks of rivers where plenty of water was available; crop products had to be dried in the open air and sun to increase the shelf-life; they knew how to carve hand tools from stones and later from iron, and so on. As agriculture evolved, so agricultural engineering was also revolutionized. Formal historic education in agricultural engineering (AE) commenced at the beginning of 20th century in the United States and India, and soon other countries followed their examples. In 1905, Jay Brownlee Davidson designed the first US professional AE curriculum at Iowa State College. Courses included agricultural machines; agricultural power sources, with an emphasis on design and operation of steam tractors; farm building design; rural road construction; and field drainage. Iowa State's Agricultural Engineering Department may be the oldest program of its kind in the United States. Davidson saw clearly the need to educate student engineers to prepare them to apply engineering technology to the solution of agricultural problems. J. B. Davidson founded the American Society of Agricultural Engineers (ASAE) on December 27, 1907, at the University of Wisconsin. Davidson is sometimes referred to as "Father of Agricultural Engineering in USA."

During the first 100 years, the American Society of Agricultural and Biological Engineers (ASABE) has broadened Davidson's scope to expand its role beyond farm production to other components of the supply and delivery

DB Davidson [Photo   http://shs.umsystem.edu/manuscripts/invent/3130.pdf
courtesy of Iowa State   Prof Mason Vaugh, Father of Agricultural Engineering in India,
University Library]   founded Agricultural Institute at Allahabad, India in 1921.

processes. The rivers of change that redirected ASABE for the first 100 years will continue to provide the opportunity for AE to not only maintain but also to improve its effectiveness in the delivery of a world-class quality level of education of the students. ASABE (asabe.org) mentions, "Providing the necessities of life: Innovators, collaborators, stewards, ASABE members are leaders in the production, transport, storage, and use of renewable resources. They put science to work to meet humanity's most fundamental needs: safe and abundant food; clean water; fiber, timber, and renewable sources of fuel; and life-enhancing and life-saving products from bio-based materials. And they do this with a constant eye toward the improved protection of the people, livestock, wildlife, and natural resources involved." For more in-depth study of AE, the reader is referred to: ASABE, 2950 Niles Road, St. Joseph, MI 49085, USA, <hq@asabe.org and asabe.org>.

My proceedings paper #047 on *Scope and Opportunities in AE* was presented at The Fourth Latin American and Caribbean Conference [LACCEI] on June 21–23, 2006 at Mayagüez, Puerto Rico. This paper summarizes what is AE today.

Carl W. Hall, ASABE Fellow and past president, has combed the literature to gather together 1332 notable landmarks and firsts in the history of machines, implements, apparatus, processes, and organizations related to agricultural and biological engineering (ABE), particularly in the United States and Canada. The resulting book on timelines help's place in perspective the growth of ABE. Potato chips were first made by George Crumb, in New York in 1853. In 1905, Charles Hart and Charles Parr established the first US factory devoted to manufacturing a traction engine powered by an internal combustion engine. Smaller and lighter than its steam-driven predecessors, it ran all day on one tank of fuel. Hart and Parr are credited with coining the term "tractor" for the traction engine. The first state license to practice engineering was issued in Wyoming in 1907. In 1954, the number of tractors on US farms exceeded the number of mules and horses for the first time. One might continue reading on more tidbits in his book. The ASABE website (http://www.asabe.org/about-us/history.aspx) outlines US historic events in ABE.

The <http://shs.umsystem.edu/manuscripts/invent/3130.pdf> narrates that Professor Mason Vaugh (June 27, 1894–October 7, 1978) was an American agriculturalist in Missouri, USA who developed the first agricultural engineering department outside North America in 1921 at Allahabad Agricultural Institute, India. In 1919, he obtained B.Sc. in Agriculture from the University of Missouri and earned the equivalent of a M.Sc. in Agricultural Engineering in 1928. In 1921, Vaugh and his wife Clara Pennington went to India as lay missionaries of the Presbyterian Church. They were assigned to the Allahabad Agricultural Institute, where Vaugh was head of the Department of Agricultural Engineering until 1953. He was the first person in India to teach a course in agricultural engineering. Vaugh utilized traditional Indian materials which he adapted for farmers to use as modern agricultural implements. Among his innovations was the "Shabash," an improved plow consisting of a plowshare, a mold board, a few bolts and a wood beam. It costed 15 Indian rupees equivalent to US $3.00 at that time. Improved implements such as the Shabash made it possible for farmers to plow larger areas than previously possible. Vaugh also introduced hoes, cultivators and wheat threshers. He was the leader of the Agricultural Development Society (ADS) that was established in Naini to manufacture and sell improved implements developed by ADS and the Institute. He retired in 1957 and returned to the United States. He has often been called the "Father of agricultural engineering in India." In his honor, the Indian Society of Agricultural Engineers (ISAE) established the Mason Vaugh Agricultural Engineering Pioneer Award. Prof. Ralph C. Hay was the first recipient of this award for his contribution in establishing agricultural engineering programs at IIT Kharagpur (1954–56) and Pantnagar University (1962–1964). The Vaughs returned to the United States in 1957. After retirement, Vaugh remained interested in Indian agricultural problems. At the Ohio State Experiment Farm, he researched and tested a thresher specifically designed for Indian crops. Professor Newmann Fernandes of wrote a detailed biography of Mason Vaugh.

How did I get interested in Agricultural Engineering? My dean Dr. C. M. Jacob, my maternal uncle Dr. K. R. Aggarwal, Dr. Gordon L. Nelson, and my elder brother Dr. K. C. Goyal are the motivators who led me into this challenging profession. Here goes the story: After passing my matriculation in 1965, my uncle and brother motivated me to apply for admission at Punjab Agricultural University (PAU)—Hisar Campus. In the meantime the first College of Agricultural Engineering was started at PAU in 1964. I applied for a transfer and was accepted. Being limited by financial sources to pursue my studies, Dr. C. M. Jacob offered me the scholarship to cover my educational expenses, including boarding and lodging. He adopted me as his professional son. When our first batch graduated in 1969, there were no jobs for them. *Fittest of fat:* Nobody will recognize our profession. Employers thought that we were agronomists and not engineers. We all went on a one-trimester strike and made good publicity to sell our profession. Finally, the strike yielded fruits. One of my college mates became Chief Conservator of Punjab and the other was named as Secretary of Agriculture for eight years by Haryana State Chief Minister, and the success story goes on. Today, agricultural engineers hold distinguished positions in government and private agencies not only in India but also throughout the world.

Finally at my graduation in 1971, I got married to the agricultural engineering profession. During 1973–75 at Haryana Agricultural University, I had the opportunity to perfect the technology of acid delinting that led to the design of cotton de-linters, cotton planters and knapsack cotton harvesters by my fellow engineers. At the Ohio State University, I defined mechanics of soil crusting and predicted optimum temperature for soybean seedling emergence that was able to increase the soybean yield by 10% in the midwest belt of the USA. In 1979, among four job offers, I accepted the lowest paying job at the University of Puerto Rico—Mayaguez Campus to work on *micro irrigation technology.* I was the first licensed profession agricultural engineer in Puerto Rico. My research, teaching, and extension activities earned me the title of "Father of Irrigation Engineering of 20th Century of Puerto Rico" and *"Hombre de Gota* (Drip Man)." Opportunities are now abundant in this budding profession for anyone who want to join this fraternity to serve the humanity. Without any farming family background in agriculture, I have been a successful agricultural engineer and have been able to defend my profession with all my heart.

Apple Academic Press Inc. published my first book *Management of Drip/ Trickle or Micro Irrigation,* and a 10-volume set under book series Research Advances in Sustainable Micro Irrigation, in addition to other books in the focus areas of Agricultural and Biological Engineering.

This book contributes to the ocean of knowledge on ABE. Agricultural and biological engineers work to better understand the complex problems on the farm. While making a call for chapters for a book volume on ABE, I mentioned to the prospective authors following focus areas:

- Academia to industry to end-user loop in agricultural engineering
- Agricultural mechanization
- Aquacultural engineering
- Biological engineering in agriculture
- Biotechnology applications in agricultural engineering
- Energy source engineering
- Food and bioprocess engineering
- Forest engineering
- Human factors engineering
- Information and electrical technologies
- Irrigation and drainage engineering
- Nanotechnology applications in agricultural engineering
- Natural resources engineering
- Nursery and greenhouse engineering
- Power systems and machinery design
- Robot engineering in agriculture
- Simulation and computer modeling
- Soil and water engineering
- Structures and environment engineering

Therefore, I conclude that scope of ABE is wide, and focus areas may overlap one another. The mission of this very book volume is to introduce the profession of ABE. I cannot guarantee the information in this book series will be enough for all situations.

At the 49th annual meeting of ISAE at PAU during February 22–25 of 2015, a group of ABEs convinced me that there is a dire need to publish book volumes on focus areas of ABE. This is how the idea was born on new book series titled Innovations in Agricultural and Biological Engineering.

The contributions by the cooperating authors to this book volume have been most valuable in the compilation. Their names are mentioned in each chapter and in the list of contributors. This book would not have been written without the valuable cooperation of these investigators, many of whom are renowned scientists who have worked in the field of ABE throughout their professional careers.

I will like to thank editorial staff, Sandy Jones Sickels, Vice President, and Ashish Kumar, Publisher and President at Apple Academic Press, Inc., for making every effort to publish the book when the diminishing water resources are a major issue worldwide. Special thanks are due to the AAP Production Staff for the quality production of this book.

I request that the reader offer his constructive suggestions that may help to improve the next edition.

I express my deep admiration to my family for understanding and collaboration during the preparation of this book. I dedicate this book volume to my maternal uncle, the late Dr. Khem Raj Aggarwal, who inspired me into the budding profession of Agricultural Engineering. He was my motivator and guru to lead me on the path of success since my childhood. Being a teacher since 1965, he made many disciples in India while teaching at Chandigarh, Hisar, and Ludhiana. After migrating to the United States in 1981, he taught at Temple University, Philadelphia; University of Wisconsin, Whitewater; and at Louisiana State University, Alexandria. His colleagues testify to his friendly attitude toward the students. He was very patient with students, and without putting them down, he took time to explain until they understood the concepts. In addition to teaching, he wrote software for practicing physicians and games for young kids. His simple life style included serenity and understanding of a human being. He was very pleasant as a human being. He always had a smile when he was among the people and made everyone felt comfortable in his company. He was a patient listener and gave very good advice as needed.

I cannot forget the help my family and I got from Khem Raj and his wife Swaran during my lifetime. Swaran has been the main driving force behind all the successes of Khem Raj, and I take this opportunity to applaud both of them for the multiple activities of Dr. Khem Raj Aggarwal. Instead of becoming a rich person monetarily, he was rich in qualities. I inherited many qualities of a successful educator from his acts.

As an educator, there is a piece of advice to one and all in the world: "Permit that our almighty God, our Creator and excellent Teacher, irrigate the life with His Grace of rain trickle by trickle, because our life must continue trickling on… and Get married to your profession…."

—Megh R. Goyal, PhD, PE
Senior Editor-in-Chief

WARNING/DISCLAIMER

Read It Carefully

The goal of this book volume, *Emerging Technologies in Agricultural Engineering*, is to guide the world community on how to manage efficiently for technology available for different processes in food science and technology. The reader must be aware that dedication, commitment, honesty, and sincerity are important factors for success. This is not a one-time reading of this compendium.

The editors, the contributing authors, the publisher and the printer have made every effort to make this book as complete and as accurate as possible. However, there still may be grammatical errors or mistakes in the content or typography. Therefore, the content in this book should be considered as a general guide and not a complete solution to address any specific situation in food engineering. For example, one type of food process technology does not fit all cases in food engineering/science/technology.

The editors, the contributing authors, the publisher and the printer shall have neither liability nor responsibility to any person, any organization or entity with respect to any loss or damage caused, or alleged to have caused, directly or indirectly, by information or advice contained in this book. Therefore, the purchaser/reader must assume full responsibility for the use of the book or the information therein.

The mention of commercial brands and trade names are only for technical purposes. No particular product is endorsed over another product or equipment not mentioned. The author, cooperating authors, educational institutions, and the publishers Apple Academic Press Inc., do not have any preference for a particular product.

All weblinks that are mentioned in this book were active on September 20, 2016. The editors, the contributing authors, the publisher and the printing company shall have neither liability nor responsibility, if any of the weblinks are inactive at the time of reading of this book.

ABOUT SENIOR EDITOR-IN-CHIEF

 Megh R. Goyal, PhD, PE, is a Retired Professor in Agricultural and Biomedical Engineering from the General Engineering Department in the College of Engineering at University of Puerto Rico–Mayaguez Campus; and Senior Acquisitions Editor and Senior Technical Editor-in-Chief in Agriculture and Biomedical Engineering for Apple Academic Press Inc.

He has worked as a Soil Conservation Inspector and as a Research Assistant at Haryana Agricultural University and Ohio State University. He was first agricultural engineer to receive the professional license in Agricultural Engineering in 1986 from College of Engineers and Surveyors of Puerto Rico. On September 16, 2005, he was proclaimed as "Father of Irrigation Engineering in Puerto Rico for the twentieth century" by the ASABE, Puerto Rico Section, for his pioneer work on micro irrigation, evapotranspiration, agroclimatology, and soil and water engineering. During his professional career of 45 years, he has received many prestigious awards. A prolific author and editor, he has written more than 200 journal articles and textbooks and has edited over 35 books. He received his BSc degree in engineering from Punjab Agricultural University, Ludhiana, India; his MSc and PhD degrees from Ohio State University, Columbus; and his Master of Divinity degree from Puerto Rico Evangelical Seminary, Hato Rey, Puerto Rico, USA.

BOOK MESSAGES

Agricultural engineers are indispensable for energy and food security. It is our ethical duty to educate the public on food security.

—Miguel A Muñoz, PhD
Ex-President of University of Puerto Rico and
Professor of Soil Scientist

I believe that this innovative book on emerging issues in energy and natural resources will aid educators throughout the world.

—A. M. Michael, PhD
Former Professor/Director, Water Technology Centre—IARI
Ex-Vice-Chancellor, Kerala Agricultural University, Trichur, Kerala

Agricultural engineers render an important service to the conservationists.

—Gajendra Singh, PhD
Ex-President (2010–2012) of ISAE
Former Deputy Director General (Engineering) of ICAR, and
Former Vice-President/Dean/Professor and Chairman Asian Institute of
Technology, Thailand

Agricultural engineers need to provide leadership opportunities for food, fiber, shelter, and energy stability of agriculture.

—V. M. Mayande, PhD
Former Vice Chancellor, Dr. Panjabrao Deshmukh Krishi Vidyapeeth
& MAFSU, Nagpur; Former President, Indian Society of Agricultural
Engineers

Emerging technologies have matured into reliable and practical applications over the past couple of decades. The usage of emerging technologies, especially in the energy and natural resources field, continues to increase in the world, and this book series will serve as a reference for engineers and practitioners worldwide.

—Felix B. Reinders, PhD
Professional Engineer Agricultural Research Council-
Institute for Agricultural Engineering, South Africa

OTHER BOOKS ON AGRICULTURAL & BIOLOGICAL ENGINEERING BY APPLE ACADEMIC PRESS, INC.

Management of Drip/Trickle or Micro Irrigation
Megh R. Goyal, PhD, PE, Senior Editor-in-Chief

Evapotranspiration: Principles and Applications for Water Management
Megh R. Goyal, PhD, PE, and Eric W. Harmsen, Editors

Book Series: Research Advances in Sustainable Micro Irrigation
Senior Editor-in-Chief: Megh R. Goyal, PhD, PE

Volume 1: Sustainable Micro Irrigation: Principles and Practices
Volume 2: Sustainable Practices in Surface and Subsurface Micro Irrigation
Volume 3: Sustainable Micro Irrigation Management for Trees and Vines
Volume 4: Management, Performance, and Applications of Micro Irrigation Systems
Volume 5: Applications of Furrow and Micro Irrigation in Arid and Semi-Arid Regions
Volume 6: Best Management Practices for Drip Irrigated Crops
Volume 7: Closed Circuit Micro Irrigation Design: Theory and Applications
Volume 8: Wastewater Management for Irrigation: Principles and Practices
Volume 9: Water and Fertigation Management in Micro Irrigation
Volume 10: Innovation in Micro Irrigation Technology

Book Series: Innovations and Challenges in Micro Irrigation
Senior Editor-in-Chief: Megh R. Goyal, PhD, PE

Volume 1: Principles and Management of Clogging in Micro Irrigation
Volume 2: Sustainable Micro Irrigation Design Systems for Agricultural Crops: Methods and Practices
Volume 3: Performance Evaluation of Micro Irrigation Management: Principles and Practices
Volume 4: Potential of Solar Energy and Emerging Technologies in Sustainable Micro Irrigation

Volume 5: Micro Irrigation Management Technological Advances and
   Their Applications
Volume 6: Micro Irrigation Engineering for Horticultural Crops: Policy
   Options, Scheduling, and Design
Volume 7: Micro Irrigation Scheduling and Practices
Volume 8: Engineering Interventions in Sustainable Trickle Irrigation

Book Series: Innovations in Agricultural and Biological Engineering
Senior Editor-in-Chief: Megh R. Goyal, PhD, PE

- Modeling Methods and Practices in Soil and Water Engineering
- Food Engineering: Modeling, Emerging issues and Applications.
- Emerging Technologies in Agricultural Engineering
- Dairy Engineering: Advanced Technologies and Their Applications
- Food Process Engineering: Emerging Trends in Research and Their
  Applications
- Soil and Water Engineering: Principles and Applications of Modeling
- Developing Technologies in Food Science: Status, Applications, and
  Challenges
- Agricultural and Biological Engineering Practices
- Soil Salinity Management in Agriculture: Emerging Technologies
  and Applications
- Engineering Practices for Agricultural Production and Water Conser-
  vation: An Interdisciplinary Approach
- Flood Assessment: Modeling and Parameterization
- Food Technology: Applied Research and Production Techniques
- Processing Technologies for Milk and Milk Products: Methods,
  Applications, and Energy Usage
- Engineering Interventions in Agricultural Processing
- Technological Interventions in Processing of Fruits and Vegetables
- Technological Interventions in Management of Irrigated Agriculture
- Engineering Interventions in Foods and Plants
- Technological Interventions in Dairy Science: Innovative Approaches
  in Processing, Preservation, and Analysis of Milk Products
- Novel Dairy Processing Technologies: Techniques, Management, and
  Energy Conservation
- Sustainable Biological Systems for Agriculture: Emerging Issues in
  Nanotechnology, Biofertilizers, Wastewater, and Farm Machines
- State-of-the-Art Technologies in Food Science: Human Health,
  Emerging Issues and Specialty Topics

EDITORIAL

Apple Academic Press Inc., (AAP) will be publishing various book volumes on the focus areas under book series titled *Innovations in Agricultural and Biological Engineering*. Over a span of 8 to 10 years, Apple Academic Press, Inc., will publish subsequent volumes in the specialty areas defined by *American Society of Agricultural and Biological Engineers* (http://asabe.org).

The mission of this series is to provide knowledge and techniques for agricultural and biological engineers (ABEs). The series aims to offer high-quality reference and academic content in Agricultural and Biological Engineering (ABE) that is accessible to academicians, researchers, scientists, university faculty, and university-level students and professionals around the world. The following material has been edited/modified and reproduced from: *"Megh R. Goyal, 2006. Agricultural and biomedical engineering: Scope and opportunities. Paper Edu_47 Presentation at the Fourth LACCEI International Latin American and Caribbean Conference for Engineering and Technology (LACCEI'2006): Breaking Frontiers and Barriers in Engineering: Education and Research by LACCEI University of Puerto Rico – Mayaguez Campus, Mayaguez, Puerto Rico, June 21–23."*

WHAT IS AGRICULTURAL AND BIOLOGICAL ENGINEERING (ABE)?

"Agricultural Engineering (AE) involves application of engineering to production, processing, preservation and handling of food, fiber, and shelter. It also includes transfer of technology for the development and welfare of rural communities," according to http://isae.in. *"ABE is the discipline of engineering that applies engineering principles and the fundamental concepts of biology to agricultural and biological systems and tools, for the safe, efficient and environmentally sensitive production, processing, and management of agricultural, biological, food, and natural resources systems,"* according to http://asabe.org. *"AE is the branch of engineering involved with the design of farm machinery, with soil management, land development, and mechanization and automation of livestock farming, and with the efficient planting,*

harvesting, storage, and processing of farm commodities," definition by:
http://dictionary.reference.com/browse/agricultural+engineering.

"AE incorporates many science disciplines and technology practices to the efficient production and processing of food, feed, fiber and fuels. It involves disciplines like mechanical engineering (agricultural machinery and automated machine systems), soil science (crop nutrient and fertilization, etc.), environmental sciences (drainage and irrigation), plant biology (seeding and plant growth management), animal science (farm animals and housing), etc.," (Source: http://www.bae.ncsu.edu/academic/agricultural-engineering.php)

According to https://en.wikipedia.org/wiki/Biological_engineering: *"Biological engineering (BE) is a science-based discipline that applies concepts and methods of biology to solve real-world problems related to the life sciences or the application thereof. In this context, while traditional engineering applies physical and mathematical sciences to analyze, design and manufacture inanimate tools, structures and processes, biological engineering uses biology to study and advance applications of living systems."*

SPECIALTY AREAS OF ABE

Agricultural and Biological Engineers (ABEs) ensure that the world has the necessities of life including safe and plentiful food, clean air and water, renewable fuel and energy, safe working conditions, and a healthy environment by employing knowledge and expertise of sciences, both pure and applied, and engineering principles. Biological engineering applies engineering practices to problems and opportunities presented by living things and the natural environment in agriculture. BA engineers understand the interrelationships between technology and living systems, have available a wide variety of employment options. The http://asabe.org indicates that *"ABE embraces a variety of following specialty areas."* As new technology and information emerge, specialty areas are created, and many overlap with one or more other areas.

1. **Aqua Cultural Engineering**: ABEs help design farm systems for raising fish and shellfish, as well as ornamental and bait fish. They specialize in water quality, biotechnology, machinery, natural resources, feeding and ventilation systems, and sanitation. They seek

ways to reduce pollution from aqua cultural discharges, to reduce excess water use, and to improve farm systems. They also work with aquatic animal harvesting, sorting, and processing.

2. **Biological Engineering** applies engineering practices to problems and opportunities presented by living things and the natural environment.

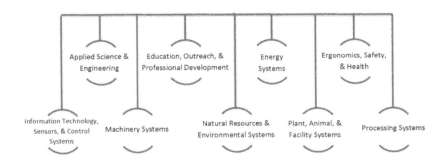

3. **Energy:** ABEs identify and develop viable energy sources – biomass, methane, and vegetable oil, to name a few – and to make these and other systems cleaner and more efficient. These specialists also develop energy conservation strategies to reduce costs and protect the environment, and they design traditional and alternative energy systems to meet the needs of agricultural operations.

4. **Farm Machinery and Power Engineering**: ABEs in this specialty focus on designing advanced equipment, making it more efficient and less demanding of our natural resources. They develop equipment for food processing, highly precise crop spraying, agricultural commodity and waste transport, and turf and landscape maintenance, as well as equipment for such specialized tasks as removing seaweed from beaches. This is in addition to the tractors, tillage equipment, irrigation equipment, and harvest equipment that have done so much to reduce the drudgery of farming.

5. **Food and Process Engineering:** Food and process engineers combine design expertise with manufacturing methods to develop economical and responsible processing solutions for industry. Also food and process engineers look for ways to reduce waste by devising alternatives for treatment, disposal and utilization.

6. **Forest Engineering**: ABEs apply engineering to solve natural resource and environment problems in forest production systems and related manufacturing industries. Engineering skills and expertise are needed to address problems related to equipment design and manufacturing, forest access systems design and construction; machine-soil interaction and erosion control; forest operations analysis and improvement; decision modeling; and wood product design and manufacturing.

7. **Information and Electrical Technologies Engineering** is one of the most versatile areas of the ABE specialty areas, because it is applied to virtually all the others, from machinery design to soil testing to food quality and safety control. Geographic information systems, global positioning systems, machine instrumentation and controls, electromagnetics, bioinformatics, biorobotics, machine vision, sensors, spectroscopy: These are some of the exciting information and electrical technologies being used today and being developed for the future.

8. **Natural Resources:** ABEs with environmental expertise work to better understand the complex mechanics of these resources, so that they can be used efficiently and without degradation. ABEs determine crop water requirements and design irrigation systems. They are experts in agricultural hydrology principles, such as controlling drainage, and they implement ways to control soil erosion and study the environmental effects of sediment on stream quality. Natural resources engineers design, build, operate and maintain water control structures for reservoirs, floodways and channels. They also work on water treatment systems, wetlands protection, and other water issues.

9. **Nursery and Greenhouse Engineering**: In many ways, nursery and greenhouse operations are microcosms of large-scale production agriculture, with many similar needs – irrigation, mechanization, disease and pest control, and nutrient application. However, other engineering needs also present themselves in nursery and greenhouse operations: equipment for transplantation; control systems for temperature, humidity, and ventilation; and plant biology issues, such as hydroponics, tissue culture, and seedling propagation methods. And sometimes the challenges are extraterrestrial: ABEs at NASA are designing greenhouse systems to support a manned expedition to Mars!

10. **Safety and Health:** ABEs analyze health and injury data, the use and possible misuse of machines, and equipment compliance with standards and regulation. They constantly look for ways in which the safety of equipment, materials and agricultural practices can be improved and for ways in which safety and health issues can be communicated to the public.

11. **Structures and Environment:** ABEs with expertise in structures and environment design animal housing, storage structures, and greenhouses, with ventilation systems, temperature and humidity controls, and structural strength appropriate for their climate and purpose. They also devise better practices and systems for storing, recovering, reusing, and transporting waste products.

CAREERS IN AGRICULTURAL AND BIOLOGICAL ENGINEERING

One will find that university ABE programs have many names, such as biological systems engineering, bioresource engineering, environmental engineering, forest engineering, or food and process engineering. Whatever the title, the typical curriculum begins with courses in writing, social sciences, and economics, along with mathematics (calculus and statistics), chemistry, physics, and biology. Student gains a fundamental knowledge of the life sciences and how biological systems interact with their environment. One also takes engineering courses, such as thermodynamics, mechanics, instrumentation and controls, electronics and electrical circuits, and engineering design. Then student adds courses related to particular interests, perhaps including mechanization, soil and water resource management, food and process engineering, industrial microbiology, biological engineering or pest management. As seniors, engineering students' work in a team to design, build, and test new processes or products.

For more information on this series, readers may contact:

Ashish Kumar, Publisher and President
Sandy Sickels, Vice President
Apple Academic Press, Inc.,
Fax: 866-222-9549
E-mail: ashish@appleacademicpress.com
http://www.appleacademicpress.com/
publishwithus.php

Megh R. Goyal, PhD, PE
Book Series Senior Editor-in-Chief
Innovations in Agricultural and Biological Engineering
E-mail: goyalmegh@gmail.com

PART I
Emerging Technologies

CHAPTER 1

DRAINAGE CONGESTION IN HIRAKUD COMMAND, ODISHA: ISSUES AND OPPORTUNITIES TO AUGMENT CROP PRODUCTION

BALRAM PANIGRAHI[1*], JAGADISH CHANDRA PAUL[1], and KAJAL PANIGRAHI[2]

[1]*Department of Soil & Water Conservation Engineering, College of Agricultural Engineering & Technology, Orissa University of Agriculture & Technology, Bhubaneswar, Odisha, India. E-mail: kajal_bp@yahoo.co.in, jcpaul66@gmail.com*

[2]*Department of Civil Engineering, National Institute of Technology, Rourkela, Odisha, India. E-mail: kajalpanigrahi@yahoo.in*

Corresponding author.

CONTENTS

Abstract 4
1.1 Introduction 4
1.2 Waterlogging in Hirakud Command 6
1.3 Canal Operation Schedule in the Command 9
1.4 Causes of Waterlogging in the Command 9
1.5 Spatial Variation of Groundwater Table in the Command 11
1.6 Opportunities to Reduce Drainage Congestion and Waterlogging ... 14
1.7 Surface and Subsurface Drainage System 15
1.8 Conclusions 17
Keywords 17
References 18

This chapter is based on the article Managing Drainage Congestion to Increase Crop Production and Productivity in Hirakud Command, India by Balram Panigrahi and Jagadish Chandra Paul, in Journal of Agricultural Engineering and Biotechnology, http://www.bowenpublishing.com/jaeb/paperInfo. aspx?PaperID=16789.

ABSTRACT

Hirakud command of Odisha is one of the biggest and oldest multipurpose irrigation projects in India. It caters the irrigation demand to 1.59 lakh ha in *kharif* and 1.12 lakh ha in *rabi*. The average groundwater table in the irrigated areas has come up by more than 6.0 m since commensurate of the project. Commensuration of the project has caused about 20% of the total cultivated area waterlogged in the command where crop diversification is impossible. The present paper discusses the causes of rising groundwater table that creates drainage congestion and waterlogging in the Hirakud command. The study reveals that in the head reach, the rise of groundwater table is more as compared to the lower reach. The main factors that contribute to the drainage congestion are (1) faulty irrigation and water management practice, (2) intensive rice–rice cultivation, (3) plot to plot irrigation in rice field instead of using field channel, (4) continuous canal flow, (5) excessive seepage losses in the canal systems, etc. Opportunities to improve the drainage system and reclaim the waterlogged areas and thereby decreasing the groundwater table are also mentioned in this paper. Construction of parallel field surface drain at 10 m drain spacing is recommended in the command to decline the rising groundwater table which may facilitate the farmers to grow non-rice crops in the command and increase crop production and productivity.

1.1   INTRODUCTION

The populations of the world which was about 2.50 billion in half a century ago has already crossed the 7.0-billion mark and is likely to cross the 9.0-billion mark in the next 25 years. India has already touched 1.26 billion population and is expected to touch 1.64 billion by 2050 AD. To feed such a great population, about 450 million tons of food grain production is needed which requires about 95% increase in the total food grain production considering the present production of 255 million tons. This will necessitate enhancing the natural resources including both soil and water optimally in a very judicious manner. Irrigation development in India during the post-independence era has greatly facilitated enhancement of agricultural production by increasing the irrigation potential. However, in the wake of such phenomenal strides of irrigation development from 23 M ha (million hectare) in the early 1950s to nearly 109 M ha at present, twin problems of waterlogging and salinity have also come up in the irrigation commands due to hydrologic disturbances beyond the capacity of natural drainage systems. The abundant loss of water in the form of seepage in the conveyance systems is because of earthen canals/channel sections, badly damaged canal outlets, etc. coupled with low water application efficiency in the crop field due to improper water management practice. This has resulted in waterlogging and salinization thus rendering vast areas to be unproductive. They not only threaten the capital investment but also the sustainability of irrigated agriculture and have become an environmental concern too.

The waterlogged saline soils are found to occur all over the country. About 8.5 M ha area has been salt-affected; of this, 5.6 M ha is waterlogged saline area in the commands of irrigation projects and is commonly referred to as man-made or wet deserts.[2] Based on the extrapolation of the data on individual schemes, it is estimated that irrigation induces salinization and waterlogging on an average of 10% of the net irrigated areas.[4]

Recently in India, waterlogging in canal irrigated areas is increasing at an alarming rate. In Tungabhadra Project in Karnataka, more than 33,000 ha has been reported to suffer from waterlogging and salinity and these areas are increasing at the rate of 600 ha/year. Large areas in Haryana have been rendered waterlogged after introduction of Bhakra canal irrigation. Rising water table in Gujarat has also been observed due to the flow of water in Mahi command.

Rice is the most important crop of India and the second most important crop of the world and is the staple food for nearly half of the world population. Among the rice growing countries, India has largest area under rice in the world (about 43.2 M ha) and ranks second in production, next to China. Rice in India consumes about 66% of irrigation water in the country. The water requirement of rice invariably is put 1000–2000 mm (for the irrigated transplanted crop) depending on soil type and climate. Rice is considered to be an inefficient water consumer. Unlike in other arable crops, seepage and percolation losses (SP losses) in rice greatly exceed the evapotranspiration demand. The SP losses are dependent on soil texture, irrigation practices, and crop duration and could vary within 52–93% of the total water expenses.[5,6] The SP losses in the rice field along with inadequate drainage facility are the major factor to cause waterlogging by raising the groundwater table. Research on drainage requirement of rice or management of the drainage water in India is a neglected component in water management, knowing fully well that when irrigation comes, drainage cannot be left behind. Drainage is an integral part of irrigation water management which provides desirable environment in the crop root zone for healthy growth of crops.

1.2 WATERLOGGING IN HIRAKUD COMMAND

After the construction of the Hirakud dam (Fig. 1.1) across the river Mahanadi in Odisha, India, there has been a gradual increase in irrigation potential in the pre-divided districts of Sambalpur and Bolangir of Odisha. This project now commands an area of 1.59 lakh ha in *kharif* and 1.12 lakh ha in *rabi*. The command area of Hirakud covers five blocks in Sambalpur, six blocks in Bargarh, two blocks in Sonepur, and one block in Bolangir through four canal systems, that is, Baragarh main canal, Sason main canal, Sambalpur distributary, and Hirakud distributary. Basic details of canal systems of the command are presented in Table 1.1.[1]

FIGURE 1.1   Views of the Hirakud dam across the river Mahanadi, Odisha.

Because of the introduction of the project, the average cropping intensity of the command has increased from 110% to 187% during the last 55 years of initiation of the project. The watershed of the command can be divided into six major streams which discharge water into the Mahanadi. The sub-watersheds are Ong, Jira, JhanJhor, KulerJhor, Harda, and Malati. The details of the watersheds are presented in Figure 1.2.[9] The topography of the command is undulating marked by ridges and valleys. The land slope of the command goes as high as 10%. The topography of the command is classified

as att (unbunded upland), mal (bunded upland), berna (medium land), and
bahal (low land). The geological formation of the command belongs to the
Areean group. The major soil texture of the command area is sandy loam and
clay loam. Soils are broadly classified in to Alfisol (63%), Entisol (19%),
Inceptisol (11%), and Vertisol (7%). Average physiological properties of the
surface soil (0–30 cm) are summarized in Table 1.2.

TABLE 1.1   Basic Details of Hirakud Canal System.

Types of canal	Length (km)	Area irrigated (*kharif*) (ha)	Discharge at canal head (cusec)
Main canal (2)			
Bargarh main canal	88	6900	3800
Sason main canal	23	1900	680
Branch canal (2)	35	4070	506–906
Distributaries (35)	444.44		7.5–38.1
Minor canal (84)	293.64		5.6–96.8
Sub-minors (32)	74.88	146,091	4.0–50.0
Sub sub-minors (4)	12.60		3.0–10.0
Water courses (2985)	2433.70		1.0–3.0
Total	**3406.66**	**158,961**	

FIGURE 1.2   Topographic map of the Hirakud command.

TABLE 1.2 Physical and Chemical Properties of Surface Soil of the Hirakud Command.

Physical properties		Chemical properties	
Bulk density	1.66–1.72 g/cm³	Organic carbon	0.30–0.50%
Field capacity	19.6–20% (weight basis)	Mg²⁺	1.8–6.2 me/100 g soil
Hydraulic conductivity	1.5–6.2 cm/h	Available K₂O	100–112 kg/ha
Infiltration capacity	0.3 cm/h	Ca²⁺	6.1–8.0 me/100 g soil
Lower plastic limit	9.3–26.8% (weight basis)	Available P₂O₅	9–25 kg/ha
Upper plastic limit	16.8–40% (weight basis)	Total nitrogen	0.03–0.05%
Particle density	2.50–2.70 g/cm³	EC (at 25°C)	0.06–0.25 mmhos/cm
Permanent wilting point	7.0–10.4% (weight basis)	K⁺	2.1–6.2 me/100 g soil
Saturation moisture	24–27% (weight basis)	Na⁺	0.1–0.2 me/100 g soil
Texture	Sandy loam to clay loam	pH	5.3–7.5

Climate in the region is warm, subhumid characterized by hot dry summer and short and mild winter. The average rainfall of the command is 1419 mm. Average temperature varies from a minimum of 9°C to a maximum of 45°C. Relative humidity varies from 43.1% to 92.6%. Overall irrigation water quality is suitable for crops. Table 1.3 provides the water quality values of both canal and pond in the command.

TABLE 1.3 Quality of Irrigation Water Available from Canal and Pond in the Command.

Particulars	Canal water	Pond water
pH	8.60	8.30
EC (at 25°C) (mmhos/cm)	0.14	0.16
Anions (me/l)		
CO_3	0.12	0.15
HCO_3	2.50	2.30
Cl	4.56	3.20
NO_3	0.06	0.06
SO_3	Trace	Trace
Cations (me/l)		
Ca^{2+}	1.23	1.70
Mg^{2+}	1.30	1.75
K^+	0.09	0.05
Na^+	0.19	0.16

The average groundwater table in the irrigated areas has come up by more than 6.0 m during the last 50 years. There are some pockets in the command where it lies only 0.10 m below the ground level. Commensuration of the project has caused about 20% of the total cultivated area waterlogged in the command where crop diversification is impossible or not cost effective.[3]

1.3 CANAL OPERATION SCHEDULE IN THE COMMAND

Water is supplied to the command area almost throughout the year. Periodically canals are closed for about 22–24 days in the month of November and December. Moreover, for some time, the canal is closed in May and June for repair works. The canal operation schedule (mean of last 24 years, 1990–2013) is presented in Table 1.4. From the table, it is observed that the average duration of canal opening period is 278 days and only for 87 days, the canal remains closed.

TABLE 1.4   Canal Schedule Data of the Hirakud Command (Mean of 24 Years).

Sl. No.	Month	Canal opening period	Days
1	Jan	1st Jan–31st Jan	31
2	Feb	1st Feb–28th Feb	28
3	Mar	1st Mar–31st Mar	31
4	Apr	1st Apr–30th Apr	30
5	May	1st May–11th May	11
6	June	21st Jun–30th Jun	10
7	Jul	1st Jul–31st Jul	31
8	Aug	1st Aug–31st Aug	31
9	Sep	1st Sep–30th Sep	30
10	Oct	1st Oct–31st Oct	31
11	Nov	1st Nov–8th Nov	8
12	Dec	25th Dec–31st Dec	6
Total			278

1.4 CAUSES OF WATERLOGGING IN THE COMMAND

Farmers in the command, especially in the head reach of the canal system cultivate rice–rice and grow them under shallow to deep submerged condition. About 97% of the total irrigated area in *kharif* season is cultivated with rice where 98% of the total irrigation water is consumed by rice. Similarly about 70% of the total irrigated area in *rabi* is grown with rice consuming about 91% irrigation water. Most of the farmers in the command, especially in head end maintain 15–25 cm ponded waster in the rice field. This has caused serious problems of waterlogging in the command especially in the head and mid reach of the command. In general, the main causes of water-logging in the command are as follows:

- Faulty water management practice
- Intensive rice–rice cultivation
- Absence of field channels to irrigate the fields
- No rotational irrigation system
- No control over flow, that is, absence of volumetric measurement device to measure and supply irrigation water
- No crop diversification
- Drainage congestion in outlets
- Unlined/badly maintained channels causing major seepage losses in canal system
- Mismatch between the demand and supply system
- Lack of land development and
- Inadequate drainage facilities

One of the most important problems of waterlogging and drainage in the command is the land locked topo-system. The canal that flows in the upland called locally as att land suffers from severe seepage loss. This seepage water continuously flow down and also laterally from the att land to the bahal land (low land) causing stagnation of water in the bahal land (Fig. 1.3). Thus, the att land suffers from moisture stress, whereas the bahal land suffers from severe waterlogging. It is needless to mention here that both the deficit and excess moisture in the crop field are not congenial to crop growth.

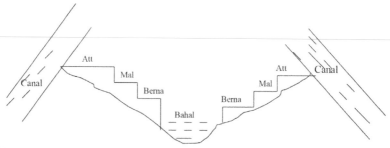

FIGURE 1.3   Land lay in Hirakud ayacut.

1.5 SPATIAL VARIATION OF GROUNDWATER TABLE IN THE COMMAND

Data on rise of groundwater table in the command were collected from different sources for 30 years (1985–2014). The four different sources from which data were collected are

1. irrigated area,
2. areas near to canal,
3. uncultivated areas, and
4. dry farming areas.

The depth of water table below the ground level was measured at weekly interval throughout the year for 30 years as mentioned above and the data are presented in Table 1.5.

TABLE 1.5 Fluctuation of Groundwater Table in Different Areas (Average of 30 years, 1985–2014).

Week No.	Depth of groundwater table below ground level (m)			
	Near canal	Irrigated area	Uncultivated area	Dry farming area
1	3.95	0.39	3.94	2.44
2	3.60	0.37	3.95	2.33
3	3.90	0.35	3.91	2.25
4	3.91	0.33	3.85	2.30
5	3.95	0.40	3.81	2.31
6	3.99	0.42	3.75	2.18
7	3.86	0.45	3.54	2.28
8	3.95	0.51	3.60	2.34
9	3.93	0.33	3.53	2.36
10	3.87	0.32	3.48	2.20
11	3.92	0.41	3.48	2.23
12	4.03	0.45	3.36	2.40
13	3.90	0.53	3.42	2.45
14	3.65	0.55	3.48	2.50
15	3.77	0.33	3.36	2.34
16	3.89	0.30	3.42	2.52
17	3.99	0.30	3.48	2.65
18	4.05	0.35	3.81	2.28

TABLE 1.5   *(Continued)*

Week No.	Depth of groundwater table below ground level (m)			
	Near canal	Irrigated area	Uncultivated area	Dry farming area
19	4.19	0.33	3.96	2.35
20	4.28	0.41	4.18	2.36
21	4.39	0.42	4.35	2.41
22	4.55	0.40	4.46	2.45
23	4.60	0.45	4.47	2.40
24	4.28	0.50	4.21	2.15
25	4.05	0.47	3.96	2.02
26	3.81	0.41	3.80	2.00
27	2.25	0.40	1.85	1.85
28	2.28	0.41	1.75	1.65
29	1.98	0.45	1.55	1.54
30	1.95	0.42	1.49	1.50
31	1.65	0.28	1.35	1.32
32	1.87	0.22	1.30	1.25
33	1.58	0.23	1.28	1.22
34	1.55	0.25	1.20	1.20
35	1.50	0.31	1.15	1.24
36	1.53	0.46	1.12	1.32
37	1.58	0.50	1.25	1.30
38	1.60	0.51	1.31	1.25
39	1.75	0.52	1.26	1.20
40	1.80	0.54	1.29	1.46
41	1.99	0.47	1.52	1.50
42	2.00	0.44	1.48	1.52
43	2.02	0.38	1.82	1.54
44	2.10	0.45	1.80	1.45
45	2.65	0.44	1.85	1.50
46	2.85	0.51	1.88	1.88
47	3.10	0.55	2.25	1.90
48	3.05	0.39	2.30	2.02
49	3.25	0.46	3.10	2.10
50	3.18	0.47	3.23	2.25
51	3.83	0.45	3.56	2.35
52	3.88	0.41	3.96	2.32

Note: There are 52 weeks in 1 year.

From the observed data, it is revealed that the water table data in the *kharif* (standard meteorological weeks 27–47 are higher, that is, near to the ground surface than the *rabi* season in all the places. The depth of water table near the canal varies from 1.50 to 4.60 m, whereas the water table in the irrigated areas, uncultivated areas and dry farming areas varies from 0.22 to 0.55 m, 1.12 to 4.47 m, and 1.20 to 2.65 m, respectively. The rise of water table in the irrigated areas is found to be significant compared to other areas. The entire command areas receive water almost throughout the year. The rise of water table causing water logging and drainage problem in the command areas are mainly attributed to three points. They are (1) intensive rice–rice cultivation, (2) faulty irrigation and water management practice especially maintaining deep submergence in the rice field, and (3) high rainfall in the rainy season in conjunction with continuous canal flow. The water table remains at a shallow depth as compared to other areas in the irrigated areas due to continuous infiltration, seepage, and deep percolation from the rice field and seepage from the unlined canal systems. The rise of water table is less in the summer months of May to June because of less rainfall in these months and also due to less number of days of canal opening periods. In dry farming areas, the rise of water table is found to be higher than the uncultivated areas because of field bunds raised around the fields to conserve maximum amount of rainfall. This has hastened up the infiltration of conserved rainwater in the field causing rise of the ground water table. It is interesting to note that in the dry farming areas, the rise of water table is higher in the months of rainy season than the winter and summer. The reason may be attributed to the high quantum of rainfall that is received in the rainy season as compared to winter and summer.

The rise of water table is found to be different for different reaches along a canal. In the upend (head reach) of the canal the rise of water table is found to be higher than that in the middle and tail end of the canal. Observations were taken to record the ground water table in five places along the canal reach of Baragarh distributary of Hirakud command. Out of these five places, two were in the head reach, two in the middle, one was in the tail reach of the canal. The average groundwater table position in head, middle, and tail reaches were recorded for last 14 years (2000–2013) and are presented in Figure 1.4.

Data of Figure 1.4 reveal that because of extensive irrigation and continuous water supply in the canal head end, the ground water table has risen up more compared to other two reaches. In the last 13 years, the water table in the head end has come up from 9.66 to 1.29 m, whereas in the middle and tail reaches, the water table has come up from 9.05 to 2.05 m and 10.2 to

4.80 m. Thus, the data reveal that the drainage and waterlogging problems are becoming severe in the head reach than the middle and tail reaches in the canal in the command. It is apprehended that if the present trend of the irrigation and cropping system (rice–rice) continues, then by 2030 AD, about 45% of the areas of the command would become waterlogged consequently affecting the crop yield severely. It, is hence, felt imperative to study the techno-economically feasible drainage system to reclaim the waterlogged areas or to arrest the upcoming waterlogging problem in canal commands.

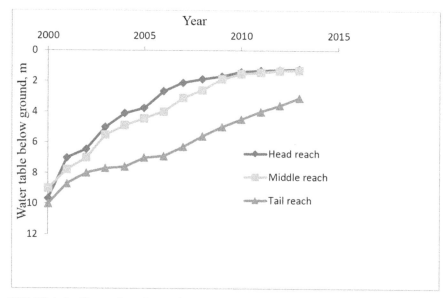

FIGURE 1.4   Fluctuation of groundwater table in different reaches.

1.6 OPPORTUNITIES TO REDUCE DRAINAGE CONGESTION AND WATERLOGGING

Waterlogging in the Hirakud command is an alarming situation and unless some tangible measures are taken then the situation will go beyond control and the fertile land will turn into fallow after some decades. Negative effects of waterlogging has caused the problems like (1) decreasing the soil health, (2) affecting the growth and yield of crops, (3) increasing the flood hazards, (4) affecting the soil temperature, (5) raising soil salinity and alkalinity, (6) causing severe environmental pollutions, and (7) posing a severe problem to crop diversification. Some of the remedial measures to reduce the drainage and waterlogging condition in the command are:

1. Irrigating rice fields through field channels instead of field to field to irrigation.
2. Irrigating the rice by intermittent irrigation method instead of maintaining standing water in field always.
3. Encouraging crop diversification instead of rice–rice cultivation.
4. Using accurate irrigation scheduling including proper methods of irrigation.
5. Adopting water-saving irrigation techniques in the crops.
6. Proper land smoothing and grading using appropriate farm machineries.
7. Using high tech irrigations like sprinkler and drip especially in fruits, orchards and vegetables which has high application efficiency.
8. Reduction of SP in crop fields.
9. Using cost effective mulching in crops fields to check soil evaporation and reduce the frequency of irrigation in crop fields.
10. Decreasing the conveyance losses of canals by lining the canal systems.
11. Volumetric supply of water in canal commands.
12. Charging water rates on volumetric basis which will enforce the farmers to use right amount of water in the crop fields.

1.7 SURFACE AND SUBSURFACE DRAINAGE SYSTEM

Out of different drainage systems, surface drainage is the simplest and cheapest drainage system. Figure 1.5 shows a surface drainage system for evacuation of excess water from crop field.

FIGURE 1.5 An open-surface drain used to evacuate excess surface water.

Parallel field surface drains are the widely adopted surface drain systems. In the design of parallel field surface drains, one of the most important design parameter is deciding the optimum drain spacing. The drain spacing would be optimum when it lowers down the groundwater table to the maximum amount and thereby facilitating to grow non-rice crops and at the same time it is economically viable. Experiments in the loamy soil at Regional Research and Technology Transfer Station, Chiplima in the command reveal that out of various drainage spacing, 10-m drain spacing works better in lowering down the water table and enables the farmers to grow rice-based cropping systems like rice–wheat–mung and rice–mustard–sesamum in chronically waterlogged ecosystem. The 6-year experiments in the field revealed that the ground water level could be lowered down by 35.3, 32.6, and 29.4 cm below the ground level with 10, 15, and 20 m drain spacings, respectively.[7,8] Figure 1.6 shows the fluctuation of groundwater table by different drain spacings by parallel filed surface drains. From the experimental data, it is recommended that in chronically waterlogged ecosystem, excess water from the soil surface as well as the groundwater from the soil subsurface should be evacuated by constructing 10-m parallel field surface drains.

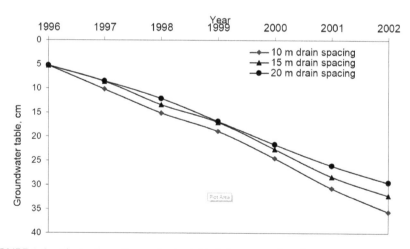

FIGURE 1.6   Fluctuation of groundwater table below ground surface.

Groundwater table in the sub-soil region can also be reduced by installing subsurface drains like tile and mole drains at suitable spacing and at suitable depth below the ground level. However, installation of subsurface drains is more costly than the surface drain methods, but it has several advantages also. It does not hamper in conveyance of machineries and in farm operation.

The costly agricultural lands are also not wasted toward construction of the drains.

1.8 CONCLUSIONS

The present study deals with finding out the causes of waterloging in the Hirakud command of the state of Odisha, India. The study reveals that during the 55 years of construction of the project, there has been more than 5 m rise in the groundwater table in the command. The rise of groundwater table is found to be more in the upper reaches of the canal command compared to the lower reaches. The main causes of waterlogging are faulty water management practice, intensive rice–rice cultivation, continuous canal flowing, maintenance of deep submergence in rice field, irrigation from plot to plot without field channel and congestion of drainage channels. Some remedies to reclaim the waterlogging problems are suggested in this paper including surface and subsurface drainage system. It is suggested that construction of parallel field surface drains at 10 m spacing can lower down the rising groundwater table and can facilitate in cultivation of non-rice crops in the ayacut.

KEYWORDS

- canal
- drain spacing
- drainage congestion
- groundwater table
- Hirakud command
- irrigation and water management
- lining
- percolation
- salinization
- seepage loss
- subsurface drains
- surface drains

- **water productivity**
- **water-saving irrigation**
- **waterlogging**

REFERENCES

1. Anonymous. Annual Report of All India Co-ordinated Research Project on Water Management, Chiplima Center, Orissa University of Agriculture and Technology, Bhubaneswar, 2005; p 106.
2. Balakrishnan, P.; Lingappa, S.; Shirahatti, M. S.; Kuligod, V. B.; Kulkarni, G. N. Reclamation of Waterlogged Saline Areas in Upper Krishna Project Command of Karnataka state, India. First Asian Regional Conference, Seoul, Korea (ICID), 2001; pp 55–61.
3. Behera, B. P.; Panigrahi, B.; Samantray, S. K.; Sahu, N. N. Ground Water Quality and Fluctuations in Hirakud Command—A Case Study. *Indian J. Power River Valley Develop.* **2001,** July–August Issue, 158–160.
4. Jain, A. B. Improper Technologies can Jeopardize Further. *Hindu Surv. Indian Agric.* **2002,** 183–186.
5. Pande, H. K.; Mitra, B. N. Water Management for Paddy. Seminar on Irrigation Water Management by Water Management Forum, New Delhi, India, 1992, pp 425–445.
6. Panigrahi, B. *Water Balance Simulation for Optimum Design of On-farm Reservoir in Rainfed Farming System.* Unpublished Ph.D. Thesis submitted to Indian Institute of Technology, Kharagpur, India, 2001.
7. Panigrahi, B.; Behera, B. P. Conjunctive use of Surface and Groundwater for Augmenting Crop Production in Hirakud Command. Souvenir of 53rd Annual Session of Orissa Engineering Congress, Jan. 21st, 2008; pp 9–22.
8. Panigrahi, B.; Behera, B. P.; Samantrai, S. K. Effect of Drain Spacing on Reclamation of Waterlogged Areas. *J. Agric. Eng.* **2007,** *44*(1), 1–7.
9. Tyagi, N. K.; Lenka, D.; Acharya, N. Waterlogging in Hirakud Ayacut. Souvenir of 23rd Annual Session of Orissa Engineering Congress, Jan. 21st, 1988; pp 9–22.

CHAPTER 2

IRRIGATION TECHNOLOGY OPTIONS FOR SELECTED AGRICULTURAL CROPS

VISHAL K. CHAVAN

AICRP for Dryland Agriculture, Dr. Panjabrao Deshmukh Krishi Vidyapeeth, Krishinagar PO, Akola 444104, MS, India.
E-mail: vchavan2@gmail.com

CONTENTS

Abstract 20
2.1   Irrigation Technology Options for Selected Agricultural
      Crops: Farmer's Practice 20
2.2   The Problems of Several Nonsystem Tank Irrigation System 25
2.3   Mulching 26
Keywords 30
References 30

ABSTRACT

In this chapter, technology options are presented for sorghum, millet, maize, pulses, sugarcane, tomato, banana, acid lime, cotton, coconut, ginger, soybean, groundnut, coconut, tank irrigation system, mulching.

2.1 IRRIGATION TECHNOLOGY OPTIONS FOR SELECTED AGRICULTURAL CROPS: FARMER'S PRACTICE

2.1.1 SORGHUM—FARMER'S PRACTICE

- Flat-bed system—irrigation based on prevailing weather and eye judgment.

2.1.1.1 TECHNOLOGY OPTIONS

- Furrow irrigation once in 15–16 days during first 20 days of sowing and six irrigations with an interval of 6 days during the rest of the crop period.
- Raton sorghum: six irrigation, namely, at ratooning, 4–5 leaf stage, milking, soft dough, and hard dough.
- Surge irrigation is feasible in long furrow (>100 m) in level lands.

2.1.2 PEARL MILLET—FARMER'S PRACTICE

- Beds and channel irrigation.

2.1.2.1 TECHNOLOGY OPTIONS

- Irrigating with IW/CPE ratio of 0.75 at 4 cm depth was found to be optimum.

2.1.3 FINGER MILLET—FARMER'S PRACTICE

- Flat-bed system irrigation based on prevailing weather and eye judgment.

2.1.3.1 TECHNOLOGY OPTIONS

- Irrigating with IW/CPE ratio of 0.75 at 4 cm depth of water.

2.1.4 MAIZE—FARMER'S PRACTICE

- Flat-bed system—irrigation based on prevailing weather and eye judgment.

2.1.4.1 TECHNOLOGY OPTIONS

- Irrigating the field at 10 days interval.

2.1.5 PULSES—FARMER'S PRACTICE

- Beds and channels and excess irrigation.

2.1.5.1 TECHNOLOGY OPTIONS

- Black gram and green gram irrigation at critical stage, that is, one at sowing, second at flowering, and third at pod formation with 4 cm depth.
- Irrigation once in 18 days was optimum.
- Soybean irrigation at 80 mm, CPE once in 11–12 days interval.

2.1.6 GROUNDNUT—FARMER'S PRACTICE

- Beds and channel-irrigation based on prevailing weather and eye judgment.

2.1.6.1 TECHNOLOGY OPTIONS

- Irrigation at sowing and establishment stages and 25 days after sowing.
- Irrigation once in 7–9 days found to be optimum.

2.1.7 GINGELLY—FARMER'S PRACTICE

- Flat-bed system and copious irrigation.

2.1.7.1 TECHNOLOGY OPTIONS

- Irrigation at flowering stage and capsule formation.

2.1.8 SUNFLOWER—FARMER'S PRACTICE

- Flat-bed system.

2.1.8.1 TECHNOLOGY OPTIONS

- Irrigation at IW/CPE ratio of 0.75 with 20:30:20 kg of NPK/ha.
- Surge irrigation under long furrow in level lands.

2.1.9 COCONUT—FARMER'S PRACTICE

- Check basin and copious irrigation.

2.1.9.1 TECHNOLOGY OPTIONS

- Irrigation through drip system@100 L of water/tree/day.
- For stress management, the palm basins to be opened to a radius of 1.8 m with receipt of late showers and mulching can be done.
- Husk mulching can be done to absorb rain water and making available to palm.

2.1.9.2 APPLICATION OF GREEN MANURE AND FYM IN THE BASIN

- Spreading dried coconut leaves and other organic residues.
- Additional tank silt to the basin increase the water retaining capacity.
- Under drought situation lower senescent leaves may be removed.
- Pitcher irrigation can be followed where a little water is available.

2.8.10   COTTON—FARMER'S PRACTICE

- Beds and channels.

2.1.10.1   TECHNOLOGY OPTIONS

- Sowing of seeds in ridges and furrows.
- Irrigation at IW/CPE ratio of 0.75.
- Mulching with sugarcane trash@5 t/ha
- Spraying of Folicot or paraffin wax 10 g or kaolin 50 g in a liter of water.
- Sprinkler irrigation is feasible.
- Drip irrigation can be adopted.

2.1.11   BANANA—FARMER'S PRACTICE

- Trench and mounds method of irrigation.

2.1.11.1   TECHNOLOGY OPTIONS

- Irrigation at 0.75–0.9 IW/CPE ratio.
- Chain basin method can be adopted.
- Basins are formed around the suckers and the basins are connected through channels.
- Drip irrigation with high density planting, fertigation is preferred in favorable locations (well-irrigated lands).
- Gradual widening of furrows with stage of crops (Fig. 2.1)

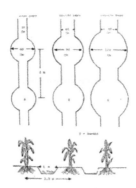

FIGURE 2.1   Planting methods in banana cultivation.

2.1.12   ACID LIME—FARMER'S PRACTICE

*   Basin irrigation.

2.1.12.1   TECHNOLOGY OPTIONS

*   Irrigation through drippers at 75% of water supplied through basins.
*   Drip irrigation is preferred.

2.1.13   TOMATO—FARMER'S PRACTICE

*   Beds and channels.

2.1.13.1   TECHNOLOGY OPTIONS

*   Furrow irrigation may be recommended.
*   Drip irrigation especially with micro sprinklers may also be recommended.
*   Irrigation at IW/CPE ratio of 1.00 during fruit formation and ripening.
*   Sprinkler irrigation in tomato with 1760 m^3 gave significantly higher water use efficiency than any other irrigation method.

2.1.14   SUGARCANE—FARMER'S PRACTICE

*   Excess irrigation through ridges and furrows.

2.1.14.1   TECHNOLOGY OPTIONS

*   Irrigation at IW/PE ratio of 0.9.
*   Mulching with sugarcane trash in garden land situation reduce evaporation loss.
*   Foliar application of kaolin@12.5 kg in 750 L of water per hectare reduce the transpiration loss.
*   Removal of old dried leaves in 5–7 months old crop.
*   Alternate or skip furrow irrigation can be followed (Fig. 2.2).

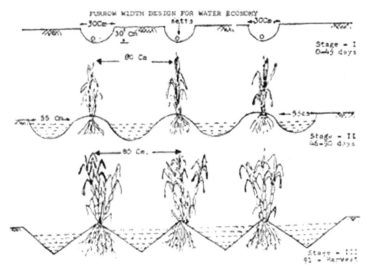

FIGURE 2.2   Corrugated furrows for sugarcane cultivation.

- Irrigate the field based on sheath moisture percentage.
- In deep trench system of planting 30 cm deep trenches are opened at 80 cm apart, sets are planted in trenches.
- Drip irrigation with fertigation is highly suitable.[1-4]
- Surge irrigation may be adopted in long fields in light-textured soils.

2.2   THE PROBLEMS OF SEVERAL NONSYSTEM TANK IRRIGATION SYSTEM

- Encroachment, siltation, soaking of supply channels resulting in poorer/no-inflow of water, pollution of tank water by tannery, and dying factory influence (Coimbatore, Erode, Salem Districts).
- Tank chains almost disappear and their hydrologically interlinking, any improvement can revive the tank will have benefit of exploiting full tank irrigation through appropriate or selective modernization benefits.
- Owing to vagaries of man only 50–60% of supply is realized a crop diversification strategy with non-rice crops is suggested.
- De-silting for reviving the original capacity, tank fore shore, plantation to arrest the silt flow, feasibility of connecting small different tanks into the percolation pond for ground water recharge, rehabilitation of tank structure, and inward channels are the solutions emanated from the tank system researches.

- On-farm development structures have to be strengthened the any tank command areas for equitable water distribution from head to tail end along with farmers' participation.
- Other technology options for poor quality water.
- Conjunctive use of relatively fresh surface water and poor quality ground water with proper proportions are recommended.
- Growing of salt tolerant crops in the saline water, irrigation belt along with proper drainage facility.
- Community bore wells during the period of erratic water supply in canal command areas enhanced the crop water availability and there by their yields and net returns.

2.3 MULCHING[4]

2.3.1 BENEFITS OF MULCHING

Mulch is a layer of material applied to the surface of an area of soil. Mulch prevents rain from hitting the soil directly, reducing the impact of the water drops. Water soaks into the soil gradually instead washing the soil away.

Soil that is high in organic matter is very much alive. In just one pinch of soil, there are about a billion individual living organisms, perhaps 10,000 distinct species of microbes. As the microbes decompose organic matter, it supplies nutrients needed by growing plants. The beauty of this natural nutrient cycle is that nutrients are released in harmony with the needs of the plants. When environmental conditions are favorable for rapid plant growth, the same conditions favor a rapid release of nutrients from the organic matter.

When a mulch of organic matter is added to the surface of the soil, it decays producing slimes and gums that help to form and stabilize soil structure. The extra organic matter is food for soil creatures. These burrow their way through the soil, mixing the organic matter in and creating passage ways within the soil through which air and water can infiltrate. In this way, mulching can help loosen up heavy clay soils, making it easier for the farmer to work, and making it easier for plant roots and shoots to push their way through. Some people call earthworms "a farmer's best friends." As earthworms multiply, the soil becomes looser and more porous—a better place for plant roots to grow. Mulching also prevents the soil from getting a hard crust. When rain drops hit bare soil in a heavy rainstorm, it breaks into smaller pieces. These pieces stick together and form a hard crust when the soil dries. This crust makes it difficult for water to soak into the ground. It also makes it hard for young plants to push through the soil crust. Mulch is a protective cover for the soil, sheltering the soil from hard-hitting rain.

Mulch around crop plants mimics the litter layer of a forest floor. The nutrients in the mulch are gradually released and taken up by the crops. Mulching is cheaper than chemical fertilizers, and because it also improves soil structure, the nutrients will not be washed away or leached from the soil by heavy rain.

Mulch reflects a lot of the sun that otherwise beats down on the soil. This keeps the soil cooler and helps prevent evaporation. This is especially important in hot, dry climates. Also, by slowing down rainwater run-off, mulch increases the amount of water that soaks into the soil. The loose soil structure, created by the soil life fed by the mulch, helps hold water in the soil. And more water in the soil means more water for your crops. With mulch, it may be possible to grow crops like tomatoes and cucumbers where only drought tolerant crops grew before. For many crops, mulching can increase yields, prevent erosion, and ensure that soil stays fertile for the future.

2.3.2   POTENTIAL DRAWBACKS TO MULCHING

For one thing, mulches may provide a good environment for pests. For example, slugs sometimes cause serious losses to crops such as beans when mulched. Harmful insects, mice, rats, rabbits, and snakes may also find thick mulches an attractive habitat. Also, mulching can sometimes lead to a lack of nitrogen for crop plants. The nitrogen can be locked up in the bacteria that are decomposing the fresh organic matter in the mulch. Nitrogen deficiency may make the crop plants more susceptible to disease.

However, as a long-term strategy for soil improvement, the benefits of mulching far outweigh the potential disadvantages. Mulching costs are extra.

2.3.3   MATERIALS FOR MULCHING

Farmer can use whatever organic matter is readily available and transportable. Common materials include compost, manure, straw (crop stems and stalks), dry grass clippings, sawdust, leaves, and other left-over crop residues. Alternative mulching materials include black plastic sheeting, newspaper, or cardboard. However these materials do not add nutrients to the soil or improve its structure.

It is better not to use plant material from the same type of crops that are grown in the field. For example, maize residue should not be used as mulch for maize as it might still be carrying insects or diseases of maize. Green vegetation is not normally used as it can take a long time to decompose and can attract pests and fungal diseases. One must try which mulches in the area last the longest. The longer the mulch lasts, the less often it needs to be applied.

2.3.4   MULCHING METHODS

For large plants spread the mulch between the rows and around each plant. For small plants or seedlings apply it between the rows, not directly around the plants. In this way one will not encourage diseases, but one will still reduce weeds and add organic matter to the soil. Try different thicknesses of mulch to see which works best for a crop. Always apply mulches to a warm, wet soil. Mulch applied to a dry soil will keep the soil dry.

If one puts mulch too thickly, it might shade seedlings, so they will grow tall and spindly. Too much mulch can also prevent airflow and encourage diseases. This can be a especial problem in areas with a lot of rain. To allow the germination of planted seeds through the mulch, a layer of less than 10 cm should be used. To clear an area of land of persistent weeds a layer of 10 cm or more can be used.

2.3.5   LABOR INPUT FOR MULCHING[1-6]

Actually, by improving the soil and helping to fight weeds, mulching can save gardeners time and work in the long run. One will spend less time weeding and because a soil with lots of organic matter is looser, those weeds that do grow are lot easier to pull out. Digging in a looser soil is also a lot easier. Plus, as mulching prevents water from evaporating from the surface of the soil, less watering is necessary. However, it is a lot of work to carry in enough mulching material to cover an entire field. Sometimes farmers simply cut the weeds in their fields, right before planting, to form a mulch. Other farmers grow cover crops, which they chop up and leave as a mulch just before food crops are planted. Plastic mulching costs extra.

2.3.6   VERTICAL MULCHING

Vertical mulching consists of creating of holes around the base of a tree or shrub that is stressed. The holes are filled with a mixture of organic material

(Fig. 2.3). The back fill and aeration can dramatically improve root growth and reduce or eliminate stress caused in lawn environment or construction damage. Some salient features of this technique are

- to alleviate soil compaction and allow air to flow more easily to the roots,
- to add organic rich compost to poor soil,
- to expand the root system by the addition of Mychorizae, and
- to allow faster percolation of water in wet areas by changing soil structure and aerating.

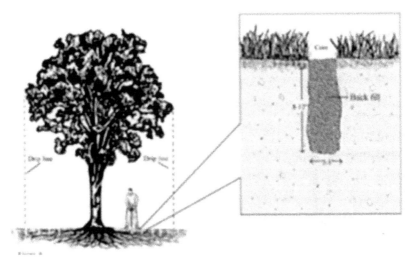

FIGURE 2.3   Technology of vertical mulching.

It is a relatively simple and unobtrusive process. The holes are created with an air spade or soil auger. Starting a few feet out from the trunk and continued in a grid pattern (approximately 2′ apart) beyond the drip line. They are usually back filled with a mixture of compost, organic fertilizer, and Mychorizae. The actual fill may vary according to the individual needs of the plant treated. Some of the benefits are

- reduced compaction,
- increased drainage,
- increased oxygen to the root system,
- introduction of organic matter to existing soil,
- increased soil organisms, and
- increased water retention.

Vertical mulching is one of the most effective ways to treat stressed trees and shrubs. The aeration and soil treatment create an environment that mimics nature and allows for greater nutrient uptake, increased root growth and the potential regeneration of a damaged root system. In vertical mulching, one may use compost tea and humate applications to further help the root system repair and provide for itself.

KEYWORDS

- check basin
- drip irrigation
- furrow irrigation
- mulching
- organic matter
- plastic mulching
- technology
- vertical mulching

REFERENCES

1. Goyal, M. R. *Research Advances in Sustainable Micro Irrigation, Volumes 1 to 10.* Apple Academic Press Inc.: Oakville, ON, 2015.
2. Goyal, M. R. *Innovations and Challenges in Micro Irrigation, Volumes 1 to 4.* Apple Academic Press Inc.: Oakville, ON, 2016.
3. Goyal, M. R. *Innovations in Agricultural & Biological Engineering.* Apple Academic Press Inc.: Oakville, ON, 2016.
4. Goyal, M. R. *Management of Drip/Trickle or Micro Irrigation.* Apple Academic Press Inc.: Oakville, ON, 2012.
5. Goyal, M. R.; Harmsen, E. W. *Evapotranspiration: Principles and Applications for Water Management.* Apple Academic Press Inc.: Oakville, ON, 2014.
6. http://www.paceproject.net/Userfiles/File/Soils/mulching.pdf.

PART II
Energy Management in Agriculture

CHAPTER 3

SOLAR ENERGY: PRINCIPLES AND APPLICATIONS

M. K. GHOSAL

Department of Farm Machinery and Power, College of Agricultural Engineering and Technology, Orissa University of Agriculture and Technology, Bhubaneswar 751003, Odisha, India.
E-mail: mkghosal1@rdiffmail.com

CONTENTS

Abstract 34
3.1   Introduction 34
3.2   Energy Resources and Their Classification 36
3.3   Principles of Solar Energy 51
3.4   Solar Energy and Its Applications 65
Keywords 103
References 105

ABSTRACT

The ever-increasing demand of energy for development of the society is fulfilled by a variety of energy sources. Large-scale energy utilization has led to a better quality of life and faster all-round development; it has also generated many critical problems. The most prominent of these is the harmful effect on the environment in various forms leading to global warming and climate change. At the same time, the fossil fuel resources are also fast depleting due to over exploitation. Therefore, it is worth to explore the alternative energy sources, systems, and technologies for sustainable development, if not fully but at least to substitute an appreciable amount of conventional energy to mitigate the harmful effect to some extent.

Other than fossil fuels, nuclear and large hydropower, there are a number of sources of energy which have started contribution in a small way to the world's present energy demand and supply scenarios. These include energy sources like wind energy, small hydro, photovoltaic conversion, biomass, tidal, geothermal energy, and solar thermal power plants. Among the renewable energy sources, solar energy plays a vital role both for its thermal and photovoltaic applications. In this chapter, the authors have focused on the thermal conversion route of solar energy with respect to solar energy collection devices, working fluids, solar thermal energy storage, etc.

3.1 INTRODUCTION

The sun is a source of infinite energy that can be used directly or indirectly and the energy harnessed from the sun is known as solar energy. The energy from the sun, amounting to nearly 4000 trillion kW h every day in the form of electromagnetic radiation[7] exceeds the current primary energy supply used by mankind, providing 10,000 times more than the annual global energy consumption[4] and which is also significantly more than other energy sources available at the ground level, such as geothermic or tidal energy, nuclear power and fossil fuel burning.[8]

India lies in the sunny belt of the world.[2] The scope for generating power and thermal applications using solar energy is huge. Most parts of India get 300 days of sunshine a year, which makes the country a very promising place for solar energy utilization.[6] The daily average solar energy incident over India varies from 4 to 7 kW h/m[2] with the sunshine hours ranging between 2300 and 3200 per year, depending upon location.[9]

The technical potential of solar energy in India is huge. The country receives enough solar energy to generate more than 500,000 TW h/year of electricity, assuming 10% conversion efficiency for PV modules.[1,3] The use of solar energy in recent years has reached a remarkable edge. The continuous research for an alternative power source due to the perceived scarcity of fuel fossils is its driving force. It has become even more popular as the cost of fossil fuel continues to rise. The earth receives in just 1 h more energy from the sun than what we consume in the whole world for 1 year.[10] Its application has proven to be most economical, as most systems in individual uses require a few kilowatt of power. Limited fossil resources and environmental problems associated with them have emphasized the need for new sustainable energy supply options that use renewable energies. Development and promotion of nonconventional/alternate/new and renewable sources of energy such as solar, wind and bioenergy, etc., are also getting sustained attention. Of all the renewable sources of energy available, solar thermal energy is the most abundant one and is available in both direct as well as indirect forms.

The sun emits energy at a rate of 3.8×10^{23} kW, of which, approximately 1.8×10^{14} kW is intercepted by the earth, which is located about 150 million km from the sun. About 60% of this amount (1.08×10^{14}) reaches the surface of the earth. The rest is reflected back into space and absorbed by the atmosphere. About 0.1% of this energy, when converted at an efficiency of 10%, would generate four times the world's total generating capacity of about 3000 GW.[3] However, 80% of the present worldwide energy used is based on fossil fuels. Risks associated with their use are that they are all potentially vulnerable to adverse weather conditions or human acts. World demand for fossil fuels (starting with oil) is expected to exceed annual production, probably within the next two decades. International economic and political crisis and conflicts can also be initiated by shortages of oil or gas. Moreover, burning fossil fuels release harmful emissions such as carbon dioxide, nitrogen oxides, aerosols, etc. which affect the local, regional and global environment.[5]

This chapter describes the present day solar thermal technologies. Performance analyses of existing designs and fabrication of innovative designs with suggested improvements (development) have been discussed in this chapter. A detailed description of major solar thermal technologies comprising of solar flat plate collector, solar water heaters, solar cookers, solar driers, and solar stills has been made in this chapter.

3.2 ENERGY RESOURCES AND THEIR CLASSIFICATION

Energy resources are main sources of energy from which the energy can be extracted and utilized for mankind. Energy is a key input in economic growth. The growth of a nation largely depends on the availability of energy resources. Energy resources can be classified into various categories based on certain criteria.

3.2.1 CLASSIFICATION OF ENERGY RESOURCES

3.2.1.1 BASED ON USABILITY

The sources of energy which are obtained directly from the environment are called primary sources of energy. Secondary sources of energy are not directly obtained from the environment (Fig. 3.1).

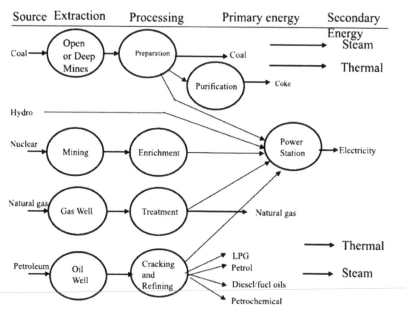

FIGURE 3.1   Major primary and secondary sources of energy.

Primary energy sources are those that are either found or stored in nature. Common primary energy sources are coal, oil, natural gas, biomass, etc. Other primary sources include nuclear energy from radioactive substances, thermal energy stored in earth's interior, and potential energy due to the earth's gravity. The primary energy sources are mostly processed for getting final form of energy sources supplied to the consumer for use.

The sources of energy after processing are called **secondary sources of energy**. Examples are coal, oil, or gas converted into steam and electricity for direct use. Primary energy sources can also sometimes be used directly like biomass, coal, etc.

3.2.1.2 BASED ON COMPARATIVE ECONOMIC VALUE

3.2.1.2.1 Commercial Energy

The energy sources that are available in the market for a definite price are known as commercial energy. These are capital intensive. Important forms of commercial energy are electricity, coal, refined petroleum products. Commercial energy forms basis of industrial, agricultural, transport, and commercial development in the modern world. In the industrialized countries, commercial fuels are predominant source not only for economic production, but also for many household tasks of most of the population of our country.

3.2.1.2.2 Noncommercial Energy

The energy sources that are not available in the commercial market for a price or available very cheaply are classified as noncommercial energy. Noncommercial energy sources include fuels such as firewood, cattle dung, and agricultural wastes, which are traditionally gathered and not bought at a price used especially in rural households.

3.2.1.3 BASED ON RENEWABILITY

3.2.1.3.1 Renewable Energy Sources

Renewable energy is the source of energy that can be regenerated or renewed after its use and is virtually inexhaustible that is found in the natural environment. This source of energy is subsequently replenished after its use. This is also referred to as nonconventional source of energy. This source of energy lasts indefinitely and is called infinite source of energy. Examples are solar, biomass, wind, ocean thermal, wave, tidal, hydro energy, etc.

3.2.1.3.2 Nonrenewable Energy

Nonrenewable energy is the source of energy that cannot be regenerated or renewed after its use and is virtually exhaustible in nature. It is obtained from static stores of energy that remain unutilized if not released by human interaction. This source of energy is called finite source of energy and cannot be replenished after its use. This is also referred as conventional sources of energy. Examples are fossil fuels (coal, oil, natural gas).

3.2.1.4   BASED ON MUSCULAR ENERGY

3.2.1.4.1   Animate Energy Sources

Animate energy is the source of energy, which is obtained from living organism. It is also called as energy source derived from muscles.

3.2.1.4.2   Inanimate Energy Sources

Inanimate energy is the source of energy, which is obtained from nonliving sources. It is not derived from the muscular energy. Examples are electricity, fossil fuels, solar energy, etc.

3.2.2   FORMS OF ENERGY

Forms of energy are basically of stored form of energy (potential) or working form of energy (kinetic). For example, the food we eat contains chemical energy and our body stores this energy until we release it when we work or play.

3.2.2.1   POTENTIAL FORM OF ENERGY

Potential energy is stored energy and the energy of position (gravitational). It exists in various forms.

   a.   **Chemical energy** is the energy stored in the bonds of atom and molecules. Biomass, petroleum, natural gas, propane, and coal are example of stored chemical energy.

b. **Nuclear energy** is the energy stored in the nucleus of an atom—the energy that holds the nucleus together. The nucleus of a uranium atom is an example of stored energy.

c. **Stored mechanical energy** is the energy stored in objects by the application of a force. Compressed springs and stretched rubber bands are example of stored mechanical energy.

d. **Gravitational energy** is the energy of place or position. Water in a reservoir behind a hydropower dam is an example of gravitational energy. When the water is released to spin the turbines, it becomes motion energy.

3.2.2.2   KINETIC ENERGY

It is the energy in motion: the motion of waves, electrons, atoms, molecules, and substances. It exists in various forms.

a. **Radiation energy** is the electromagnetic energy that travels in transverse waves. Radiant energy includes visible lights, X-rays, gamma rays, and radio waves. Solar energy is an example of radiant energy.

b. **Thermal energy** is the internal energy in the substances—the vibration and movement of atoms and molecules within the substances. Geothermal energy is the example of thermal energy.

c. **Electrical energy** is due to the movement of electrons.

d. **Mechanical energy** is due to motion of shaft.

Energy exists in different forms. Energy is changed from one form to another, but the total amount in any closed system remains constant. This is known as the conservation of phenomena occurring in nature. Energy can also be changed from some useful form to some other form that for all practical purposes is useless, even though formally energy is conserved in the process. Different forms of energy are summarized below.

1. Muscular energy—Energy spent by muscle to do work.
2. Kinetic energy—Energy that causes an object to move.
3. Potential energy—The energy of position of an object.
4. Heat energy—Energy that brings warmness or coldness.
5. Light energy—Energy getting the vision of objects.
6. Chemical energy—Energy due to chemical reaction.

7. Nuclear energy—Energy due to nuclear reactions like fission and fusion.
8. Solar energy—Energy from sun.
9. Wind energy—Energy from moving wind.
10. Hydel energy—Energy from flowing water.
11. Tidal energy—Energy from tides in sea-water.
12. Ocean thermal energy (OTE)—Energy from temperature gradients in ocean water.
13. Geothermal energy—Energy from hot water springs found under the earth.

3.2.3 ADVANTAGES OF RENEWABLE ENERGY

1. Renewable energy sources are available in considerable quantities and also available continuously in the nature.
2. Renewable energy sources are financially and economically competitive for certain applications such as in remote location where the cost of transmitting electrical power or transporting conventional fuels are high. They favor power system decentralization.
3. Renewable energy sources cannot be depleted unlike fossil fuels. They can provide a reliable and sustainable supply of energy indefinitely. In contrast, the nonrenewable sources of energy are finite and can be diminished by extraction and consumption.
4. Renewable energy sources provide less environmental impacts as compared to other sources of energy. The nonrenewable energy sources cause water/air pollution, acid rain, global warming, climate change, ozone depletion which affect human and animal life and also vegetation in the earth.

3.2.4 DISADVANTAGES OF RENEWABLE ENERGY

1. The availability of renewable energy is very intermittent, irregular and most erratic.
2. Introduction or use of energy storage system makes the energy generation system costly.
3. They are site and location specific.
4. Area required for considerable energy extraction from the renewable energy sources is vast as compared to nonrenewable energy sources.

5. The efficiency of renewable energy conversion device is very less as compared to nonrenewable systems.
6. All renewable energy devices are installed far away from residence.
7. Implementation of renewable energy devices is not so easy due to lack of efficient devices, public awareness and sufficient information.

3.2.5 COMPARISON BETWEEN RENEWABLE AND NONRENEWABLE ENERGY SOURCES

TABLE 3.1   Comparison between Renewable and Nonrenewable Energy Sources.

Feature of comparison	Renewable energy sources	Nonrenewable energy sources
Source	Solar, wind, biomass, tidal, etc.	Coal, oil, natural gas, etc.
Availability	Natural local environment	Concentrated stock in the earth
Life time of supply	Infinite	Finite
Cost at source	Free	Increasingly expensive
Location for use	Site and society specific	General and public use
Scale of use	Small scale and economic	Increased scale
Skill	Interdisciplinary and varied wide range of skill	Strong links with electrical and mechanical engineering, specific range of skill
Area of use	Rural and decentralized applications	Urban and centralized uses
Pollution and environmental damage	Less	More
Safety	Less dangerous	Most dangerous when faulty

3.2.6 TYPES OF ENERGY RESOURCES

3.2.6.1 THERMAL ENERGY

Thermal energy is stored as heat energy in the fossil fuels. Fossil fuels are the fuels obtained from the earth that have been accumulated over thousands of years due to decaying of plants. These fuels produce heat energy when they are burnt. Heat energy is mainly used for transportation and electric power generation in thermal power plants. The fossil fuels can be classified as shown in Figure 3.2.

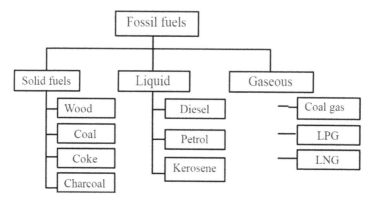

FIGURE 3.2   Types of fossil fuels.

3.2.6.2   HYDEL ENERGY

Hydel energy is the potential energy of water created due to the storage of water at a higher level. A dam is built across the river to store water at a higher level. When this stored water in dam flows from higher level under pressure to the lower level, it can run the turbine to generate electrical power. A layout of hydroelectric power plant is shown in Figure 3.3. It consists of (1) reservoir, (2) penstock to carry water from reservoir to turbine, (3) turbine to convert water energy into mechanical energy, (4) generator to convert mechanical energy to electrical energy, and (5) power transmission system.

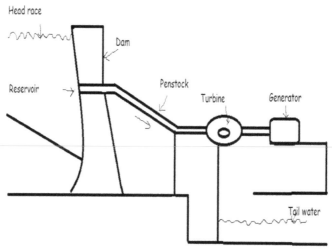

FIGURE 3.3   A typical hydroelectric power station.

3.2.6.3   NUCLEAR ENERGY

The nuclear energy is released when atoms of certain unstable materials split during the process of fission. A small mass of nuclear fuel such as uranium can release an enormous amount of heat energy when it undergoes fission process. One kilogram of uranium-235 can give heat energy on fission process, which is equal to the heat which can be obtained by burning 4000 t of high grade petroleum. The uranium can be made to undergo fission process inside a nuclear reactor. The nuclear fission is a chain reaction as shown in Figure 3.4.

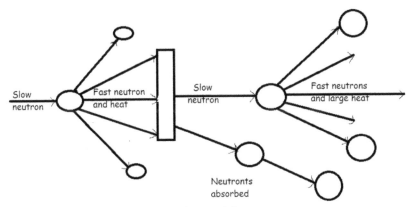

FIGURE 3.4   Fission process in the nuclear model.

3.2.6.4   SOLAR ENERGY

The sun is a continuous fusion reactor, in which hydrogen combines to form helium and liberates large amount of heat during the process. The sun rays contain a large amount of energy in the form of electromagnetic radiation due to the continuous nuclear fusion reaction taking place in the sun. The energy is released at the rate of 3.7×10^{20} MW. This heat energy contained in the sun rays can be utilized to generate electrical power. The sun rays are focused on solar collector to heat butane water to generate butane gas in the butane boiler. The butane gas under high pressure from the boiler is taken to butane turbine to perform mechanical work. A generator is coupled to the turbine to generate electrical power as shown in Figure 3.5. The potential of power generation by solar energy can be in the order of 1.75×10^{11} MW. The solar collectors are mechanical devices, which help to collect the solar radiations so as to convert them into heat energy.

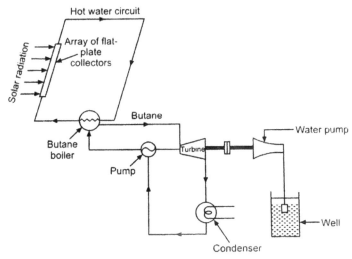

FIGURE 3.5   Solar power plant.

Photovoltaic conversion is a direct electricity generation method, in which sunlight is converted into electricity using solar cells. The most common solar cells are manufactured from a highly refined silicon material. A single solar cell can produce electric power of 1 W at a voltage of 0.5 V. Several solar cells are at intermittent energy source and generally used with batteries to store generated electricity, thereby providing a more economical power generating system.

3.2.6.4.1   Advantages of Solar Energy

1. Solar energy is available freely in nature.
2. It is a renewable energy resource.
3. It does not pollute the environment
4. It can be directly converted into electricity by employing photovoltaic cells.

3.2.6.4.2   Disadvantages of Solar Energy

1. It is available only during daytime and on clear days.
2. Solar energy obtainable also depends on seasonal variations.
3. It requires a large area to entrap appreciable solar energy for the generation of an economical amount of electricity.

3.2.6.5 BIOMASS ENERGY

Biomass refers to the mass of biological material produced from the chemical processes. This includes the materials derived from plants as well as from animals. Chemically biomass refers to hydrocarbons containing hydrogen, carbon and oxygen, which can be represented in the form of $C_{6n}(H_2O)_{6n}$. We extract biomass from numerous sources like plants, trees, agricultural crops, raw material from the forest, household waste, and wood. The contribution of the biomass to our energy requirement comes in the form of food and fuel. There are several other requirements that are being served by the use of biomass, namely shelter preparation, fodder for animals, nutrient for soil, etc. The earliest inhabitants on the earth burned wood in their campfires for heat, and since then it has been a source of energy for meeting human needs. Biomass is solar energy stored in organic matter. As trees and plants grow, the process of photosynthesis uses energy from the sun to convert CO_2 of the atmosphere into carbohydrates (sugar, starch, and cellulose). The process of photosynthesis can be written as follows:

$$6CO_2 \text{ (gas)} + 6H_2O \text{ (liquid)} + \text{light} \rightarrow C_6H_{12}O_6 \text{ (solid)} + 6O_2 \text{ (gas)} \qquad (1)$$

Carbohydrates are the organic compounds that make up biomass. When plants die, the process of decay converts the organic matter into fossil fuels like coal, gas, and oil. Nowadays, we are using fossil fuels for our energy requirements. The consumption of fossil fuels results in emission of CO_2 back to atmosphere. In this way, one full cycle of CO_2 is completed.

Biomass is a renewable energy source because the growth of new plants and trees replenishes the supply. It is found in almost all regions of the world. The biomass has been used extensively in the development of societies since the beginning of civilizations. It has played very important role in the development of societies. Many people in the developing countries still depend on the use of biomass for their daily livelihood. They use it for food, fuel as well as a source of income. It is estimated that biomass contributes to about 14% of the world's total energy requirement. In many developing countries, the contribution to total energy requirement is as high as 90%. In addition to being useful for human purposes, the biomass acts as an essential medium for sustaining earth's ecological balance. Through photosynthesis process, biomass helps in balancing the CO_2 in the atmosphere, it enriches and conserves the natural vegetation and soil and at the same time provides long-term secondary energy. As a renewable energy source, biomass should be able to meet all our requirements in various forms, provided the balance between production and consumption of biomass is maintained.

But during the last several decades, human activities have resulted in the generation of CO_2 (massive burning of fossil fuels) which is much more than what biomass can absorb. This is causing global warming. Also, there is imbalance in production and consumption of biomass. Therefore, any future energy production based on biomass needs to be designed carefully. Advantages and disadvantages of biomass energy are listed below:

Advantages	Disadvantages
Available almost everywhere	Energy from biomass is more expensive than conventional fossil fuels
Municipal, agricultural and industrial waste can be utilized for useful purposes	It is a less concentrated form of energy
It is an inexhaustible fuel source if there is a balance in production and consumption	Releases some emissions, especially if burnt improperly
Sulfur, nitrogen oxides, and carbon emissions can be significantly reduced if biomass is used as source of energy	

3.2.6.6   WIND ENERGY

Wind is induced in atmosphere by uneven heating of earth's surface by the sun. The wind energy is associated with the movement of large masses of air from cold to hot regions. The movement results from uneven heating of atmosphere by sun, thereby creating differences in temperature, density, and pressure. The wind energy can be used to run windmill, which in turn will drive a generator to produce electric power or run water pumps. Windmill is a device which converts the kinetic energy of the moving mass of air or wind into mechanical energy. The windmills can be classified depending on the orientation of axis of rotation as horizontal axis and vertical axis windmills as shown in Figure 3.6. The windmills can also be classified based on the number of blades as single-bladed windmill, double-bladed windmill, three-bladed windmill and multibladed windmill (Fig. 3.7). The blades of the windmills are generally made of composite materials such as fiber-reinforced plastic because this material is less costly, easy to use for manufacturing and possesses high strength to weight ratio.

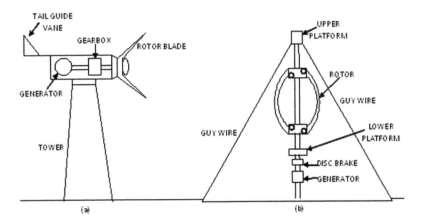

HORIZONTAL AND VERTICAL AXIS WIND MILLS
(a) HORIZONTAL AXIS WIND MILL (b) VERTICAL AXIS WIND MILL

FIGURE 3.6   Types of wind mills: (a) horizontal axis and (b) vertical axis.

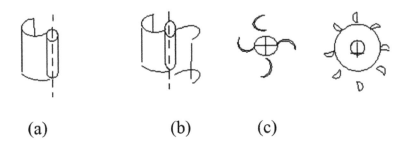

(a)   **(b)**   **(c)**

FIGURE 3.7   Types of windmills: (a) single-bladed rotor, (b) double-bladed rotor, and (c) multibladed rotor.

3.2.6.6.1   *Advantages of Wind Energy*

1.   It is freely and abundantly available in nature.
2.   It is a renewable energy source.
3.   It does not cause pollution to environment.
4.   Windmills require minimal maintenance operating cost.

3.2.6.6.2   *Disadvantages of Wind Energy*

1.   It cannot produce steady and consistent power.
2.   It can generate only low power.

3.2.6.7   TIDAL ENERGY

Ocean waves and tides contain a large amount of both potential and kinetic energies, which can be utilized for power generation. A tide is the periodical rise and fall of sea water caused principally by the interaction of gravitational fields of sun and the moon. The highest level of tidal water is called flood or high tide and the lowest level is called low or ebb tide. The level difference between the high and the low tide is called tidal range. The up and down movement of the tide is used for filling and emptying the tidal basin of the plant. The typical tidal plant is shown in Figure 3.8. The tidal basin is filled up during high tide, and it is emptied out during low tide. The flowing in and flowing out water between sea and tidal basin is used to run a turbine and generate electricity.

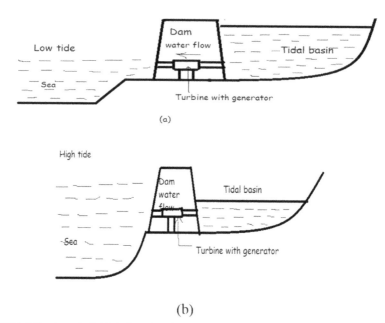

(a)

(b)

FIGURE 3.8   A typical tidal plant showing low and high tides.

3.2.6.7.1   Advantages of Tidal Energy

1. It is free from pollution.
2. It is superior to hydel energy as it does not depend on rain.
3. The tidal basin can also be used for fish farming.
4. It is best suited to meet peak power demands.

3.2.6.7.2   *Disadvantages of Tidal Energy*

1. Tidal power plant is costly compared to thermal and hydel power plants.
2. Limited locations are available for the construction of tidal power stations.
3. Power generation is not continuous and depends on the capacity of tidal basin.

3.2.6.8   *GEOTHERMAL ENERGY*

The word "geothermal" is a Greek word, meaning the heat of the earth. The temperature at earth's core is in the order of 4000°C. The internal heat energy available at a considerable depth below the surface of the earth is called as geothermal energy. It is the heat source in the form of molten rock within the earth which is called magma with temperature of about 3000°C. A geothermal power plant is shown in Figure 3.9. The water is made to flow down through a porous layer to magma heat source, where the water is converted into steam by the heat available at magma. The steam comes out through the vents of the earth surface. This steam is used to vaporize certain low-boiling refrigerant. This high-pressure refrigerant steam is used to run the turbine. The turbine runs a generator to produce electric power.

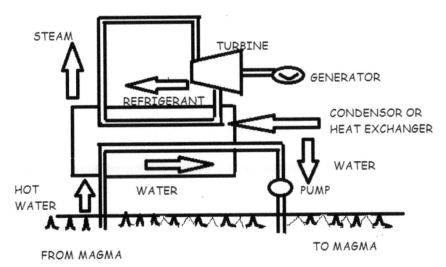

FIGURE 3.9   A typical geothermal power plant.

3.2.6.8.1 Advantages of Geothermal Energy

1. Energy is continuously available. It is more reliable.
2. It has a good potential to meet the power requirement.
3. Capital cost is low in comparison to nuclear and thermal power plants.

3.2.6.8.2 Disadvantages of Geothermal Energy

1. Components of the plants are liable to be corroded.
2. Gaseous effluent creates nuisance at the site for workers.
3. Gaseous effluent also creates thermal pollution to the environment.
4. Groundwater is likely to be polluted from gaseous effluents.

3.2.6.9 OCEAN ENERGY

The various types of energy resources which ocean can provide are the tides of the ocean can be used to generate electricity, the wind produces large waves in the ocean having high kinetic energy which can be converted into electrical power, and the temperature gradient from the surface of ocean to the great depth inside the ocean can be used to provide thermal energy to generate electricity.

The water at the ocean surface is around 25°C, while it is about 5°C at a depth of 100–200 m. Hence, there is a temperature gradient of about 20°C between these two levels and this can be used for generation of electricity by ocean thermal eddy currents (OTEC). A low boiling point liquid such as ammonia, propane, or Freon can be vaporized into high-pressure vapor using the heat of warm water available at the ocean surface into a boiler as shown in Figure 3.10. The liquid vapor is then used to run a turbine coupled with a generator to produce electricity. After expansion in the turbine, the liquid vapor is condensed into liquid in the condenser using cold water from the deep ocean at a temperature of about 5°C. The condensed liquid is pumped back to the boiler so as to be heated by warm water from the ocean surface. The cycle is repeated.

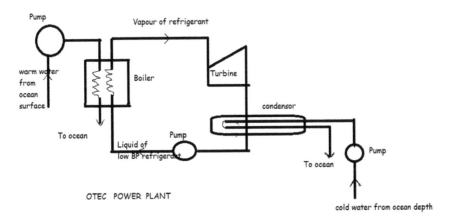

FIGURE 3.10   OTEC power plant.

3.2.6.9.1   Advantages of OTEC

1.   Power generation is continuous throughout the year.
2.   Energy is available from nature at no cost.

3.2.6.9.2   Disadvantages of OTEC

1.   It has a small temperature gradient, which gives a small thermody-
     namic efficiency.
2.   Capital cost is high due to necessity of heat exchanger, boiler and
     condenser.

3.3   PRINCIPLES OF SOLAR ENERGY

The sun is a source of enormous energy. The energy from the sun in the form
of radiation is called solar energy. All substances, solids, liquid, and gases,
above the absolute zero of temperature emit energy in the form of electromagnetic waves. This energy is called radiation. Heat transfer by radiation
can be thought as transport of energy by photons (packages or bundled of
energy) being released from excited molecules of atoms and traveling on
straight paths until they are absorbed or scattered by some other atoms. Heat
transfer by radiation is distinguished from the other forms of heat transfer
(conduction and convection) by its velocity of propagation which in a

vacuum is independent of frequency and has the value of 3.0×10^8 m/s and by the fact that no intervening medium is required for its transmission. The radiations from the sun lie within the ultraviolet, visible, and infrared (IR) spectral regions.

Radiant energy of sun (solar energy) is a virtually inexhaustible energy source, potentially capable of meeting a significant portion of the nation's future energy needs with a minimum of adverse environmental consequences. The incoming solar radiation is absorbed by a number of different processes. Some of it is absorbed by green plants for the growth of organic matter. Part of the energy absorbed by the atmosphere and the earth's surface is converted into wind, because the heating rate is not equal over the globe. Another part provides energy for the evaporation of water which consequently returns to the earth as participation and thus gives us flowing water energy. The remainder of the absorbed sunshine goes into the heating of land and water masses. Thus solar energy is the most promising of the nonconventional energy sources.

In view of this encouraging potential of solar energy, one might wonder why solar energy has not replaced conventional energy systems, not even to a significant fraction. The primary reason is that the efficient extraction of energy from the sun is difficult. Solar energy is a very dilute source, and it is available only at intermittent intervals. The most favorable sites for collecting and exploiting solar energy are confined to areas between the latitude 35° north and south of the equator. These areas receive about 2500–3500 h of sunshine per year, and the amount of solar energy incident on a horizontal surface ranges from 4 to 7 kW h/m²/day. Solar research and technology aims at finding the most efficient ways of capturing the dispersed solar energy and the developing systems to store the collected heat for use at night or during overcast period.

Another important reason is the cost of solar energy based technology, especially in comparison to the lower priced commercial fuels. Even though the amount of solar radiation reaching the earth has high energy density, yet only a small fraction of it gets captured by the collecting devices. There are also problems of storage of captured energy, and the devices needed for it are of low efficiency and high costs.

Due to these and other technological reasons, the addition of solar technologies to a residence to provide heating, cooling, or electricity can cost millions of rupees. The use of solar technology in a new home will raise the price of that home by 5% or more, which might be unacceptable to a potential buyer even though the savings on heating/cooling bills might yield payback of 5–10 years. If technological breakthroughs can bring down the cost of tapping and converting solar energy to electrical power, it would be a major boon because sun will be there and shining till long after all the fossil fuels are exhausted. Further solar energy is considered economically and politically less risky than many conventional supplies (especially oil), costs of which are subject to wide and unpredictable fluctuations. Also with solar energy, it would be possible to rapidly increase energy production to meet its rising demand because it is flexible and modular and most of the materials required for making solar collectors and controls, etc. are easily available and are not very complex to design.

The sun is the largest member of our solar system. It is a sphere of extremely hot gaseous matter with a diameter of 1.39×10^9 m, and it maintains a distance of 1.495×10^{11} m from the earth as shown in Figure 3.11. Sun's high temperature is maintained by enormous nuclear energy being released by the continuous fusion reaction. The fusion reaction involves four hydrogen atoms combining to form one helium atom and releasing 26.7 MeV of energy [$4 \times {}_1H^1_2 = He^4 + 26.7$ MeV]. The sun radiates heat energy uniformly in all directions. The radiated heat energy moves out as electromagnetic waves and increases the temperature of a body on its interception and absorption. This radiated heat energy from the sun is called **solar energy**, and it provides the energy needed to sustain life on our solar system.

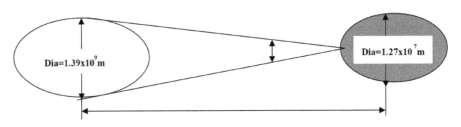

FIGURE 3.11   Radiation of heat energy from one source to other.

Sun can be considered for all practical purposes as a black body having high surface temperature of 6000 K. The radiation spectrum consists of emission at various wavelengths but more at shorter wavelengths (Fig. 3.12). The maximum emissive power of the radiation takes place at wavelength of 0.48 μm. Similarly, the earth can also be considered as a black body at temperature of 288 K. The radiation spectrum from the earth consists of emission generally of longer wavelengths. The maximum emission is taking place at wavelength of about 10 μm.

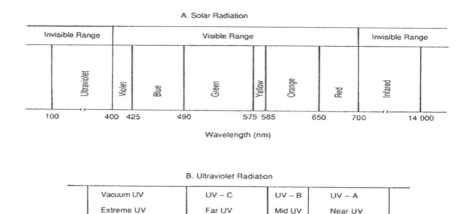

FIGURE 3.12   Wave length versus type of radiation energy.

Solar radiation is radiant energy emitted by a sun as a result of its nuclear fusion reactions.

Radiant energy is the energy of electromagnetic waves.

Irradiance is the rate at which, energy is incident on a unit surface area. It is the measure of power density of sunlight falling per unit area and time. It is measured in watts per square meter. Heat energy is measured in joules and while watt (or joules per second) is a unit of power.

Irradiation is the solar energy per unit surface area which is striking a body over a specified time. Hence, it is integration of solar illumination or irradiance over a specified time (usually an hour or kilowatt a day). It is measured in kilowatt hour or kilowatt day per square meter. For example, if irradiance is 20 kW/m² for 5 h, then irradiation is 20 × 5 = 100 kW h/m².

Solar constant is the energy from the sun per unit time, received on a unit area of surface perpendicular to the direction of propagation of the solar radiation at the mean sun–earth distance outside the earth's atmosphere. The value of solar constant based on experimental studies is 1366 W/m².

WIEN'S DISPLACEMENT LAW

The wavelength at which maximum radiation occurs for a specified temperature is given by the relation

$$\lambda_{max} T = 2897 \ \mu m \ K \qquad (3.1)$$

3.3.1 SPECTRAL DISTRIBUTION OF SOLAR RADIATION (FIG. 3.13)

Radiation type	Wavelength	Percentage of energy (%)
Ultraviolet range	0.2–0.38 μm (1 μm = 10^{-6} m)	7
Visible range	0.38–0.78 μm	47.3
Infrared radiations	0.78–3.0 μm	45.7

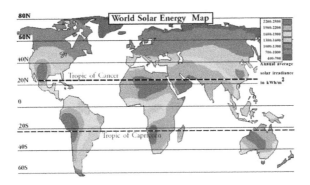

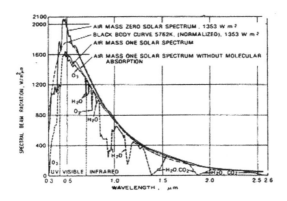

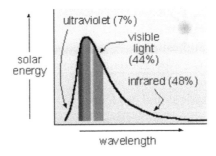

FIGURE 3.13 Spectral distribution of solar radiation.

Irradiance is given in W/m² and is represented by the symbol G. It is the rate at which radiant energy is incident on a surface per unit area of surface. Irradiation is given in J/m² and is the incident energy per unit area on a surface—determined by integration of irradiance over a specified time, usually an hour or a day. Insolation is a term specifically used to solar energy irradiation.

Of the above electromagnetic waves, we can sense only those wavelengths between approximately 0.1 and 100 μm. Electromagnetic waves falling within these limits cause our body to heat up. Hence these waves are called thermal radiation. Solar energy utilization is schematically shown in Figure 3.14.

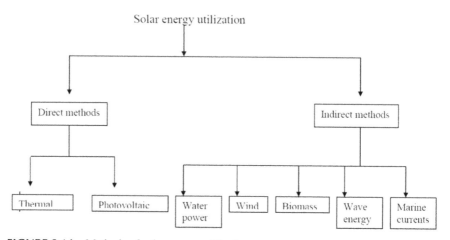

FIGURE 3.14   Methods of solar energy utilization.

3.3.2 APPLICATIONS OF SOLAR ENERGY

The applications of solar energy are as follows:

1. Heating and cooling of buildings.
2. Solar water and air heating.
3. Salt production by evaporation of seawater.
4. Solar distillation.
5. Solar drying of agricultural products.
6. Solar cookers.
7. Solar water pumping.
8. Solar refrigeration.

9. Electricity generation through photovoltaic cells.
10. Solar furnaces.
11. Industrial process heat.
12. Solar thermal power generation.

3.3.3 *PROPAGATION OF SOLAR RADIATION IN ATMOSPHERE*

While solar radiation passes through the earth's atmosphere, the following three mechanisms occur:

1. **Reflection**: This happens when sun's rays fall on bigger size particle like cloud.
2. **Absorption**: This happens due to various gases present in the atmosphere and fraction of it also absorbed by the earth's surface.
3. **Scattering**: Sun's rays are thrown in all directions.

Solar radiations while passing through the earth's atmosphere are subjected to the mechanisms of atmospheric absorption and scattering. A fraction of the radiation reaching the earth surface is reflected back into the atmosphere and is subjected to the atmospheric phenomenon like absorption and scattering again (Figs. 3.15 and 3.16). The remainder is absorbed by the earth's surface. Atmospheric absorption is due to ozone, oxygen, nitrogen, carbon monoxide, carbon dioxide, and water vapor. Scattering is due to air molecules, dust, and water droplets. The X-rays and extreme ultraviolet radiations of sun are absorbed highly by nitrogen, oxygen, and ozone. IR long wave radiations ($\lambda > 2.3$ μm) are absorbed by carbon dioxide and water vapors in atmosphere. Scattering is by air molecules, clouds and dust particles results in attenuation of radiation. In fact most of the terrestrial solar energy (i.e., energy received by earth) lies within the range of 0.29–2.3 μm.

Solar radiation is absorbed, scattered, and reflected by components of the atmosphere, including ozone, carbon dioxide, and water vapor, as well as other gases and particles. Cloud cover and local conditions such as dust storms, air pollution, and volcanic eruptions can also greatly reduce the amount of radiation reaching the surface of Earth. The two major types of radiation reaching the ground are direct radiation and diffuse radiation. *Total global radiation* is all of the solar radiation reaching Earth's surface and is the sum of direct and diffuse radiations.

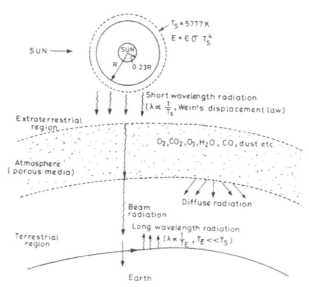

FIGURE 3.15   Position of terrestrial and extraterrestrial regions.

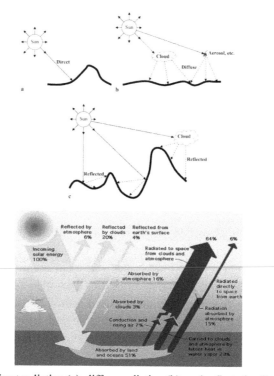

FIGURE 3.16   Direct radiation (a), diffuse radiation (b), and reflected radiation (c).

Direct or beam radiation (W/m²) is the solar radiation propagating along the line joining the receiving surface and the sun without change of direction.

Diffuse radiation (W/m²) is the solar radiation received from the sun after its direction has been changed by reflection and scattering in the atmosphere.

Total or global solar radiation (W/m²) is the sum of the beam and diffuse radiations intercepted at the surface of the earth per unit area.

Sun at zenith refers to a position of sun directly overhead. The sun's ray makes 90° angle with the horizontal surface.

Air mass characterizes atmospheric attenuation of solar radiation. It is the ratio of the length of path of the sun's ray through the atmosphere to the length of path of the sun's ray when the sun is at overhead (Fig. 3.17). It is equal to the cosecant of the solar altitude angle. Large value of air mass indicates that the solar radiation travels a longer distance at the atmosphere and hence prone to attenuation.

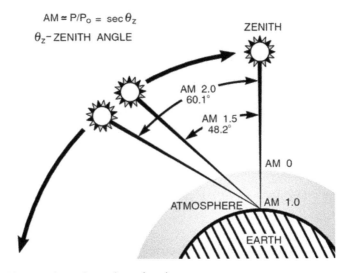

FIGURE 3.17 Air mass above the surface of earth.

The amount of solar radiation that is absorbed or scattered in the atmosphere depends on how much atmosphere it passes through before reaching earth's surface. When the sun is at zenith, the amount of atmosphere that the sun's rays have to pass through to reach Earth's surface is at a minimum. *Zenith* is the point in the sky directly overhead a particular location. The *zenith angle* is the angle between the sun and the zenith. As the zenith angle increases (the sun approaches the horizon), the sun's rays must pass through a greater amount of atmosphere to reach Earth's surface. This reduces the quantity of solar radiation and also changes its wavelength composition. Differences between beam and diffuse radiation are given below:

Beam radiation	Diffuse radiation
It is the radiation reaching directly from sun	It is the radiation reaching after attenuation, scattering and redirection
It has indirection propagation	It has uniform propagation in all directions
It is a very intense radiation and produces sun burns and shadow	It is a mild radiation
Its value is usually highest during noon	It has usually highest value during cloudy condition

3.3.4 INSTRUMENTS TO MEASURE SOLAR RADIATION AND SUNSHINE HOURS

For utilization of solar energy through solar collectors, measurements of solar radiation and recording of sunshine hours in a day are important to predict the performance of the solar energy devices. It is important to measure solar radiation owing to the increasing number of solar heating and cooling applications and necessity for accurate solar radiation data to predict performance.

3.3.4.1 WORKING PRINCIPLE OF SOLAR RADIATION MEASURING INSTRUMENTS

- Presence of thermo-electric sensing elements (thermocouple/ thermopile).
  Thermo-emf produced is proportional to the intensity of solar radiation.
- Presence of photovoltaic sensing elements (silicon solar cell).
  Electricity produced in solar cell is proportional to intensity of solar radiation.

3.3.4.2   ADVANTAGES OF PHOTOVOLTAIC SENSING ELEMENTS

1. They are less expensive than thermocouples.
2. Response is very small.
3. High current output.
4. Stable to environmental condition.

3.3.4.3   DISADVANTAGES OF PHOTOVOLTAIC SENSING ELEMENTS

It is sensitive to red and near IR components of solar radiations and insensitive to the blue and violet radiations and IR radiations of wavelengths longer than 1.2 µm.

3.3.4.4   TYPES OF SOLAR RADIATION AND SUNSHINE HOURS MEASURING INSTRUMENTS

3.3.4.1   Eppley Normal Incidence Pyrheliometer

Pyrheliometer (Greek words: 'pyr' means fire, 'helio' sun or light, 'meter' measurement) includes Eppley Normal Incidence Pyrheliometer (by Eppley Laboratories in USA) and Angstrom Pyrheliometer (after the name of Scientist K. Angstrom in 1893). Both instruments use thermoelectric-type sensing element. It is an instrument for measuring intensity of beam radiation at normal incidence (Fig. 3.18). It uses thin silver disc as absorber coated with parson optical black lacquer placed at the base of the tube. Fifteen junctions of fine bismuth–silver thermocouples are used. The cold junctions of thermopile are in contact with the brass or copper tube. A series of diaphragm inside the tube limits the aperture to a circular cone of full angle of around 5.7°. The voltage produced from the thermopile is a function of the incident radiation.

3.3.4.2   Angstrom Pyrheliometer

In this instrument, A and B, two identical blackened manganin strips, are arranged in such a way that either of them is exposed to solar radiation at the base of the collimating tube by operating a reversible shutter (Fig. 3.19). One strip is exposed to the solar radiation and the other is shaded, and

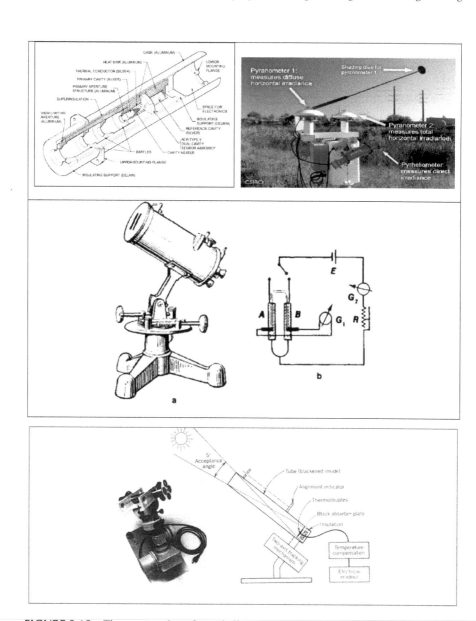

FIGURE 3.18 The construction of a pyrheliometer.

current is passed through it to heat it to same temperature as the exposed strip. Thermocouples on the back of each strip are connected in opposition through a sensitive galvanometer are used to test the equality of temperature. When there is no difference in temperature between two strips, the electrical energy absorbed by the shaded strip must be equal to the solar energy absorbed by the exposed strip.

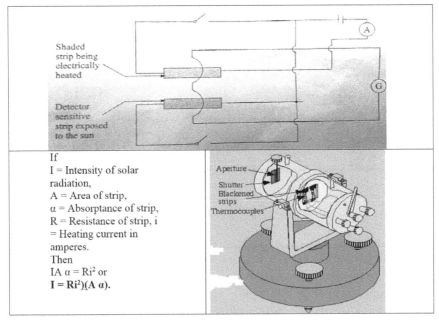

If
I = Intensity of solar radiation,
A = Area of strip,
α = Absorptance of strip,
R = Resistance of strip, i
= Heating current in amperes.
Then
IA α = Ri² or
I = Ri²)(A α).

FIGURE 3.19   Electrical circuit for Angstrom compensation pyrheliometer.

3.3.4.3   *Pyranometer*

A pyranometer is a radiation energy measuring device which is designed to measure global or total radiation, usually on a horizontal plane. When instruments shaded using a shading ring to prevent beam radiation reaching its detector, the pyranometer in this condition can measure only diffuse radiation. The instrument consists of a thermopile whose sensitive surface forms its hot junction (Fig. 3.20). This surface is blackened circular in shape and exposed to radiation. The temperature of the hot junction increases depending on amount of the incident radiation energy absorbed by it. The cold junction of the thermopile is completely shaded to prevent the radiation reaching (shown as white). The sensing element is covered by two concentric hemispherical glass domes to shield it from wind and rain. This also helps to reduce or prevent heat inflow to the detector by the convection heat transfer from air. The instrument is protected from direct solar radiation by mounting a circular guard plate at the level of detector which is also painted white to prevent the absorption of any solar radiation.

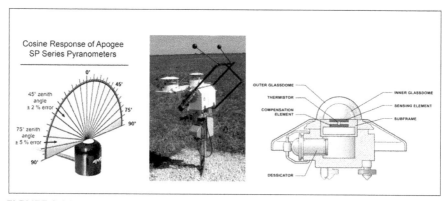

FIGURE 3.20   Construction and use of a pyranometer.

The instrument can be accurately leveled by means of three leveling screens provided at its base. Inside the instrument, a tube containing silica gel is provided to keep the interior of the instrument dry without the adverse effect of moisture on the inside of a glass of the domes. The temperature difference between the hot and cold junction is the function of radiation falling on the sensitive surface (black junction). The terms solarimeter and actinometer also mean pyranometer and pyrheliometer, respectively.

3.3.4.4   Sunshine Recorder

It is an instrument to measure the duration in hours of bright sunshine in a day. It consists of a glass sphere installed in a section of spherical metal bowl having grooves for holding a recorder card strip. The glass sphere acts as a convex sphere that focuses sun's ray to a point on the card strip held in a groove in the spherical bowl mounted concentrically with the sphere (Fig. 3.21).

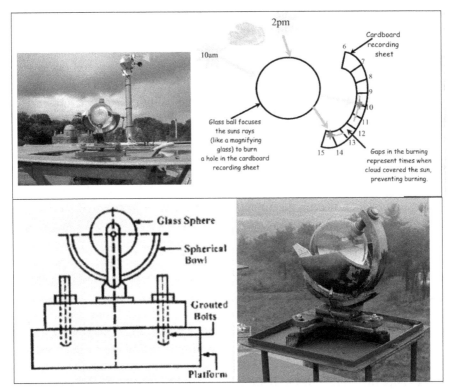

FIGURE 3.21   Sunshine recorder.

Whenever there is a bright sunshine, the image formed is intense enough to burn a spot on the card strip. As sun moves across the sky in a day, the images (burning spot) moves along the strip. Thus a burnt space, whose length is proportional to the duration of sunshine, is obtained in the strip. The burnt spots are formed in the card strip bearing a time scale.

3.4   SOLAR ENERGY AND ITS APPLICATIONS

The radiation continuously showered on earth by the sun represents the most basic and inexhaustible source of energy, which is the prime source of all forms of energy: conventional or nonconventional, renewable or nonrenewable (the only exceptions being geothermal energy and nuclear energy). Plants use solar energy to cause photosynthesis process to occur, converting the carbon derived from atmospheric carbon dioxide to plant tissues. This gives rise to plant biomass. Animals including humans use plant biomass to get food energy. The animal tissues generated via consumption of food become animal biomass. This animal biomass becomes food energy for carnivores. We also derive energy from animal biomass in terms of work we do or the work done for us by animals such as donkeys, bullocks, and horses. Besides providing food energy, plant biomass is also a major energy source: it is either directly converted into heat by burning, or is converted into chemicals (methanol, ethanol, methane, coke) which in turn, become sources of energy.

The heat from the sun also causes continuous evaporation of water from the oceans, lakes, rivers, plants, and soil. The sun also heats up the air. Due to differences in the nature of terrains, altitudes, and distances from the sun, this heating is not uniform: the air acquires different temperatures at different points horizontally as well as vertically. This leads to winds (slow and fast) providing wind energy. The evaporation of water coupled with action of winds drives what we call hydrologic cycle leading to rains and changes in weather. Water stored at an elevation when allowed to flow by gravity is a major source of hydropower or hydel energy. The waves generated in oceans by winds, and the gravitational pull of sun and moon cause wave energy. The sun heats up the top layer of oceans (up to the depths to which its light penetrates which in turn depends on how turbid the ocean water at a place is). The bottom, darker layers in oceans remain cooler than the top layers. This difference in temperatures is used to generate *OTE*. Thus, all renewable energy sources, with the exception of nuclear energy, are solar in origin. The nonrenewable sources of energy such as petroleum, coal, and lignite, are also solar in origin, because generated by the action of heat, pressure, and time on forests and animals that existed some million years ago. In this section, the utilization of direct solar energy is discussed. Solar energy is thus a clean, cheap, and abundantly available renewable energy. Energy cycle is shown in Figure 3.22. Solar energy can be used in two ways (1) directly as thermal energy and (2) indirectly using solar photovoltaic cells to convert it to electricity.

3.4.1 SOLAR COLLECTOR

Solar energy can be converted into **thermal energy** by using solar collector. It can be converted into electricity by using photovoltaic cell. Solar collector is a device for collecting solar radiation and then transferring the absorbed energy to a fluid passing through it. A solar collector absorbs solar energy in the form of heat and simultaneously transfers this to heat exchanger so that it

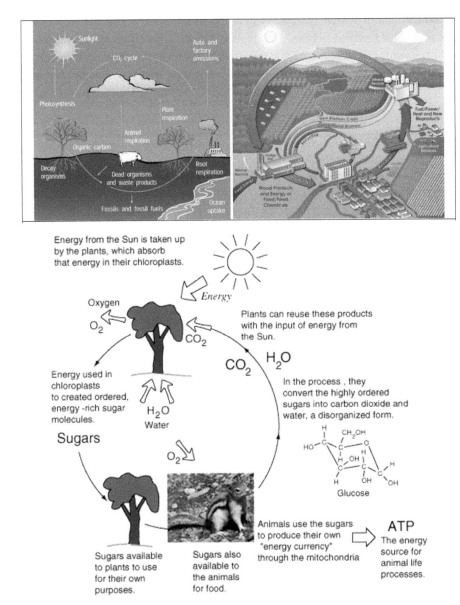

FIGURE 3.22   Energy cycle showing importance of solar energy.

can be utilized in a solar thermal system. Solar collector forms the first basic unit in a solar thermal system. Hence, it is a device for converting, trapping, or harnessing the energy present in solar radiation into more usable form. The flat plate collector is basically a heat exchanger that transfers the radiant energy of incident sunlight to sensible heat of a working fluid (liquid or air).

3.4.1.1   TYPES OF COLLECTORS

1. In **non-concentrating or flat plate type solar collector**, the area of collector to trap the solar radiation is equal to area of an absorber plate with a concentration ratio of 1. If $(T_c - T_a)$ = temperature difference between water temperature in the collector and the ambient temperature, then as $(T_c - T_a)$ increases the flat plate collector loses efficiency more rapidly. This means the flat plate collector is generally less efficient in producing water higher than 25°C above ambient, for example, for running steam turbines, etc.
2. **Concentrating or focusing type solar collector**: In such collectors, the area of collector to trap the solar radiation is more than area of an absorber plate. The ratio of effective area of the aperture to the surface area of the absorber (**concentration ratio**) is high.

3.4.1.2   ADVANTAGES, DISADVANTAGES, AND APPLICATIONS OF FLAT PLATE COLLECTOR

3.4.1.2.1   Advantages

- Both beam and diffuse solar radiations are used.
- Requires least maintenance.
- No tracking device needed.
- Mechanically simpler than focusing collectors.

3.4.1.2.2   Disadvantages

- Low temperature is achieved.
- Heavy in weight.
- Large heat losses by conduction due to large area.

3.4.1.2.3   Applications

- Domestic solar water heating.
- Suitable for low temperature power generation.

3.4.1.3   ADVANTAGES, DISADVANTAGES OF CONCENTRATING COLLECTORS

3.4.1.3.1   Advantages

- High concentration ratio.
- High fluid temperature can be allowed.
- Less thermal losses.
- Efficiency of system increases at high temperature.

3.4.1.3.2   Disadvantages

- Collect only beam radiation. Because for diffuse radiation coming from all directions, only a small portion is from the direction for which focusing occurs and rest all is lost.
- Requires costly tracking device.
- High initial cost.
- Need maintenance to retain the quality of reflecting surface against dirt and oxidation.

3.4.1.4   COLLECTORS IN VARIOUS RANGES AND APPLICATIONS (FIGS. 3.23 AND 3.24)

Low temperature, $<100°C$
   Flat plate collector: Water heating, space heating, drying.

Medium temperature, $<100–200°C$
   Cylindrical parabolic concentrator: cooking, process heating, and refrigeration.

High temperature, $>200°C$
   Parabolic mirror arrays: steam engine, Stirling engine.

3.4.1.5 DIFFERENCES BETWEEN FLAT AND FOCUSING COLLECTORS

Flat collector	Focusing collector
The absorber area is large	Absorber area is small
Concentration ratio is 1	Concentration ratio is high varying from more than 1 to even 3000
Temperature range is low, generally not more than 70°C	
	Temperature range is high, which is up to 3000°C
It uses both beam and diffuse radiation	It uses mainly beam radiation
Simple in construction and maintenance, no tracking system is required	More complicated design and difficult maintenance, tracking system is required
Less costly	
Application limited to low temperature uses	More costly
Suitable for all places as it can work in clear and cloudy days	High-temperature application such as power generation
	Suitable where there are more clear days in a year

3.4.1.6 COMPONENTS OF FLAT PLATE COLLECTOR (FIG. 3.23)

A flat plate collector consists of following essential components (Fig. 3.23):

1. **Absorber plate** is meant to intercept and absorb incident solar radiation. It is primarily a blackened heat absorbing plate. It may also be given a coating to minimize the emission of heat from its surface. It is usually made from copper, aluminum, steel, brass, silver, etc.
2. **Transparent cover** is made of one or more transparent sheets of glass or plastic. It is placed above the absorber plate. The cover allows radiation to reach the absorber plate, but it prevents any re-radiation and heat losses due to convection.
3. **Fluid tubes or channels** are arranged in thermal contact with the absorber plate so that heat can be transferred from the absorber plate to the fluid in the tubes or channels.
4. **Thermal insulation** is provided under the absorber plate and fluid tubes to minimize any heat loss by transmission or convection from the absorber plate and fluid tubes. Insulation material is Crown white wool, Glass wool, Expanded polystyrene, or foam, etc.
5. **Tight Container or box** protects all the above components of the collector. Cover plate is made from glass or Teflon, etc.

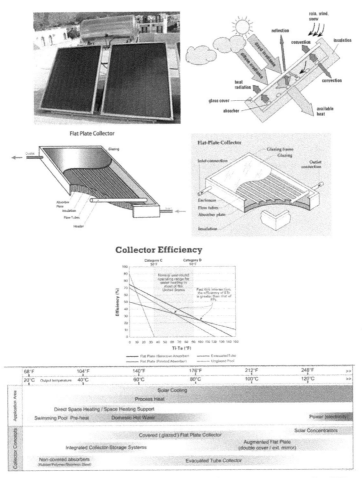

FIGURE 3.23   A typical flat plate collector (top) and temperature range for different types of collectors (bottom).

3.4.1.7   CONCENTRATING COLLECTOR

Concentrated solar power (also called concentrating solar power, concentrated solar thermal, and CSP) systems generate solar power by using mirrors or lenses to concentrate a large area of sunlight, or solar thermal energy, onto a small area (Fig. 3.24). Electricity is generated when the concentrated light is converted to heat, which drives a heat engine (usually a steam turbine) connected to an electrical power generator or powers a thermochemical reaction (experimental as of 2013). Figure 3.25 shows how does the solar energy is absorbed by an evacuated tube.

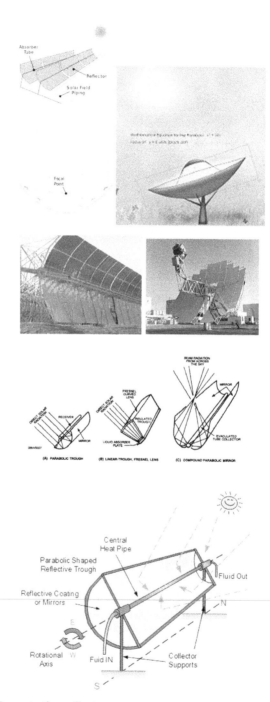

FIGURE 3.24   Concentrating collector.

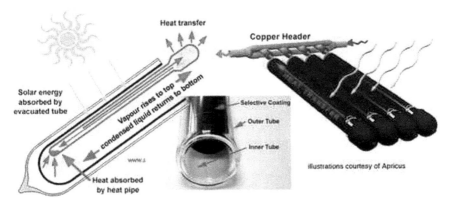

FIGURE 3.25   Solar energy absorbed by an evacuated tube.

3.4.2   SOLAR WATER HEATER

Solar water heater is a device, in which direct solar energy is utilized for sensible heating of cold water. Solar water heater is a practical application of direct solar energy utilization to heat water for a low temperature up to 60–80°C.

3.4.2.1   ADVANTAGES

- It is simple to construct and install.
- Little maintenance and no running cost.
- Retro-fittable to existing building.

3.4.2.2   OBJECTIVES

The main objective of solar water heater is to convert as much solar radiation as possible into heat at the highest attainable temperature and for lowest investment in materials and labor.

3.4.2.3   COMPONENTS OF SOLAR WATER HEATER

- Cold water overhead tank.
- Flat plate collector.

- Storage tank.
- Circulation system and auxiliary heating system.
- Control of system.

3.4.2.4   CLASSIFICATION OF SOLAR WATER HEATER (FIG. 3.26)

- Natural circulation or thermosiphoning solar water heater.
- Forced circulation type (use of water pump for forced circulation).

PRINCIPLE OF OPERATION

In natural circulation systems, the natural tendency of a less dense fluid to rise above a denser fluid can be used in a simple solar water heater to cause fluid motion through a collector. The density difference is created within the solar collector where heat is added to the liquid. In this system, as water gets heated in the collector, it rises to the storage tank and the cold water from the tank moves to the bottom of the collector, setting up a natural circulation loop. It is also called as thermosyphon water heater. Since this water heater does not use a pump, it is a passive water heater. For thermosyphon water heater, the storage tank must be located higher than the collector. Thermo-siphoning is used for circulation of liquids and volatile gases in heating and cooling applications, such as heat pumps, water heaters, boilers, and furnaces. Thermosiphoning also occurs across air temperature gradients such as those utilized in a wood fire chimney, or solar chimney.

Thermosyphon systems are simple, relatively inexpensive, and require little maintenance due to the absence of moving parts (Fig. 3.26). They are however not suitable in freezing climates, because water remains in collector always and may freeze on very cold days, resulting in the bursting of the pipes. However a valve can be added to drain the collectors when freezing temperatures occur. To provide heat during long cloudy periods, an electrical immersion heater can be used as a backup for the solar water heater system. The immersion heater is located near the top of the tank to enhance stratifica-tion and so that the heated fluid is at the required delivery temperature.

In **forced circulation solar water heater** (Fig. 3.26), the driving force for the circulation of water is not density difference of the fluid but a pump to maintain a forced circulation. A pumped system due to higher flow rates has higher efficiency. The higher is the rate of heat removal from the collector, the greater will be the absorption of energy by the collector. When a large amount of hot water is required, a natural circulation system is not suitable. Large arrays of flat-plate collectors are then used and forced circulation is maintained with a water pump. The restriction, that the storage tank should be at a higher level, is thus removed. All pumped systems have at least four basic solar components: the collector, the storage tank, the circulation pump, and the control system. The pump in forced circulation system is generally controlled by electronic differential controller which operates by turning on and turning off the pump at a predetermine temperature difference between the water tank and collector outlet temperature. In a simple direct-pumped system, there is a bank of solar collectors connected to a single hot-water storage tank.

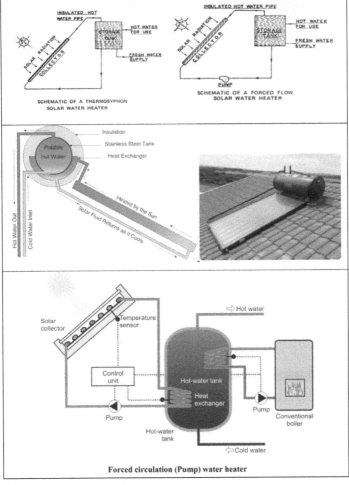

FIGURE 3.26   Types of solar water heaters.

3.4.2.1   DESIGN AND COSTING OF SOLAR HEATING SYSTEMS

The design of a solar water heater system concerns: How much hot water is required (quantity in liters), at what temperature the hot water is required, what kind of heating should be used, what should be the size of collectors, etc. The information about how much water is required depends on the end use application. For a family of 3–5 people, about 100 L of hot water is required for bathing, cooking, washing, etc. For domestic applications, water at 60°C is good enough. From the industrial point of view, temperature and amount of water requirements may be different. One has to know these two values before proceeding for the design. The basic concern in designing a solar water heater system is the quantity of water required, its temperature, size of collectors, and the type of heating. Following considerations should be taken into account while designing:

- Collector should face south.
- Ratio of height to diameter of tank = 2
- Nonreturn valve is required to prevent reverse circulation and resultant nighttime heat losses from the collector.
- Insulation thickness of storage tank >10 cm.
- In cold climates: Antifreeze liquids (ethylene glycol, glycerin, etc.) are used freezing of water in the pipes.

3.4.2.1.1   Heat Capacity of Water

If water is to be heated to a certain temperature, then we need to estimate how much heat should be added to water to reach the desired temperature. Water has a heat capacity of 4.18 kJ/kg°C. Therefore, if we need to raise the temperature of 1 kg water (=1 L of water) at some temperature by 1° (say, from 25 to 26°C), a heat of 4.18 kJ should be added to water. Based on this information we can calculate how much heat should be added to raise the temperature of m kg water by ΔT temperature (where ΔT = required temperature − ambient water temperature):

Therefore, heat requirement, $Q = 4.18 \times m \times \Delta T$, in kJ

If fluid other than water is used, then corresponding heat capacity should be used. Assumptions for designing the system are

- Hot water requirement of 250 L.

- Water (or other fluid) required at a temperature of 80°C (may be 120°C for other than solar water heating requirement).
- Ambient water temperature of 30°C.
- Daily solar radiations of 5.5 kW h/m² day.
- Efficiency of solar water heater of 0.40 (efficiency can range between 40% and 60% for flat plate collector).
- Heat capacity of fluid (water in this case) = 4.18 kJ/kg K.

STEP 1: FIND OUT USEFUL SOLAR ENERGY

Daily available solar radiation = 5.5 kW h/m² day
    Useful solar heat energy (which is actually used in raising temperature of water)

> = Daily solar radiation × solar water heating efficiency = 5.5× 0.4 = 2.2 kW h/m² day; = 7920 kJ/m² day using 1 W h = 3600 J

STEP 2: HEAT REQUIRED BRINGING THE WATER FROM THE AMBIENT TO DESIRED TEMPERATURE

The ΔT or rise in temperature required = Required temperature − ambient water temperature = 80 − 30 = 50°C. The mass of water (1 L = 1 kg) = 250 kg
    Heat energy required to raise water temperature to desired level, everyday

$$= 4.18 \times m \times \Delta T, KJ = 4.18 \times 250 \times 50 = 52250 \text{ kJ/day}$$

STEP 3: TOTAL AREA OF THE COLLECTOR REQUIRED TO FULFILL THE REQUIREMENT

Collector area required = required heat energy/useful heat energy = 52,250/7920 = 6.60 ≈ 7 m²

3.4.2.2   COST OF SOLAR HOT WATER HEATER

In order to estimate the cost of solar water heaters, following thumb rules can be used:

- The typical price of a collector is about Rs. 5000 per square meter.

- Tank and other installation cost is about 50% of the collectors cost.
- 50 L of hot water per day requires 1 square meter of collector area.
- Thus, cost of a 100-L hot water per day system = 100/50 (5000 + 1/2 5000) = Rs. 15,000.

3.4.2.2.1 Simple Payback Period

The life of a water heating system is between 15 and 20 years. During this period significant amount of electricity can be saved, which in turn is saving of money. Let us consider 100 L of daily hot water requirement (at 60°C). As per above calculations, the cost of the water heating system will be Rs. 15,000. How much electricity is required per day to heat 100 L of water to 60°C (ambient water temperature is 30°C)?

The heat energy requirement = 4.18 × 100(60 − 30) = 12,540 kJ.

Electrical units required per day = 12,540/3600 = 3.48 kW h, 1 W h = 3600 J.

Efficiency of electric water heater = 0.90 (90%).

Actual electrical energy = 3.48/0.90 = 3.87 kW h.

Consider cost of an electrical energy unit (kW h) is Rs. 4, thus

Total cost of electricity per day = 4 × 3.87 = Rs. 15.20

Let us assume, there are 300 sunny days in a year, thus yearly savings will be

= 15.2 × 300 = Rs. 4650

Number of years in which the investment cost is earned/saved = Investment/Savings =15,000/4650 = 3.28 years is the payback period = within 3–4 years for a hot water system.

3.4.3 SOLAR COOKER

Solar cooker is used to cook food by utilizing the heat radiations coming from the sun. It is also an important thermal application of solar energy on small scale particularly for domestic cooking because: Firewood used for cooking causes deforestation, commercial fuels are not available in sufficient quantities, dry cow dung and agricultural wastes used for cooking are good fertilizer, and labor used for collecting fuel can be diverted for some other useful jobs.

ADVANTAGES OF SOLAR COOKER

- No orientation to sun is needed.
- No attention is needed during cooking.
- No fuel, maintenance and recurring cost.
- Simple to use and fabricate.
- No pollution.
- No loss of vitamins in the food.

DISADVANTAGES OF SOLAR COOKER

- Cooking can be done only when there is sunshine.
- Quick cooking is not possible.
- Comparatively it takes more time.
- All types of foods cannot be cooked.

3.4.3.1 TYPES OF SOLAR COOKER

3.4.3.1.1 Box-type Solar Cooker

It consists of an outer box, a blackened aluminum tray, a double glass lid, a reflector, insulation, and cooking pot (Fig. 3.27). The blackened aluminum tray is fixed inside the box and the sides are covered with an insulation cover to prevent heat loss. The reflecting mirror provided on the box cover increases the solar energy input. Metal pots are painted black on the outside. Food to be cooked is placed in cooking pots and cooker is kept facing the sun to cook the food. A typical box-type solar cooker can have

- The solar rays penetrate through the glass covers and absorbed by blackened metal trays (boxes) kept inside the cooker.
- The upper cover has two glass sheets each 3 mm thick fixed in the wooden frame with 20 mm distance between them.
- The outer box size 60 × 60 × 20 cm. Inner box 50 × 50 × 14 cm.
- Overall dimensions of the latest model are 60 × 60 × 20 cm.

- Insulation thickness: 5 cm.
- Cost Rs. 2500 per piece of box-type cooker.
- Dish cooker cost Rs. 6000–8000.
- Can be used for boiling food (rice and pulses).
- Time taken for cooking = 1.5–3.0 h.
- Temperature achievable 100–120°C.
- No frying of food possible as temperature is not so high.
- This type of cooker is termed as family solar cooker as it cooks sufficient dry food materials for a family of 5–7 people.

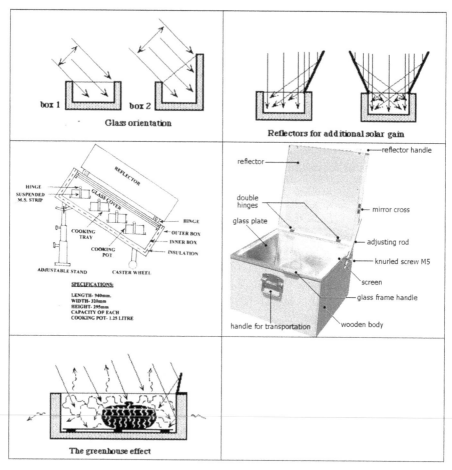

FIGURE 3.27   Box type solar cooker.

3.4.3.1.2   Dish Solar Cooker (Fig. 3.28)

Instead of a flat collector, parabolic dish is used to concentrate the incident solar radiation

3.4.3.1.3   Community Solar Cooker for Indoor Cooking (Fig. 3.28)

A community solar cooker is a parabolic reflector cooker. It has large reflector ranging from 7 to 12 square meter of aperture area. The reflector is outside the cooking space so as to reflect the solar rays into the cooking area. A secondary reflector further concentrates the rays on to the bottom of the cooking pot that is painted black. Temperature can reach up to 400°C and food can be cooked for >50 persons.

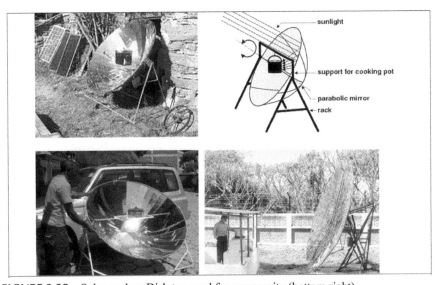

FIGURE 3.28   Solar cooker: Dish type and for community (bottom right).

3.4.3.2   PARTS OF SOLAR COOKER (FIG. 3.27)

The important parts of a hot box solar cooker are the outer box, inner cooking box or tray, the double glass lid, thermal insulator, mirror, and cooking containers.

Outer box of a solar cooker is generally made of GI or aluminum sheet or fiber reinforced plastic.

Inner cooking box (Tray) is made from aluminum sheet. The inner cooking box is slightly smaller than the outer box. It is coated with black paint so as to easily absorb solar radiation and transfer the heat to the cooking pots.

Double glass lid covers the inner box or tray. This cover is slightly larger than the inner box. The two glass sheets are fixed in an aluminum frame with a spacing of 2 cm between the two glasses. This space contains air which insulates and prevents heat escaping from inside. A rubber strip is affixed on the edges of the frame to prevent any heat leakage.

3.4.3.3 TYPICAL MATERIALS USED FOR MAKING A BOX-TYPE SOLAR COOKER

System components	Specifications		Comments
	Material	**Properties**	
Reflector Flat mirror or metal foil	Glass, Aluminum	Reflectivity should be high to ensure the increase in temperature of cooking utensil	Mirrors are more reflective but costlier and fragile
Box a. Outer box b. Inner box	GI, Aluminum sheet, FRP Aluminum or copper (high heat conductivity)	Enough to maintain stability and insulation Coated with nontoxic black paint to absorb the heat	Low cost materials like cardboard can be used. Paint must be nontoxic when dry
Transparent cover	Glass or plastics	Glass traps the infrared radiation, prevents heat loss from top	Glass cover should be well sealed to prevent heat loss from the gap
Insulation	Wool, cotton, feathers, or even crumpled newspaper	Space between the outer and inner box including bottom of the tray must be insulated to reduce heat losses from the cooker	Must be free from volatile materials
Cooking utensil	Aluminum , copper or stainless steel	Must be coated with black material to black absorb more heat	Lightweight, shallow pot must be used

3.4.3.4   PRINCIPLE OF OPERATION

A solar cooker consists of (1) an insulated box of blackened aluminum in which utensils with food materials can be kept; (2) reflector mirror hinged to one side of the box so that the angle of reflector can be adjusted; and (3) a glass cover consisting of two layers of clear window glass sheets which also serves as the box door as shown in Figure 3.27. The box is kept in such a way that solar radiation falls directly on the glass cover and reflector mirror is also adjusted in such a way that additional solar radiation after mirror reflection is also incident on glass cover. The glass cover traps heat owing to the greenhouse effect, that is short wavelengths radiation can pass inside the box but long-wavelengths radiation coming out from the box is entrapped in the enclosure, thereby providing more heating effect. The air temperature obtained inside the box ranges from 140 to 160°C. This provides sufficient heat for boiling and cooking purposes.

Cookers with reflectors on all four edges can also be fitted to enhance the temperature rise inside the box. With parabolic disc concentrator type solar cooker, temperatures of the order of 450°C can be obtained in which solar radiations are concentrated on to a focal point. A box cooker can be made out of any type of material, such as wood, plastic, cardboard, etc. Usually two boxes of varying size are needed in order to be able to fit the smaller one inside of the larger one, thus creating a gap or space around the smaller box which can then be filled with paper, hay, fiberglass, etc. to form an insulating barrier between the two boxes to prevent the escape of heat through conduction.

3.4.2.5   ENERGY REQUIRED FOR COOKING

3.4.2.5.1   Estimation of How Much Energy is Required to Cook Half a Kg of Rice

First step is to find out how much mass needs to be heated? It should include mass of vessel, mass of water, and mass of rice.

- Let us say, vessel is made of aluminum (heat capacity 0.9 kJ/kg°C) and weigh 0.5 kg. Therefore, heat energy is required to heat vessel for per degree rise in temperature = mass × heat capacity = 0.5 × 0.9 = 0.45 kJ/°C.

- Water used for cooking rice is 0.5 kg (heat capacity 4.18 kJ/kg°C). Therefore, energy required for per degree rise of water temperature is = 0.5 × 4.18 = 2.09 kJ/°C
- Half-a-kg rice needs to be cooked. Considering the heat capacity of rice same as water, the heat energy required for per degree rise of rice temperature is = 0.5 × 4.18 = 2.09 kJ°C.
- Total heat energy required for per degree rise of whole mass (vessel + water + rice) = 0.45 + 2.09 + 2.09 = 4.63 kJ/°C (=this is total $m \times c_p$ of the system).
- Total heat energy required to heat the whole mass to the boiling point = $m \times c_p \times$ (final temperature − ambient temperature) = 4.63 × (100 − 30) = 324.1 kJ.
- After reaching the boiling temperature, heat is needed to maintain the temperature against heat losses, till the food is cooked. This energy requirement is not much and can be assumed as 10–30% of heat supplied for raising the temperature of whole mass to boiling temperature. Let us consider it as 25%. Including this heat, the total energy required now = 324.1 + 0.25 × 324.1 = 405.12 kJ.

In this way, energy required for cooking of any amount of food can be estimated. This estimated value can be used to find out the required size of the cooker or to find out how much heat is needed to cook the food in a given cooker.

3.4.3.6   ESTIMATING THE TIME REQUIRED TO COOK FOOD

3.4.3.6.1   Assumptions

- Solar radiation at a given instant: 800 W/m² (actual instantaneous solar radiation varies between 100 (in the early morning and evening) to 1000 W/m² (at noon).
- Solar cooker cross-section area = 0.5 × 0.5 = 0.25 m².
- Area of the reflector (bringing extra light to box) = 0.5 × 0.5 = 0.25 m².
- Efficiency of solar cooker = 0.20 (efficiency can vary between 20% and 50%).
- Total energy required to cook food (half kg rice and aluminum vessel as estimated in previous section) = 405.12 kJ.

Step 1: Find out energy required to cook food
        Total energy required to cook food = 405.12 kJ = 405,120 J.

Step 2: Find out actual solar radiation available for heating
        Solar radiation at a given instant = 800 W/m².
        Area of the cooker + area of the reflector = 0.25 + 0.25 = 0.5 m².
        Efficiency of the solar cooker = 0.20.
        Actual wattage available inside cooker for cooking = radiation intensity per unit area × cooker area × cooker efficiency = 800× 0.50 × 0.20 = 80 W.
        (Thus, a box-type solar cooker will work, if there is an 80-W electrical heating coil kept inside the box).

Step 3: Total time required to cook food
        Total time required for cooking = (total energy required to cook food)/(actual wattage available) = (405,120)/(80 × 3600) = 1.40 h (using 1 W h = 3600 J).

Note: It is important to use reflector in a box-type cooker properly, as without the effect of reflector, the time required to cook food would be 2.8 h. This is due to half the light collection area. Proper use of reflector requires correct placing of cooker with respect to sun's position. Reflector is adjusted so that it reflects light into the box. It may require a few adjustments because of the motion of the sun.

3.4.3.7 *SIMPLE PAYBACK PERIOD*

Use of solar cooker can replace use of firewood, kerosene, LPG, and electric cooking. Depending on which fuel it is replacing, payback period varies. Step-by-step calculation for replacing LPG cylinders is shown below:

3.4.3.7.1 *Energy Content of an LPG Cylinder*

The calorific value of LPG (mixture of propane and butane) is 12.6 kW h/kg. A LPG cylinder typically will have 14 kg of gas, thus the total heating capacity of a full cylinder = 12.6 × 14 = 176.4 kW h/cylinder = 635,040 kJ/cylinder.

- But only 60% of this energy is used (other part goes as waste), therefore, useful energy = 635,040 × 0.6 = 381,024 kJ/cylinder.
- Cost of an LPG cylinder is Rs. 400.

3.4.3.7.2 Energy Produced by a Box-type Solar Cooker

- Heat energy required for cooking for single person in a day = 900 kJ.
- A box-type solar cooker typically can cook for about three people, but it can cook for, say, only about 40% of the meal (other part of the meal is considered as frying based food). Considering cooking two meals per day, the heat energy produced by solar cooker per day = 3 × 2 × 0.4 × 900 = 2160 kJ/day (as per the calculation shown under time of cooking estimation section. (This value is equivalent to about 5.5 h of use of solar cooker.)
- Typical cost of a box-type cooker is about Rs. 2000.
- Number of days required by solar cooker to produce energy equivalent of a LPG cylinder = Cylinder's total useful energy/Energy produced by cooker per day
- = 381,024 kJ/2160 kJ/day = 176.4 days (no. of cylinders saved in year = 2.06).
- Payback period = Cost of solar cooker/Cost energy saved per year = 2000/828 = 2.4 years.

It can be seen from above calculations that the payback period for solar cooker is about 2–3 years. Though, a very simple analysis of payback period calculation is done, that is, without considering inflation, interest rate, inflation in energy cost, maintenance cost, etc. But it is a good approximation. Payback period of a solar cooker depends on the type of fuel it is replacing. As compared to LPG cylinder, the payback period is less than 2.5 years.

3.4.4 SOLAR STILL

The process to convert saline water into pure water using solar radiation is called solar distillation. A solar device used for this purpose is called solar still.

Water plays a key role in the development of an economy and in turn for the welfare of a nation. Nonavailability of drinking water is one of the major problems faced by both the underdeveloped and developing countries all over the world. Around 97% of the water in the world is in the ocean, approximately 2% of the water in the world is at present stored as ice in polar region, and 1% is fresh water available for the need of the plants, animals, and human life. Today, majority of the health issues are owing to the nonavailability of clean drinking water. In the recent decades, most parts of the world receive insufficient rainfall resulting in increase in the water salinity. The pollution of water resources is increasing drastically due to a number of factors including growth in the population, industrialization, urbanization, etc. These activities adversely affect the water quality in rural areas and agriculture. A lot of time is spent each day, mostly by females, to collect water from distant, often polluted sources. Similarly, a large number of people die each year from water-related diseases. Majority of the rural people are still unaware of the consequences of drinking untreated water.

Desalination is the oldest technology used by people for water purification. Solar energy is available in abundant in most of the rural areas, and hence solar distillation is the best solution for rural areas and has many advantages of using freely available solar energy. Distillation has long been considered a way of making salt water drinkable and purifying water in remote locations.

A **solar still** operates similar to the natural hydrologic cycle of evaporation and condensation. The basin of the solar still is filled with impure water and the sun rays are passed through the glass cover to heat the water in the basin and the water gets evaporated. As the water inside the solar still evaporates, it leaves all contaminates and microbes in the basin. The purified water vapor condenses on the inner side of the glass, runs through the lower side of the still, and then gets collected in a closed container.

3.4.4.1 CLASSIFICATION OF SOLAR STILL

The solar distillation systems are mainly classified as passive solar still and active solar still. The numerous parameters, affecting the performance of the still, are water depth in the basin, material of the basin, wind velocity, solar radiation, ambient temperature, and inclination angle. The productivity of any type of solar still will be determined by the temperature difference between the water in the basin and inner surface glass cover.

In a **passive solar still**, the solar radiation is received directly by the basin water and is the only source of energy for raising the water temperature and consequently, the evaporation leading to a lower productivity. Passive system in which solar energy collected by structure elements (basin liner) itself for evaporation of saline water. The simple solar still is shown in Fig. 3.29. The sun's energy in the form of short electromagnetic waves passes through a clear glazing surface such as glass. Upon striking a darkened surface, this light changes wavelength, becoming long waves of heat, which is added to the water in a shallow basin below the glazing. As the water heats up, it begins to evaporate. The warmed vapor rises to a cooler area. Almost all impurities are left behind on the basin. The vapor condenses onto the underside of the cooler glazing and accumulates in to water droplets or sheets of water. The combination of gravity and the tilted glazing surfaces allows the water to run down the cover and into a collection trough, where it is channeled in to storage. The drawback of the lower yield of purified water in a passive solar still is solved to some extent by developing various design of active solar still (Fig. 3.29).

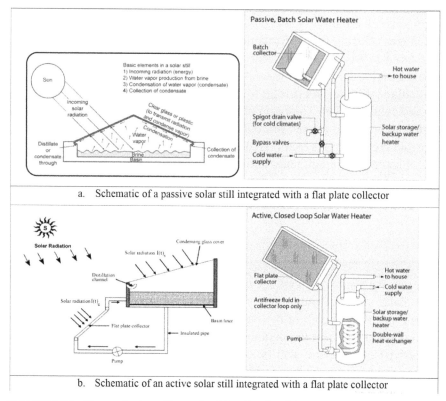

FIGURE 3.29   Types of solar still: passive (a) and active (b).

In **active solar still**, an extra thermal energy is supplied to the basin through an external mode to increase the evaporation rate and in turn improve its productivity. The active solar distillation is mainly classified as follows:

- High-temperature distillation: Hot water will be fed into the basin from a solar collector panel.

- Preheated water application: Hot water will be fed into the basin at a constant flow rate.
- Nocturnal production: Hot water will be fed into the basin once in a day.
- The differences between passive and active solar stills are shown in Figure 3.30.

3.4.4.2 PRINCIPLE OF OPERATION OF A SOLAR STILL

A solar still consists of a shallow blackened basin filled with saline or brackish water to be distilled. It is covered with sloping transparent roof as shown in Figures 3.29 and 3.30. The sun rays can pass through transparent roof, and these rays are absorbed by the blackened surface of the basin thereby increasing the temperature of water. The water in basin evaporates due to solar heat and rises to the roof. The water vapor cools down and condenses at the undersurface of the roof. The water drops or condensed water slip down along the sloping roof. The condensed water is collected by the condensate channel and drained out from the solar still and used as drinkable water.

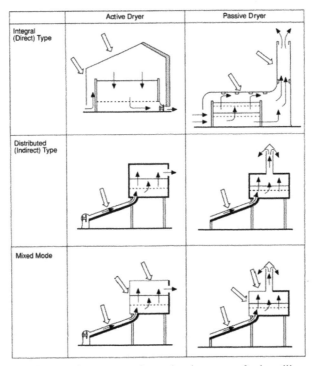

FIGURE 3.30 Differences between passive and active types of solar stills.

3.4.4.3 COMPONENTS OF A SOLAR STILL

System components	Specification	
	Material	**Purpose**
Water basin/Tank	Cement concrete or fiberglass	Container of saline water (brick-cement can be used for basin, tank should be insulated at the base to avoid thermal loss)
Insulation	PUF, putty, tars, silicon sealant.	To prevent heat losses (should not become brittle, cracks may from resulting in vapor leakage)
Transparent cover	Glass or polyethylene	To transmit solar energy (glass is heavy, prone to mechanical damage, polyethylene has low transmissivity and requires a special layer to make it water wettable
Absorber	Black butyl rubber, black polyethylene, or ink or dye	To absorb the heat (should be able to withstand temperature up to 100°C)
Condensate channel	Aluminum, galvanized iron	To collect droplets of water
Make-up water inlet	PVC pipe	To supply the saline water
Excess make-up water outlet	PVC pipe	To drain the water
Storage pot	Glass, plastic	To store distilled water

PUF: Polyurethane foam.

3.4.4.4 ENERGY REQUIREMENT AND EFFICIENCY OF A SOLAR STILL

The energy required to evaporate water is the latent heat of vaporization of water. *The latent is the heat that is required by a substance to change its phase, from solid to liquid or liquid to gas or vice-versa.* The latent heat of evaporation of water is about 2260 kJ/kg. This implies that in order to produce 1 L of water through distillation, heat input of 2260 kJ is required (assuming water density as 1 kg/L). In practice, one has to supply much higher amount of heat energy because the efficiency of the still will not be 100%.

There are several possible heat loss paths in solar still, such as heat could be lost due to vapor leakage, heat loss at the bottom and sidewalls of the still through conduction, heat loss due to radiation, heat loss due to convection from the glass cover, heat absorbed by the cover, etc. The overall heat loss can be 40–60%. Based on this information, the efficiency of solar still can vary from 30% to 45%.

3.4.4.5 SIZING OF THE SOLAR STILL

Sizing of the solar still is very simple. It involves finding out how much distilled water we will get per day from one square meter area and how much total area is required to fulfill our requirement. Following assumptions are made:

- Latent heat of water evaporation: 2260 kJ/kg (heat required to changes water from liquid to steam phase).
- Density of water: 1 kg/L.
- Efficiency of solar still: 40%.
- Average daily solar radiation on a given location: 5 kW h/m² day (data varies from one location to another and one season to another).
- Amount of distilled water required per day: 20 L.

Step 1: *Find out useful solar radiation*
Daily available solar radiation = 5 kW h/W m² day
Useful solar radiation (which is actually converting radiation heat energy to distilled water) = Daily solar radiation × solar distill efficiency = 5 × 0.4 = 2.0 kW h/m² day
= 7200 kJ/m² day (conversion factor: 1 W h = 3600 J)

Step 2: *Liters of distilled water produced per day per square meter*
Latent heat of water evaporation = 2260 kJ/kg
Number of liters of distilled water produced per square meter per day = useful solar radiation/Latent heat of water evaporation = 7200/2260 = 3.18 L/m² day

Step 3: *Total area of the solar distill to fulfill the requirement*
Total distilled water requirement per day = 20 L/day
Total area of the solar distill to fulfill family requirement = Total daily requirement/Number of liters produced per day per square meter = 20/3.18 = 6.28 m²
Alternatively, the total solar distill area required to fulfill daily requirement can also be calculated using following formula: $A = Q \times 2.26/G \times E$. where, A = total distill area required; Q = total distilled water; G = daily solar radiation in MJ/m² day; E = efficiency of solar distill in decimals.

3.4.4.6 PARAMETERS AFFECTING THE SOLAR STILL PERFORMANCE

The performance of a solar still is governed by parameters such as: climatic conditions, design parameters, and operating parameters. The climatic parameters consist of the amount of solar insolation, ambient air temperature, wind speed, sky conditions, etc. The design parameters consist of orientation of the still, thermophysical properties of materials used in the still, tilt angle of glass cover, insulation at the base, etc. The operating parameters consist of water depth in the basin, water salinity, initial water temperature, etc. Effects of some major parameters are as follows:

Solar insolation: The solar still output is a function of the amount of solar radiation falling on a given location. For a given temperature, higher solar radiation results in higher amount of distilled water. Thus, more is the solar insolation, more is the output of solar still.

Wind velocity: High wind speed will cause higher heat losses from the still, which will result in lower frequency, meaning less distilled water. Thus, lesser the wind speed more is the output of solar still.

Ambient air temperature: Increase in ambient air temperature will increase the output of solar still because heat losses will decrease. Thus, higher the ambient air temperature, more is the output of solar still.

Brine depth: As brine depth increases, the solar still productivity decreases. Thus, lesser the brine depth, more is the output of solar still.

Slope of the glass cover: Lower glass cover slope increases the output but for practical reasons it should be more than 10° because for much smaller angle the water droplets will fall in the basin. Thus, slope of solar still must be 20–30° to the horizontal.

Glass material and angle: The slope of the cover must be sufficient to ensure that the condensed water runs smoothly through the condensate channel without forming large droplets. The glass should be low absorption glazing type.

3.4.4.7 COST OF SOLAR STILL

Payback period can be calculated using the following sample calculation for distilled water. A solar still with 40% efficiency will produce about 3.18 L/day per square meter, if the daily solar radiation is 5 kW h/m² day.

3.4.4.7.1 Estimation of Payback Period for Solar Still Unit

Daily distilled water production per unit area = 3.18 L/day.

Cost of manufacturing of solar still per one square meter area = Rs. 2000/m².

Cost of distill water in the market = Rs. 15/L.

Worth of distill water produced everyday (gain) = 3.18 × 15 = Rs. 47.7/day.

The number of days required to recover the cost of solar still = cost of manufacturing/cost gain per day = 2000/47.7 = 52.31= 42 days.

Thus, it can be seen that the payback period for solar still is within one and half months.

3.4.4.8 GENERAL MAINTENANCE

1. *Cleaning the basin and flushing:* As water evaporates from the solar still basin, salts and other contaminants are left behind. Over time, these salts can build up to the point of saturation if the still is not properly maintained and flushed on a regular basis. Adequate operation of a still requires about three times as much make-up water as the distillate produced each day. If the still produces 2 L of water, then 6 L of make-up water should be added. The additional 4-L water leaves the still as excess make-up water. The excess make-up water flushes the still basin through the overflow to prevent salt buildup.

2. *Cleaning the glass and condensate channel:* The glass cover must be cleaned periodically to remove the dust from the cover, which otherwise absorbs some part of the solar radiation reducing the efficiency of the solar still. Care must be taken to avoid the breakage of glass, which may lead to vapor leak. Condensate channel must also be cleaned periodically while draining out the basin water.

3.4.5 SOLAR DRYER

Drying is a process of moisture removal due to simultaneous heat and mass transfer. Drying by exposure to the sun is one of the oldest methods using solar energy, for food preservation, as vegetables, fruits, fish, meat, etc. Since the prehistoric times, mankind used the solar radiation as the only available thermal energy source to dry and preserve all necessary foodstuffs for future time, to dry soil bricks for homes and animal skins for dressing. Reducing moisture content of foodstuff down to a certain level slows down the action of enzymes, bacteria, yeasts, and molds. Thus food can be stored and preserved for long time without spoilage. In direct solar drying called "sun drying," the product is heated directly by the sun's rays and moisture is removed by natural circulation of air due to density differences. Two basic moisture transfer mechanisms are involved in drying:

    a.   Migration of moisture from the mass inside to the surface.
    b.   Transfer of the moisture from the surface to the surrounding air, in the form of water vapor.

Artificial mechanical drying is energy intensive and expensive, and ultimately increases the product cost. Solar-drying technology offers an alternative which can process the vegetables and fruits under clean, hygienic, and sanitary conditions with lowest energy costs. It saves energy, time, occupies less area, improves product quality, makes the process more efficient and protects the environment. Solar drying can be used for the complete drying process or as a supplement to artificial drying systems, in the latter case reducing the fuel energy required. Solar dryer technology can be used in small-scale food processing industries to produce hygienic, good quality food products.

3.4.5.1   REQUIREMENTS OF SOLAR DRYER

3.4.5.1.1   Problems in Open Sun Drying

Open sun drying (OSD) though very simple, easy, and cheap, but it has following disadvantages:

    a.   Contamination of product due to dirt and insects.
    b.   Wastage by birds/mice.
    c.   Spoilage due to sudden and unpredicted rain.
    d.   No control over crop drying. Over drying results in loss of germination power, nutritional change, and sometimes complete damage.
    e.   Under drying results in deterioration of food due to growth of fungi and bacteria.

3.4.5.1.2 Postharvest Losses

In developing countries where cold storage facilities are not adequately available, postharvest losses are between 30% and 50% in perishables. These food losses can be reduced in developing countries if efficient as well as improved harvesting, storing, and handling practices are followed. Drying is an important way of preserving grain for later use. Due to the factors under A and B above, solar dryer is an easy alternative for reducing postharvest losses and improved preservation.

3.4.5.2   PARAMETERS CONSIDERED FOR A SOLAR DRYER

a. It depends on external parameters such as temperature, humidity, and velocity of the air stream.
b. It depends on drying material properties like surface characteristics (rough or smooth).
c. It depends upon chemical composition (sugar starches, etc.).
d. Physical structure (porosity, density, etc.).
e. Size and shape of the product.
f. Material characteristics (hygroscopic and non-hygroscopic in nature).

Therefore, solar dryer should be designed keeping all above drying parameters in mind. Agricultural drying process depends on the objectives, type of crop, and choice. Examples are presented for coffee and cassava shown in Figure 3.31.

3.4.5.2.1 Steps in Drying

a. Heat by convection and radiation to the surface of the product.
b. Heat enters into the interior of product.
c. Increases the temperature of the product.
d. Formation of water vapors.
e. Moisture is evaporated from the surface of product.

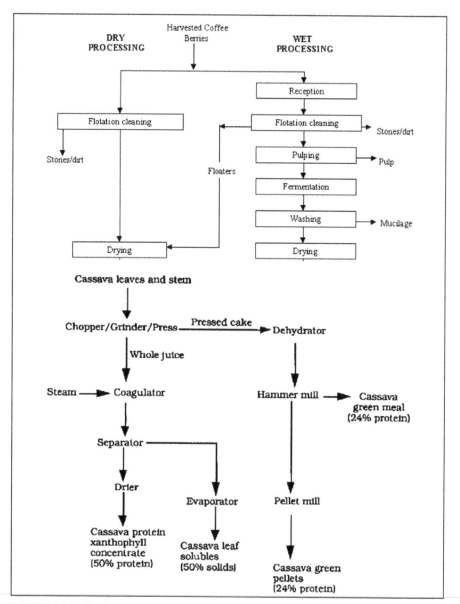

FIGURE 3.31   Drying process for coffee and cassava products.

3.4.5.3   PRINCIPLE OF OPERATION OF A SOLAR DRYER

Solar dryers can broadly be categorized into direct, indirect, and specialized solar dryers. **Direct solar dryers** have the material to be dried placed in an enclosure, with a transparent cover on it. Heat is generated by absorption of solar radiation on the product itself as well as on the internal surfaces of the drying chamber.

In **indirect solar dryers**, solar radiation is not directly incident on the material to be dried. Air is heated in a solar collector and then ducted to the drying chamber to dry the product. Specialized dryers are normally designed with a specific product in mind and may include hybrid systems, where other forms of energy are also used. Although indirect dryers are less compact when compared to direct solar dryers, yet they are generally more efficient. Hybrid solar systems allow for faster rate of drying by using other sources of heat energy to supplement solar heat. The three modes of drying are (1) open sun, (2) direct, and (3) indirect in the presence of solar energy. The working principle of these modes mainly depends upon the method of solar energy collection and its conversion to useful thermal energy.

3.4.5.3.1   Open Sun Drying

Figure 3.32 shows the principle of OSD. The short wavelength solar energy falls on the uneven crop surface. A part of this energy is reflected back and the remaining part is absorbed by the surface depending upon the color of crops. The absorbed radiation is converted into thermal energy and the temperature of crop starts increasing.

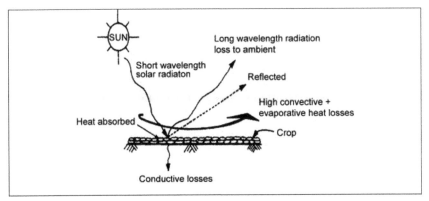

FIGURE 3.32   Principle of open sun drying.

This results in long wavelength radiation loss from the surface of crop to ambient air through moist air. In addition to long wavelength radiation loss, there is convective heat loss due to the blowing of wind through moist air over the crop surface. Evaporation of moisture takes place in the form of evaporative losses and so the crop is dried. Further a part of absorbed thermal energy is conducted into the interior of the product. This causes a rise in temperature and formation of water vapor inside the crop and then diffuses towards the surface of the crop and finally loses the thermal energy in the form of evaporation. In the initial stages, the moisture removal is rapid since the excess moisture on the surface of the product presents a wet surface to the drying air. Subsequently, drying depends upon the rate at which the moisture within the product moves to the surface by a diffusion process depending upon the type of the product.

In OSD, there is a considerable loss due to various reasons such as rodents, birds, insects, and microorganisms. The unexpected rain or storm further worsens the situation. Further, over drying, insufficient drying, contamination by foreign material like dust dirt, insects, and microorganism as well discoloring by UV radiation are the characteristics of OSD. In general, OSD does not fulfill the quality standards and therefore it cannot be sold in the international market. With the awareness of inadequacies involved in OSD, a more scientific method of solar-energy utilization for crop drying has emerged called controlled drying or solar drying.

3.4.5.3.2 *Direct Solar Drying*

The principle of direct solar crop drying is shown in Figure 3.33. This is also called cabinet dryer. A part of incidence solar radiation on the glass cover is reflected back to atmosphere and remaining is transmitted inside cabin dryer. Further, a part of transmitted radiation is reflected back from the surface of the crop. The remaining part is absorbed by the surface of the crop. Due to the absorption of solar radiation, crop temperature increases and the crop starts emitting long wavelength radiation which is not allowed to escape to atmosphere due to presence of glass cover unlike OSD. Thus, the temperature above the crop inside chamber becomes higher. The glass cover also serves as reducing direct convective losses to the ambient which further becomes beneficial for rise in crop and chamber temperatures. However, convective and evaporative losses occur inside the chamber from the heated crop. The moisture is taken away by the air entering into the chamber from below and escaping through another opening provide at the top as shown in this figure.

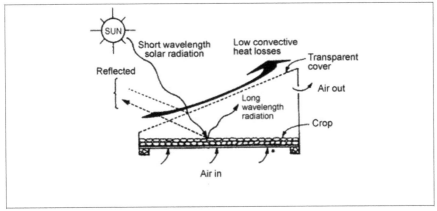

FIGURE 3.33   Working principle of direct solar drying.

A cabinet dryer has the following limitations:

1.   Due to its small capacity, its use is limited to small scale applications.
2.   Discoloration of crop due to direct exposure to solar radiation.
3.   Moisture condensation inside glass covers reducing its transmissivity.
4.   Sometimes the insufficient rise in crop temperature affecting moisture removals.
5.   Limited use of selective coatings on the absorber plate.

3.4.5.3.3   Indirect Solar Drying

The crop is not directly exposed to solar radiation to minimize discoloration and cracking on the surface of the crop. Figure 3.34 describes principle of indirect solar drying which is generally known as conventional drying. In this case, a separate unit termed as solar air heater is used for solar-energy collection for heating of entering air into this unit. The air heater is connected to a separate drying chamber where the crop is kept. The heated air is allowed to flow through wet crop. Here, the heat from moisture evaporation is provided by convective heat transfer between the hot air and the wet crop. The drying is basically by the difference in moisture concentration between the drying air and the air in the vicinity of crop surface. A better control over drying is achieved in indirect type of solar drying systems and the product obtained is of good quality.

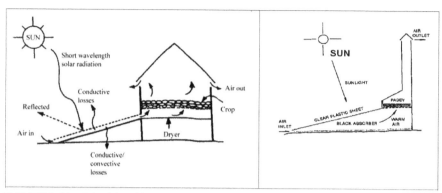

FIGURE 3.34   Principle of indirect solar drying system.

3.4.5.4   TYPES OF SOLAR DRYER

Solar dryers can generally be classified into two broad categories: active and passive. Passive dryers use only the natural movement of heated air. They can be constructed easily with inexpensive, locally available materials which make them appropriate for small farms. A direct passive dryer is one in which the food is directly exposed to the sun's rays. Direct passive dryers are best for drying small batches of fruits and vegetables such as banana, pineapple, mango, potato, carrots, and French beans. This type of dryer comprises of a drying chamber that is covered by a transparent cover made of glass or plastic (Fig. 3.35). The drying chamber is usually a shallow, insulated box with air-holes in it to allow air to enter and exit the box. The food samples are placed on a perforated tray that allows the air to flow through it and the food. The figure shows a schematic of a simple direct dryer. Solar radiation passes through the transparent cover and is converted to low-grade heat when it strikes an opaque wall. This low-grade heat is then trapped inside the box by what is known as the "greenhouse effect". Simply stated, the short wavelength solar radiation can penetrate the transparent cover. Once converted to low-grade heat, the energy radiates as a long wavelength that cannot pass back through the cover.

   Active solar dryers are designed incorporating external means, like fans or pumps, for moving the solar energy in the form of heated air from the collector area to the drying beds. Figure 3.36 shows a schematic of the major components of an active solar dryer. The collectors should be positioned at an appropriate angle to optimize solar-energy collection.

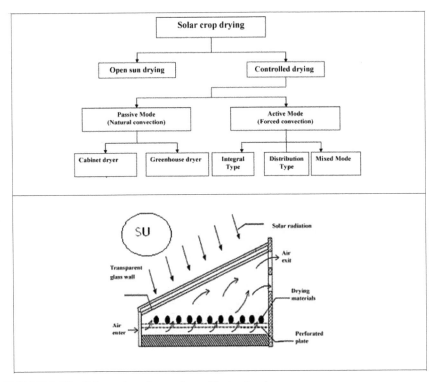

FIGURE 3.35   Schematics of a passive cabinet solar dryer.

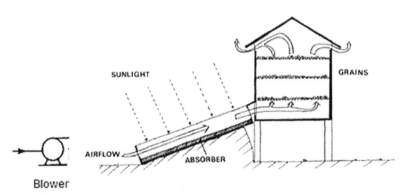

FIGURE 3.36   Schematics of an active cabinet solar dryer.

3.4.5.5   CONSTRUCTION OF CABINET-TYPE SOLAR DRYER

The cabinet type of dryer does not use fan or blower to be operated by electrical energy. It is easy to construct and operate and low in cost. It is made up of wooden material and its structure is of box type. The length of the box is generally three times its width. Its base and sides are insulated and the top is covered with a transparent roof. The inside surfaces of the box are coated with black paint and product to be dried is kept in the trays made up of wire mesh bottom. The loaded trays are kept through an openable door provided on the rear side of the dryer. Ventilation holes are provided in the bottom through which fresh outside air is sucked automatically. Holes are also provided on the upper sides of the dryer through which moist warm air escapes. When the food grain is placed in the trays and exposed to solar radiation, the temperature of the product rises resulting in evaporation of moisture. This warm moist air passes through ventilation holes by natural convection, creating partial vacuum, and drawing fresh air up through the holes provided in the base of the dryer. Temperatures as high as 90°C are achieved in this dryer. This type of dryer reduces the drying time from one half to one-third compared to OSD.

3.4.5.6   SIZING OF SOLAR DRYER (CABINET TYPE)

The dimensions of a small scale cabinet type solar dryer are about 1 m × 1 m × 0.5 m. The mass of air enclosing this size dryer is about 0.6 kg (taking density of air to be 1.2 kg/m³).

3.4.5.6.1   Heat Capacity of Air

If air is to be heated to a certain temperature, then we need to estimate how much heat should be added to air to reach the desired temperature. Air has a heat capacity of 1 kJ/kg°C. This implies that if we need to raise the temperature of 1 kg air by 1° (say, from 25 to 26°C), a heat of 1 kJ should be added to air. Based on this information, we can calculate how much heat should be added to raise the temperature of m kg air by ΔT temperature (where ΔT = required temperature − ambient air temperature). Therefore, heat requirement, $Q = 1 \times m \times \Delta T$ (in kJ). Assumptions for designing the system are

- Mass of hot air requirement = 1 kg.
- Air required at a temperature of 80°C.
- Ambient water temperature be 25°C.
- Daily solar radiation = 5.5 kW h/m² day.

- Efficiency of solar air heater is 0.40 (efficiency can range between 40% and 60% for flat plate collector).
- Heat capacity of air = 1 kJ/kg K.

Step 1: Find out useful solar energy

Daily available solar radiation = 5.5 kW h/m² day.

Useful solar heat energy (which is actually used in raising temperature of air) = Daily solar radiation × solar air heating efficiency = 5.5 × 0.4 = 2.2 kW h/m² day = 7920 kJ/m² day, using a conversion factor of 1 W h = 3600 J.

Step 2: Heat required bringing the air from the ambient to desired temperature

The ΔT or rise in temperature required = Required temperature − ambient water temperature = 80 − 25 = 55°C. The mass of air = 1 kg.

Heat energy required to raise air temperature to desired level, everyday = 1 × m × ΔT kJ = 1 × 1 × 55 = 55 kJ required per day.

Step 3: Total area of the collector required to fulfill the requirement

Collector area required = required heat energy / useful heat energy = 55/7920 = 7.0 × 10⁻³ m² = 70 cm².

KEYWORDS

- alternate energy sources
- Angstrom pyrheliometer
- atmosphere
- beam radiation
- biomass
- classification of energy resources
- climate change
- conventional energy
- diffuse radiation
- energy sources
- energy utilization
- Eppley pyrheliometer

- flat plate collector
- forms of energy
- fossil fuels
- geothermal energy
- global warming
- hydro
- hydropower
- nonrenewable sources of energy
- nuclear energy
- photovoltaic
- photovoltaic applications
- principles of solar energy
- propagation of solar radiation
- pyranometer
- renewable energy sources
- solar collector
- solar cooker
- solar dryer
- solar energy
- solar energy utilization
- solar photovoltaic system
- solar radiation
- solar still
- solar thermal energy storage
- solar water heater
- sunshine hours
- sunshine recorder
- tidal energy
- types of energy resources
- wind energy

REFERENCES

1. Anonymous. *Renewable Energy in India*. Ministry of Renewable Energy Report, Government of India, 2013. http://mnre.gov.in/.
2. Anonymous. *Solar Policy*. Government of India, 2013. http://mnre.gov.in/.
3. Anonymous. *Renewable Energy in India*. Ministry of Renewable Energy Report, Government of India, 2014. http://mnre.gov.in/.
4. IEA. *Energy Technology Perspectives*. International Energy Agency: Paris, 2010.
5. IPCC. Summary for Policymakers. In *Climate Change 2007: Mitigation. Contribution of Working Group III to the Fourth Assessment Report of the Intergovernmental Panel on Climate Change*, Metz, B., Davidson, O. R., Bosch, P. R., Dave, R., Meyer, L. A., Eds.; Cambridge University Press: Cambridge, New York, NY, 2010.
6. Mani, A.; Rangrajan, S. *Solar Radiation Over India*. Allied Publishers: New Delhi, 1982.
7. Thirugnanasambandam, M.; Iniyan, S.; Goic, R. Renewable and Sustainable Energy. *Energy Rev.* **2010,** *14*, 312–322.
8. Tiwari, G. N.; Ghosal, M. K. *Fundamentals of Renewable Energy*. Narosa Publishing House: New Delhi, India, 2007.
9. Urja, Akshay, (2012). November issue. Ministry of Renewable Energy Report, Government of India.
10. Urja, Akshay, (2013). May issue. Ministry of Renewable Energy Report, Government of India.

CHAPTER 4

ENHANCED ANAEROBIC DIGESTION BY INOCULATION WITH COMPOST

ZHIJI DING[1], NIRAKAR PRADHAN[2*], and WILLY VERSTRAETE[1]

[1]*Laboratory of Microbial Ecology and Technology (LabMET), Faculty of Bio-Engineering Science, Ghent University, Coupure Links 653, Gent 9000, Belgium. E-mail: jimmydzj2006@gmail.com*

[2]*Erasmus Mundus Joint Doctorate (ETECOS3) Program, University of Napoli Federico-II, Naples 80125, Italy. E-mail: nirakar.pradhan@ gmail.com*

*Corresponding author.

CONTENTS

Abstract 108
4.1   Introduction 108
4.2   Materials and Methods 109
4.3   Results 113
4.4   Discussions 120
4.5   Conclusions 125
Keywords 126
References 126

ABSTRACT

The increased use of the activated sludge system to treat municipal and industrial wastewater has led to a considerable production of sludge, which is potentially harmful to both the environment and human beings. Anaerobic digestion (AD) of waste-activated sludge (WAS) is a widely accepted method for sludge stabilization and biogas production. The enhancement of the AD process has been studied by implementing different pretreatments of WAS, codigestion of WAS, and other substrates and applying multiphase systems. The current study applied compost material from aerobic composting of plant litter as inoculants to the WAS digesters. The studies were carried out in two phases. First, compost with different maturations was tested. During the process of compost production, the temperature and the microbial community evolve with time. The phases of composting are indicated by temperature difference. In this test, the compost was sampled at its initial, hot and stable phases, respectively. The stable compost was found to have the strongest stimulatory effect on the biogas production. The reactor with an inoculation of 0.1 g compost DW/g COD could achieve 42% more biogas production than the control reactor at the 10th day. Subsequent experiments further investigated this effect by distinguishing it into biological and nonbiological aspects. After 60 days, the biological aspects showed a stronger stimulatory effect of biogas production and a better removal of effluent chemical oxygen demand (COD) than the control. Furthermore, an addition of 0.1 g compost DW/g COD, with a thermo-alkaline pretreatment and kitchen waste as co-substrate, would give 19% more specific biogas production (mL biogas/g COD loaded) than the control.

4.1   INTRODUCTION

Waste-activated sludge (WAS) is the major byproduct of aerobic wastewater treatment, that is, activated sludge system. The default value for sludge yield is 0.4 kg CDW/kg COD removed.[15] The production of sludge kept increasing when more and more municipal and industrial water is aerobically treated. WAS exhibits wide variation in their properties depending on its origin and previous treatment.[7] Stabilization of WAS can be achieved either aerobically or anaerobically. Aerobic stabilization requires less capital cost and maintenance but extra energy for aeration is needed. Anaerobic stabilization is an economic solution for wastewater treatment plant (WWTP) with a capacity of more than 50,000 IE.[18] Anaerobic treatment is especially suitable for developing countries where the average temperature is high. A major advantage of it is the energy recovery from its byproduct biogas, which normally contains 70–80% methane (CH_4) gas.

The anaerobic digestion (AD) is a process, which a certain group of bacteria break down organics in the absence of oxygen. In the treatment of WAS, aerobic biosolids are used as carbon source. The AD process consists of four steps, respectively, hydrolysis, fermentation, acetogenesis, and methanogenesis. The overall reaction of AD is (glucose as example):

$$C_6H_{12}O_6 + 3H_2O \rightarrow 3CH_4 + 3HCO_3^- + 3H^+ (\Delta G = -404 \text{ kJ/mol})$$

The end products are biogas and hydrophobic anaerobic biomass. The former can be flamed to generate energy and the latter goes to further dewatering and reuse/disposal. The objective of this paper is to find out if compost materials from aerobic composting of plant litter have stimulatory effect on AD of WAS in terms of biogas production and organic carbon removal.

4.2 MATERIALS AND METHODS

4.2.1 WASTE-ACTIVATED SLUDGE

Secondary sludge was transported from the Ossemeeresen WWTP located in Gent, Belgium. The Ossemeersen WWTP applies a conventional-activated sludge process treating municipal wastewater for a community of 230,000 IE. The plant is composed of pretreatment (screening and gritting), biological treatment (aeration and decant), and final clarification. The sludge is treated by thickening, AD, and centrifugal dewatering. After collection, sludge was further thickened by decanting for 2 days and stored at 4°C. The final WAS to be used as substrate has a total COD of 12 g/L and soluble COD of 770 mg/L. Analysis of parameters at different times during storage showed no significant variations.

4.2.2 COMPOST

Compost of different stages was collected from an aerobic composting site treating plant litter in Belgium. Fresh compost was collected at the starting point of the composting process, which is the same as fresh plant litter.

Active compost was collected during the thermophilic stage of composting when the temperature was at the range of 60–80°C. Stable compost was from the maturation stage of composting when the temperature cooled down to the ambient temperature. After collection, the fresh compost was stored at 4°C. The active compost was mashed manually and kept at room temperature with low humidity for less than 14 days. The stable compost was sieved (2 mm) and stored at the same condition of the active compost. The respective characteristics are listed in Table 4.1.

TABLE 4.1 Characteristics of Inoculants.

Item	Fresh compost	Hot compost	Stable compost	AD sludge
DS	43%	46%	68%	84.8 g/L
VS	20%	25%	39%	72.7 g/L
VS/DS	0.46	0.53	0.57	0.86
SMA	–	–	–	0.71 g CH_4-COD/g VSS day

DS: dry solids; VS: volatile solids; SMA: specific methane activity.

4.2.3 ANALYTICAL METHODS

The monitored parameters during the experiments were total chemical oxygen demand (TCOD), soluble COD, total suspended solids (TSS), volatile suspended solids (VSS), pH, total ammonium nitrogen, total oxidized nitrogen, volatile fatty acid, biogas production, and composition according to APHA standard methods. The pH was measured by a portable pH meter. Biogas production was noted by the headspace of the graduated gas column.

4.2.4 ANAEROBIC BATCH TEST

Four 1-L Erlenmeyer were connected to graduated gas columns by rubber stops and tubes as batch reactors. Each of them contained a mixture of WAS and anaerobic granular sludge of certain ratios depending on the different aims of the test. The setup was kept at a mesophilic temperature of 32–35°C. Biogas was collected in the gas column, which was placed above an acid solution (indicated by the red color of methyl orange) to avoid dissociation of CO_2. The acid solution was initially pulled up to the zero point of the column by a sucking pump. A buffer (2 g KH_2PO_4 + 1 g $NaHCO_3$) was used to prevent dramatic pH drop when necessary. Substrate was fed to the reactors on every 2 days with certain volumetric loading rate. Upon feeding, clear supernatant was discharged and the same amount influent was fed to the reactor. Mixture of the reactor was done manually and regularly. Cumulative biogas production as well as biogas production rate (L/day) was evaluated to compare the effect of different inoculums. Specific biogas production (L biogas/g COD loaded) was obtained by linearly fitting the figure of cumulative gas production against time over each test period. When methane content of the reactors of one test showed a significant variance, the respective methane gas production was used to evaluate the effect.

4.2.5 PHASE I EXPERIMENT: COMPARISON OF COMPOST WITH DIFFERENT AGES

The aim of Phase I experiment was to select the right type of compost that has stimulatory effect on AD of WAS. The amount of compost dry weight dosed to each reactor equal to 10% of TCOD of the substrate in the reactor. Reactors were set up as follows:

Reactor 1: No compost (control).
Reactor 2: Inoculated with well mixed fresh compost.
Reactor 3: Inoculated with mashed active compost.
Reactor 4: Inoculated with sieved stable compost.

All the reactors were followed up with the same feeding strategy. Volumetric loading rate remained the same throughout the experiment period. Upon feeding, the reactors were completely mixed. Therefore, a mixture of WAS and AD sludge was discharged and a certain amount of wash out of AD sludge was unavoidable.

Compost was dosed at 10% of the total substrate COD in dry weight. The same feeding strategy was applied to all the reactors and kept constant over the experimental period. In brief, hydraulic retention time (HRT) is kept 20 days. Sludge loading rate (SRT) kept 0.042 g COD/g VSS day. In this batch test, the volumetric loading rate (B_v) was kept at 0.6 g COD/L day pH was maintained between 7 and 8 by phosphate buffer.

4.2.6 PHASE II EXPERIMENT: EVALUATION OF BIOLOGICAL AND NONBIOLOGICAL EFFECTS

The aim of Phase II experiment was to investigate the mechanism of the effect of stable compost. Four reactors are set up as follows to investigate the biotic, abiotic, and combined biotic/abiotic effects in comparison with control:

Reactor 1: No compost (control).

Reactor 2: Compost as such, representing hypothesis that the sum of biological and nonbiological effects.

Reactor 3: The inoculated compost was incubated at 60°C for 1 h then kept in the room temperature for 2 days. This stimulated the spores in the compost thus activated the microorganisms thus represented the hypothesis of biological effect.

Reactor 4: Autoclaved compost. During autoclave, bacteria were killed and functioning enzymes were destroyed. This represented Hypothesis 2, the nonbiological effect.

Four reactors were operated in parallel, representing the control and the three postulated effects. The substrate was a mixture of WAS and kitchen waste in a COD ratio of 2:1. Here, the WAS was thermally pretreated at 52°C in alkaline condition (2.5 g NaOH/L) for 2 days to achieve a higher soluble COD. The pretreatment achieved a 35% soluble COD whereas the untreated one was only 3%.

The tested compost was stable fine compost with diameter smaller than 0.2 mm. The characteristics of the inoculants are shown in Table 4.2.

TABLE 4.2   Characteristics of Inoculants.

	Compost as such	Thermal treated compost	Autoclaved compost	AD sludge
DS	87%	97%	99%	72 g/L
VS	53%	61%	51%	62 g/L
VS/DS	0.61	0.63	0.52	0.86

DS: dry solids; VS: volatile solids.

Compost as such represented Hypothesis 1 the combined effect of biological and nonbiological aspects. Thermal treated compost was obtained by incubating the compost at 60°C for 1 h, then keeping in the room temperature for 2 days. This stimulated the spores in the compost and allowed them to grow. Autoclaved compost was free of bacteria and the functioning enzymes thus representing Hypothesis 3, nonbiological aspects.

The feeding strategy of Phase II was no longer kept constant over the whole period. Instead, the HRT, B_y, and B_x changed in response to the biogas production, pH, and effluent COD since the last feed. Upon feeding, instead of completely mixing before discharging, the clear supernatant liquor was discharged. Therefore, little biomass was lost throughout the test period.

The batch test lasted for 60 days. During the first 30 days, the volumetric loading rate was changed in response to the variation of biogas production, pH and the effluent COD of the time upon feeding. The specific biogas production level (L biogas/gCOD load) of each reactor was compared. To confirm the effect of biological aspect, during the second 30 days, the nonbiological reactors, that is, reactor 1 and 4, were mixed and redistributed. The biologically affected reactors, that is, reactor 2 and 3 were mixed and redistributed. Hence, two controls in parallel were compared with two biological reactors in parallel. The reactors continued to be operated with the feeding strategy changing over time but kept the same over the reactors. To minimize the wash out of anaerobic biomass, the clear supernatant was discharged upon feeding. Mixture of the reactors was done only after feeding.

4.3 RESULTS

4.3.1 PHASE I

4.3.1.1 BIOGAS PRODUCTION

The batch reactors were operated for a period of 20 days. The cumulative biogas production is plotted against time in Figure 4.1.

A first glance at Figure 4.1 shows that reactors inoculated with fresh and stable compost had higher biogas production than the control. The strongest effect was achieved at the 11th day when the fresh and stable compost enhanced the biogas productions by 42% over the control. The extra biogas produced in the fresh compost reactor can be explained by the dosage of biodegradable COD (fresh plant litter). An amount of 2.5 g DW/L fresh compost corresponded to 1.17 g VS/L, and therefore 1.28 g COD/L. As a rule of thumb, 1 g COD is equivalent to about 500 mL biogas. So the extra biogas was expected to be 640 mL. However, due to the low volumetric and SRT, this value was much lower.

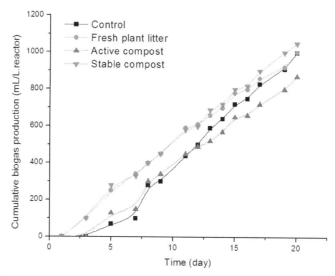

FIGURE 4.1   Cumulative biogas productions for Phase I experiment.

According to the constant feeding strategy, a linear increase in cumulative biogas production was expected. Therefore, data for daily gas production (data not shown) were used to compare the effect of the different inoculations of the sludge with that of control. A paired t-test was applied to prove the difference in average biogas production in comparison with the control. None of the three inoculated reactors showed significant difference comparing with the control on the basis of the t-test. The average daily biogas productions for all the reactors were far less than expected, that is, about 200–300 mL/d. Despite the use of buffer (NaHCO$_3$ and K$_2$HPO$_4$), significant pH drops that could be as low as 6.5–7.0, were observed throughout the experiment period. This is an indication of acidification process in the reactor. Therefore, a second batch test was set up with a higher COD concentration of the substrate.

4.3.1.2   CONFIRMATION OF THE POSITIVE EFFECTS OF STABLE COMPOST

In the second batch test, sludge of higher COD from the same source, was enriched to 25 g COD/L using glucose and starch. Here, the volumetric loading rate B_v was 2.5 g COD/L day, which was much higher than the previous one. The HRT was reduced to 10 days and pH maintained at 7.0–7.5. The cumulative biogas production was plotted against time after 10 days of experiment (Fig. 4.2). Due to the short HRT, the biomass was quickly washed out from the reactor. As a consequence, all the reactors stopped gas production after 5 days. The available numbers of samples were too small to conduct statistical tests. However, it can be observed that after 5 days, the reactor inoculated with stable compost produced 13.8% more biogas than the control.

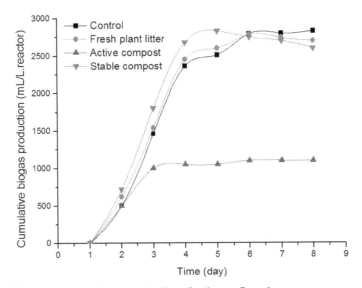

FIGURE 4.2   Cumulative biogas productions for the confirmation test.

The Phase I experiments indicated that the inoculation of stable compost in an AD reactor could stimulate the biogas production up to 42%. The problem of the test was that the reactors' overall performance of the reactors was not optimal. The influent COD was too low to be well digested and consequently they could only reach a limited biogas production.

4.3.2 PHASE II

The aim of Phase II experiments was to further investigate the mechanisms of the stimulation by stable compost, observed in Phase I.
Hypotheses:

1. Biological aspects enhanced biogas production.
2. Nonbiological aspects enhanced biogas production.

3. A combination of biological and nonbiological aspects enhanced biogas production.

4.3.2.1   BIOGAS PRODUCTION

Figure 4.3 shows the comparison of the cumulative biogas production and the cumulative COD load over the four reactors. The reactors inoculated by untreated and thermally treated compost could be loaded 4% of total COD higher than the control, while the autoclaved compost reached a lower rate. Compost as such achieved 19% more biogas than the control at the end of the experiment period.

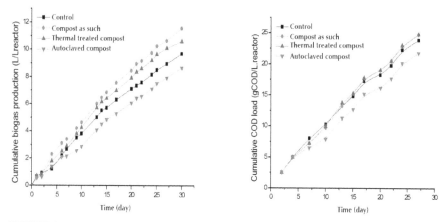

FIGURE 4.3   Comparison of biogas production (left) and COD load (right) of the reactors.

Figure 4.4 shows the individual cumulative biogas production against cumulative COD load for each reactor. The reactor with autoclaved compost has the lowest slop, which represented the lowest value of specific biogas production, that is, 362 mL biogas/g COD loaded. The value for the control was 374 mL biogas/g COD loaded. Therefore, autoclaved compost has no surplus effect. The reactor inoculated with compost as such reached the highest specific biogas production, that is, 439 mL biogas/g COD feed. This approved the effect of biological aspects. The one with the thermally treated compost had a somewhat higher biogas production than the control but inferior to the one with the compost as such. Therefore, thermal treatment of the compost could not further enhance the biological effect that compost brought about.

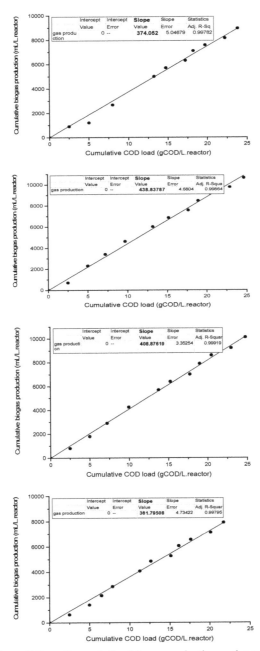

FIGURE 4.4 Linear fitting of cumulative biogas production and cumulative COD load. Upper left: control; up right: compost as such; lower left: thermal treated compost; lower right: autoclaved compost.

Figure 4.5 shows the evolution of the effluent COD concentration. The effluent from the reactor inoculated with untreated compost reached the lowest COD decrease of 73%. The control reached the highest decrease of 78%. This is probably because reactors inoculated with compost as such and thermally treated compost were loaded higher than the control (Fig. 4.3).

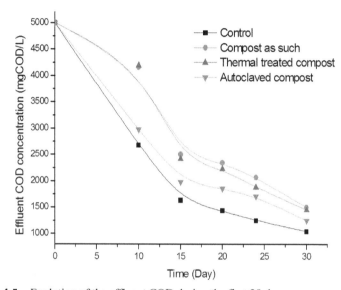

FIGURE 4.5   Evolution of the effluent COD during the first 30 days.

4.3.2.2 CONFIRMATION OF BIOLOGICAL EFFECTS OF COMPOST

To confirm the biological effect, the control reactor and the one with auto-claved compost were mixed and equally redistributed; reactors with untreated and thermally treated compost were mixed and redistributed equally. Hence, two control reactors and two reactors with biologically active stable compost were compared. All the reactors continued to be operated with the same feeding pattern. The results for biogas production and COD loading rate are shown in Figure 4.6. With the same COD load, the biologically affected reactor gave 16% more biogas than the control, complying well with the previous results. The respective specific biogas production was 422 mL biogas/g COD, while that of the control was 366 mL biogas/g COD.

The effluent COD concentration evolution is shown in Figure 4.7. Given the same COD load, the biological affected reactor achieved a higher decrease of soluble COD (36.5%) than that of the control (22%).

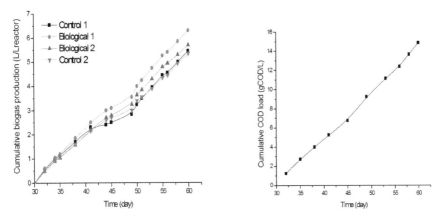

FIGURE 4.6   Comparison of biogas production (left) and COD load (right) over the reactors.

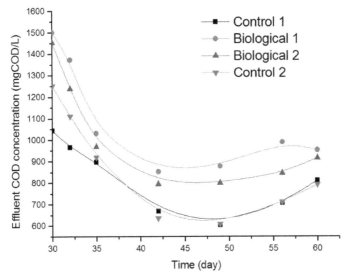

FIGURE 4.7   Effluent COD evolution of the second 30 days.

4.3.2.3   CONCLUSION OF THE PHASE II EXPERIMENTS

Biological aspects contribute to the stimulatory effect of compost toward biogas production in the AD process. The reactor inoculated with compost could be loaded higher than the noninoculated one without overloading the reactor. The inoculation also resulted in a 17% higher specific biogas production. The thermal treatment of the compost did not give an extra stimulatory effect. Autoclaved compost showed no effect probably because the treatment killed the organisms and destroyed the functioning enzymes.

4.4 DISCUSSIONS

4.4.1 EVALUATION OF PHASE I STUDY

Compost used as inoculants in anaerobic digesters has been rarely reported. Scherer et al.[22] applied plant litter compost from its hot rot phase (active compost) to the digester of fodder beet silage and reached a shorter HRT. Stable vermicompost was found to be effective in enhancing the anaerobic treatment of synthetic wastewater by increasing the methane production by 25.2%.[4] Therefore, compost material from different source and age can potentially enhance the AD process for certain substrate. In this study, plant litter compost was applied as inoculant in the WAS digester.

In the Phase I experiment, fresh, active, and stable compost were used as inoculant in the AD reactors treating WAS. The cumulative biogas production of each reactor after a period of 20 days was compared. The reactor inoculated with stable compost reached the highest biogas production, which was 42% more than the control (Fig. 4.1). Another series of batch test confirmed this stimulatory effect by the fact that the reactor inoculated with the stable compost had an increment of 13.8% more biogas than that of the control (Fig. 4.2). The reactor inoculated by fresh compost also had a higher biogas production than the control. However, this increment was most probably contributed by the dose of extra biodegradable COD from fresh compost itself. Active compost showed no stimulation effect.

During composting, organic matters are decomposed to CO_2, NH_3, H_2O, and humic substances. Veeken et al.[26] found that the content of humus decreased in the first 20 days and increased after by the way of NaOH extraction. The form of humus changed during composting from aliphatic to aromatic. The amount of humus in the stable compost could be above 10 g/kg VSS. The redox range of humus is continuous in the range from −300 mV to +400 mV. Therefore, humus molecules act as electron shuttle or buffer when other available electron acceptors, like oxygen, nitrate, are rare. Many anaerobic bacteria tend to shuttle electrons via humic acids.[17] Qi et al.[20] investigated a microbial film from landfill compost, which contained heterogeneous functional microbial groups, including fungi, yeasts, bacteria, as both aerobic and anaerobic microorganisms, and protista, primarily amoebas. It was developed to degrade humic acids in natural aquatic environments. The addition of glucose as an extra carbon source gave a synergetic effect on the degradation of both and enhanced the growth of the biofilm. Therefore, the existence of humus could be a potential abiotic aspect in enhancing the WAS digestion. In this test, the lowest humus content in active compost, according to Veeken et al.,[26] could probably be one of the reasons why it did not enhance the biogas production in WAS digestion as an inoculants. The fresh compost was composed of plant litter, which acted as extra biodegradable carbon source. However, the degradation of this carbon source was limited by its content of poorly degradable lignin. The codigestion of plant litter and WAS did not give a synergetic effect on the biodegradability of both. The stable compost contained the largest amount of humus. It also had the highest microbial diversity, which will be discussed later. The predominant existence of humus could be an abiotic contribution in the enhancement of WAS digestion.

According to the cooperative community continuum principle, there exists no single stable community; the degree of close cooperation is a key to functionality.[5] Therefore, the microbial community of a compost pile undergoes dynamic changes during the composting process. Temperature is a primary factor in determining the microbial community during the composting process. The cellulose, hemicelluloses, and lignin account for the major part of biomass to be decomposed in plant litter compost. A group of bacteria and fungi is responsible for this niche.[3]

Different methods have been used to study the compost microbiology, including the plate-count method, community level physiological profiles, phospholipids fatty acid analysis, molecular technologies, and quinone profile method. Yu et al.[27] did a detailed research on the changes of the microbial community during agricultural plant waste composting by the quinone profile method. During the thermophilic stage, the obvious increase of MK-7 (one kind of menaquinine species) indicated that the Gram-positive bacteria with a low GC content (e.g., *Bacillus* spp.), *Proteobacteria* and *Actinobacteria* may play an important role. During the maturation phase, the shift of the profile shows more cooperation between Gram-positive bacteria and fungi; the latter plays a major role in the lignin degradation.[27]

In the process of aerobic composting, the aerobic degradation takes place only at the surface of substrate particles where the available oxygen is used by aerobic microorganism. In general, it takes several weeks for this aerobic process to reach the particle center. Therefore, the majority of the substrate (up to 90%) undergoes anaerobic processes at the starting period of the composting.[15] That is the reason why methane fermentation was also observed in compost piles.[10] One of the predominant *Archaea* in compost,

Methano bacterium thermos auto trophicum, has been isolated and identified from mushroom compost.[6] Later, Thummes et al. [24] investigated the archaeal 16S rRNA gene composition in compost material at different stages by cloning.

Figure 4.8 gives the allocation of archaeal clones for different maturity of compost. The frequency of *Methanothermobacter* spp. was dominant in the 6-week-old compost material and decreased as the compost became mature. The archaeal diversity increased over the same process. The mature compost showed the highest archaeal diversity.

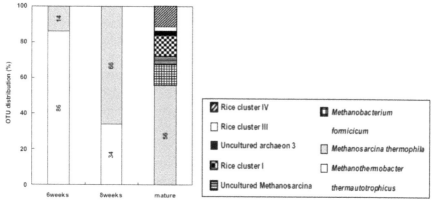

FIGURE 4.8   Allocation of the archaeal clones in 6, 8 weeks and mature compost, the frequency of their appearance as a percentage of the total number of analyzed clones.[24]

The community dynamics of methanotrophic bacteria has also been studied. Methanotrophs were considered to be the largest biological sink for greenhouse gas methane.[1] Halet et al.[8] studied the activities of the methanotrophs in compost by 16S-rDNA and 16S-rRNA. The activity of type I methanotrophs was lowest in themophilic phase and increased to peak. Finally, it decreased toward maturation, while the abundance of the methanotrophs showed no change over time. When compost was dosed into a WAS digester, new species were brought to the system. According to the "neutral theory," the biodiversity is determined by the influx, the arrival of new species.[11] In the anaerobic digester, the mesophilic methanogens were the predominant species. The introduction of the community of the active compost seemed to have negative effect on the original methane production activity (Figs. 4.1 and 4.2). One reason could probably be that the new condition favored the methanotrophs, especially type II methanotrophs, which are believed to be dominant under high methane conditions.[2] In contrast, the stable compost had the highest microbial diversity. Probably the mesophilic condition facilitated a better cooperation between the new members and the existing community and thus enhanced the overall biogas production.

4.4.2   EVALUATION OF PHASE II STUDY

Based on the findings of Phase I experiment, Phase II experiments were conducted to study the mechanism of the stimulatory effect of stable compost. The biological effect was simulated by preheating the compost at 60°C for 1 h, then keeping it at room temperature for 2 days. This allowed the spore forming organisms to form spores and grow. The nonbiological effect was simulated by autoclaving the compost, which killed all the microorganisms and destroyed the enzymes. Compost as such was used to represent the sum of biological and nonbiological effects. The substrate was a mixture of thermo-alkaline treated WAS and kitchen waste with a COD ratio of 2:1. The thermo-alkaline treatment achieved a 35% soluble COD in WAS whereas the untreated one was only 3%. During the period of the first 30 days, the reactor inoculated with compost as such had a 17.3% more specific biogas production (mL biogas/g COD loaded) than the control (Fig. 4.4). The reactor inoculated with thermally treated compost showed somewhat higher biogas production than the control but lower than the one with compost as such. The second 30 days of the test confirmed the effect of biological effect by an increment of 17% in specific biogas production (Fig. 4.6) and a higher COD removal (Fig. 4.7).

The objective of the thermal treatment of the compost was to activate the microorganisms, including some *Archaea* and bacteria, thus enhancing their biological effect when it was dosed in to the AD reactor. The stable compost has the largest biodiversity and resembles the natural soil. Some Gram-positive bacteria with a low G + C content, *Actinobacteria* and fungi were found to be ubiquitous in stable compost.[27] Some bacteria and *Archaea* are capable of producing endospores. The endospores are produced to ensure the survival of a bacterium through periods of environmental stress, including heat, radiation, etc. They are commonly found in soil and water, where they may survive for long periods of time. Endospore forming bacteria include *Bacillus*, *Thermoactinomyces*, *Clostridium*, etc. Reactivation of the endospore occurs when conditions are more favorable and be triggered by a temperature increase.[19] During the thermal treatment of the compost, the endospore were expected to be activated and subsequently induced to grow in to mature cells. Other groups, including some thermophilic bacteria, which might be under dormancy, were also expected to be reactivated. However, an enhanced biological effect of thermally treated compost over nontreated compost was not achieved, although the latter showed somewhat better performance than the control reactor. The reasons could probably be (1) Thermal treatment of 60°C might have killed some species that would potentially be important in the new biotope. (2) The condition exerted to the compost did not successfully activate the endospores. (3) The endospores might be activated but did not have enough time to grow into mature cells. In the case of (1) and (2), the thermal treatment was just another environmental stress for the compost microbial community. It could inhibit instead of stimulate the microbial activity.

The autoclave is a way of sterilization achieved by high pressure and high temperature. A holding time of 15 min at 121°C is required. In this work, the objective of the autoclaving treatment of compost was to inactivate the bacteria and fungi as well as their spores. Thus, when it was dosed into the AD reactor, few new species would be introduced. Humus is the refractory end product of aerobic composting process of plant litter waste. The structure of humus is very complex and difficult to define. Its special chemical structure is of interest in the process of WAS digestion. First, as described before, it can act as electron shuttle for many anaerobic bacteria [17]. Second, it has many active sites, which could bind some recalcitrant ions, for example, NH_4, heavy metals, etc. However, the results did not show any effect when autoclaved compost was dosed into the AD reactor, in comparison with the control. The reasons could be: (1) humification was not complete; (2) a dosage of 0.1 g compost DW/g COD was too low for the compost material to function.

In Phase II experiment, the substrate was a mixture of thermo-alkaline treated WAS and the kitchen waste with a COD ratio of 2:1. Thermo-alkaline treatment have been reported to effectively solubilize the WAS.[21,25] The process was found to be optimal at 0.1 g NaOH/g TSS at 60°C for 48 h.[16] In this test, the treatment of 0.1 g NaOH/g COD at 52°C for 48 h reached a solubilization of about 35%, whereas the soluble COD of the nontreated sludge was 3%. Codigestion of sewage sludge and kitchen waste has been well studied by Ref. [12–14]. An enhanced biogas production was found both at mesophilic and thermophilic conditions. In this test, the fresh kitchen waste was co-digested with pretreated WAS. Comparing the results from Phase I experiments, where only WAS was used as substrate, the thermo-alkaline pretreatment and codigestion brought about a much higher specific biogas production (208 mL biogas/g COD loaded for the reactor inoculated with stable compost in Phase I and 438 mL biogas/g COD loaded for the reactor inoculated with compost as such in Phase II).

When the reactor was inoculated with untreated stable compost, a 17.3% more specific biogas production and a higher loading rate could be achieved. The *Bacillus* spores are known to be able to produce autolysin enzymes that break their thick peptidoglycan layer and grow into cells. If free excess enzymes were present, they could potentially attack the thin peptidoglican layer of Gram-negative organisms. Song et al.[23] isolated a bacterial strain SY-9, which was identified as *Geobacillus* sp. from the compost of sewage sludge. It was found to have lytic activity against both intact and thermal inactivated bacterial cells. Hasegawa et al.[9] also observed that the strain *Bacillus stearothemophilus* was efficient in secreting hydrolytic enzymes such as proteases and amylases. It could enhance sludge hydrolysis under aerobic conditions. The co-substrate is mainly composed of aerobic bacterial cells from wastewater treatment and kitchen waste, which is a source of biodegradable COD. The destruction of the cell walls is the bottleneck in the overall AD process. In this test, nontreated stable compost was found to be effectively enhanced the hydrolysis (Phase III experiments), which might finally enhance the overall performance including biogas production. Since the anaerobic condition is not favorable for sporulation, only the existing extracellular enzymes could function. Other possibilities are the incoming compost microbial community could direct the hydrolysis–fermentation to a more favorable type of end product and the highly diverse *Archaea* in the compost improved the structure of the methanogenic association.

The microbial community in the compost turned out to be the major aspect that enhanced the AD of the co-substrate. The fact that the autoclaved compost showed no effect to the AD reactor could be one clue. The biological effect was not further enhanced by the thermal treatment of the compost probably because the condition chosen was not correct to activate the spores.

4.5 CONCLUSIONS

The current study found that the inoculation of fine stable compost (diameter <2 mm) from plant litter composting to the anaerobic sludge digester could enhance the biogas production by an increment of about 20%. The subsequent study on the mechanisms of the enhancement showed that biological aspect of the compost was the main contribution to the stimulatory effect on the biogas production. Compost was shown to be able to accelerate the hydrolysis in the presence of anaerobic bacteria, but the effect was insignificant to the whole process. Therefore, it could be concluded that the new arrival of the microbial community of compost material enhanced methanogenic process. It is expected that the combination of thermo-alkaline pretreatment, codigestion of sludge and kitchen waste and the inoculation of compost could give rise to the implementation of anaerobic stabilization of sludge.

KEYWORDS

- **activated sludge**
- **anaerobic digestion**
- **biogas**
- **codigestion**
- **composting**
- **nutrient removal**
- **reactor**

REFERENCES

1. Adamsen, P.; King, G. Methane Consumption in Temperate and Sub-arctic Forest Soils—Rates, Vertical Zonation, and Responses to Water and Nitrogen. *Appl. Environ. Microbiol.* **1993,** *59*(2), 485–490.
2. Amaral, J.; Knowles, R. Growth of Methanotrophs in Methane and Oxygen Counter Gradients. *FEMS Microbiol. Lett.* **1995,** *126*(3), 215–220.
3. Berg, B.; McClaugherty, C. *Plant Litter: Decomposition, Humus Formation, Carbon Sequestration Decomposition, Humus Formation, Carbon Sequestration.* Springer, 2003.
4. Chen, G.; Zheng, Z.; Yang, S. Experimental Co-digestion of Corn Stalk and Vermicompost to Improve Biogas Production. *Waste Manage.* **2010,** *30*(10), 1834–1840.
5. Curtis, T. P.; Sloan, W. T. Towards the Design of Diversity: Stochastic Models for Community Assembly in Wastewater Treatment Plants. *Water Sci. Technol.* **2004,** *54*(1), 227–236.
6. Derikx, P.; Opdencamp, H.; Vanderdrift, C.; Vangriensven, L.; Vogels, G. Isolation and Characterization of Thermophilic Methanogenic Bacteria from Mushroom Compost. *FEMS Microbiol. Ecol.* **1989,** *62*(4), 251–257.
7. EEA. *Sludge Treatment and Disposal—Management Approaches and Experiences,* 1997.
8. Halet, D.; Boon, N.; Verstraete, W. Community Dynamics of Methanotrophic Bacteria During Composting of Organic Matter. *J. Biosci. Bioeng.* **2006,** *101*(4), 297–302.
9. Hasegawa, S. Decomposition of Biological Waste Sludge by Thermophilic Bacteria— S-TE Process (Solubilization by Thermophilic Enzyme). *Japan Tappi J.* **2001,** *55*(8), 81–87.

10. Hellebrand, H. Emission of Nitrous Oxide and Other Trace Gases During Composting of Grass and Green Waste. *J. Agric. Eng. Res.* **1998,** *69*(4), 365–375.

11. Hubbell, S. *The Unified Neutral Theory of Biodiversity and Biogeography.* Princeton University Press: Princeton, 2001.

12. Kim, H. W.; Han, S. K.; Shin, H. S. The Optimisation of Food Waste Addition as a Co-substrate in Anaerobic Digestion of Sewage Sludge. *Waste Manage. Res.* **2003,** *21*(6), 515–526.

13. Kim, H. W.; Han, S. K.; Shin, H. S. Anaerobic Co-digestion of Sewage Sludge and Food Waste Using Temperature-phased Anaerobic Digestion Process. *Water Sci. Technol.* **2004,** *50*(9), 107–114.

14. Kim, H. W.; Han, S. K.; Shin, H. S. Simultaneous Treatment of Sewage Sludge and Food Waste by the Unified High-rate Anaerobic Digestion System. *Water Sci. Technol.* **2006,** *53*(6), 29–35.

15. Metcalf, A. *Wasterwater Engineering: Treatment and Reuse.* McGraw-Hill: New York, 2003.

16. Nansubuga, I. G. Investigation of Sludge Hydrolysis and Biodegradation, Master Thesis, Ghent University, 2006.

17. Nepomnyashchaya, Y. N.; Slobodkina, G. B.; Baslerov, R. V.; Chernyh, N. A.; Bonch-Osmolovskaya, E. A.; Netrusov, A. I.; Slobodkin, A. I. *Moorella humiferrea* sp. nov. a Thermophilic, Anaerobic Bacterium Capable of Growth via Electron Shuttling Between Humic Acid and Fe(III). *Int. J. Syst. Evol. Microbiol.* **2002,** *62*, 613–617.

18. Nowak, O. Optimizing the Use of Sludge Treatment Facilities at Municipal WWTPs. *J. Environ. Sci. Health, A* **2006,** *41*, 1807–1817.

19. Prescott, L. *Microbiology.* Wm. C. Brown Publishers, 1993.

20. Qi, B.; Aldrich, C.; Lorenzen, L.; Wolfaardt, G. Degradation of Humic Acids in a Microbial Film Consortium from Landfill Compost. *Ind. Eng. Chem. Res.* **2004,** *43*(20), 6309–6316.

21. Rivero, J. A. C.; Suidan, M. T. Effect of H_2O_2 Dose on the Thermo-oxidative Co-treatment with Anaerobic Digestion of Excess Municipal Sludge. *Water Sci. Technol.* **2006,** *54*(2), 253–259.

22. Scherer, P.; Schmidt, O.; Unbehauen, M.; Neumann, L. Compost as a Source of Inoculum for the Anaerobic Digestion of Renewable Biomass, Allowing Short Hydraulic Retention Time. *11th IWA World Congress on Anaerobic Digestion*, Brisbane, Australia, 2007.

23. Song, Y.; Hu, H.; Li, X. Characterization of Thermophilic Strain *Geobacillus* sp. SY-9 with Capability to Lyse Bacterial Cells. *China Environ. Sci.* **2007,** *27*(4), 456–460.

24. Thummes, K.; Kaempfer, P.; Jaeckel, U. Temporal Change of Composition and Potential Activity of the Thermophilic Archaeal Community During the Composting of Organic Material. *Syst. Appl. Microbiol.* **2007,** *30*(5), 418–429.

25. Vallom, J. K.; McLoughlin, A. J. Lysis as a Factor in Sludge Flocculation. *Water Res.* **1984,** *18*(12), 1523–1528.

26. Veeken, A.; Nierop, K. V.; Hamelers, B. Characterization of NaOH-extracted Humic Acids During Composting of a Biowaste. *Bioresour. Technol.* **2000,** *72*(1), 33–41.

27. Yu, H.; Zeng, G.; Huang, H.; Xi, X.; Wang, R.; Huang, D.; Huang, G.; Li, J. Microbial Community Succession and Lignocellulose Degradation during Agricultural Waste Composting. *Biodegradation* **2007,** *18*(6), 793–802.

CHAPTER 5

TRACTOR UTILIZATION IN VIDARBHA REGION, INDIA

D. V. TATHOD[1*], Y. V. MAHATALE[2], and V. K. CHAVAN[3]

[1]*College of Agricultural Engineering and Technology, Warwat Road, Jalgaon Jamod, District Buldhana, Maharashtra, India. E-mail: dnyanutathod@gmail.com*

Department of Farm Power & Machinery, College of Agricultural Engineering and Technology, Warwat Road, Jalgaon, District Buldhana, Maharashtra, India. E-mail: yogeshvmahatale@gmail.com

[2]*AICRP for Dryland Agriculture, Dr. Panjabrao Deshmukh Krishi Vidyapeeth, Krishinagar PO, Akola 444104, Maharashtra, India. E-mail: vchavan2@gmail.com*

*Corresponding author.

CONTENTS

Abstract 130
5.1   Introduction 130
5.2   Methodology 131
5.3   Results and Discussion 132
5.4   Conclusions 134
Keywords 135
References 135

ABSTRACT

The tractor is very popular among Indian farmers primarily because of affordability of high operational efficiency. The project titled "The utilization tractors in Jalgaon Jamod Taluka in Vidarbha Region" is a survey type project. The survey was conducted in 25 villages out of 53 villages in Jalgaon Jamod region. In each village, 10 farmers were personally contacted for collecting data.

From the study, it was concluded that 76% of tractor sales were on credit. The loan was extended by commercial banks, state land development banks, and regional rural banks. Most of farmers (24%) preferred credit from local cooperative banks that play a major role for providing the loan facility. Reduced cropping cycle due to irrigation facilities require deep tilling, which translates into higher demand for tractors. Most of the farmers bought the tractors as their status symbol by paying the amount in cash.

5.1   INTRODUCTION

The sale of agricultural tractors and other farm equipment has increased all over the world. Today more than 250,000 tractors are manufactured every year by 13 and more manufacturers. These tractors are available in different horsepower (HP) ranges of less than 25 to more than 55 HP, but most popular range is 31–35 HP.

For 2008–2009, growth of agricultural credit in India was targeted at 2.8 billion INR[1] with short-term crop loan remained to be disbursed at 7% per annum with initial provision of 16,000 million INR subvention in 2008–2009 national budget. Gross capital formation in agriculture as a promotion of GDP in the agriculture sector enhanced from low level of 10.2% in 2003–2004 to 12.5% in 2006–2007. The plan was to increase it to 16% in XI plan to achieve growth rate of 4%. A crucial factor in the development of agricultural sector in XI plan was a target of 1500 million INR on farm mechanization (Planning Commission Report of Indian Government).

As higher capital formation and subsidies are available in agriculture sector for achieving the higher growth rate, timeliness of operation and precision in the use and application of inputs are the major constraints. These constraints can be optimized through farm mechanization. Therefore, this research was undertaken to study the "Utilization of tractors and

---

[1]In this chapter, one Indian Rupee (INR) = 0.01566 US$.

implements in the Jalgaon Jamod region of Maharashtra State of India" with the following objectives.

1. To study utilization of tractors by farmers in Jalgaon Jamod region (JJR).
2. To know the availability of credit facility for the purchase of tractors and implements in JJR.
3. To judge the suitability of tractor of different sizes for the JJR.

5.2 METHODOLOGY

The study of tractors utilization in JJR was conducted through field survey to gather the data, which was analyzed to know the annual utilization of tractors, available of credit facility, average HP available in JJR, and more suitable tractor model for the farm. Out of 53 villages in JJR, 25 villages were selected for the survey. In each village, 10 farmers were personally interviewed for collecting data. The following questionnaire was prepared for the survey (Table 5.1).

TABLE 5.1 The Information in the Questionnaire for Utilization of Tractor.

1. Name of farmer and age
2. Village
3. Taluka
4. No. of family members
5. Education: illiterate/primary/higher/college
6. Family members work on farm
7. Total annual income
8. Total land holding (ha)
    - Irrigated
    - Non-irrigated
9. Opinion about the tractor
10. Name of the tractor
11. Tractor model
12. Tractor horsepower
13. Name of loan facility/money source
14. Agricultural implements

5.3   RESULTS AND DISCUSSION

The region total areas is about 110,000 ha in JJR, district Buldhana. It was found that the total tractor population is 9172. One tractor per 0.54 ha of land is available in the region and the average tractor size used in the region is 52 HP per 0.54 ha of land. The status of farm mechanization in the region is satisfactory.

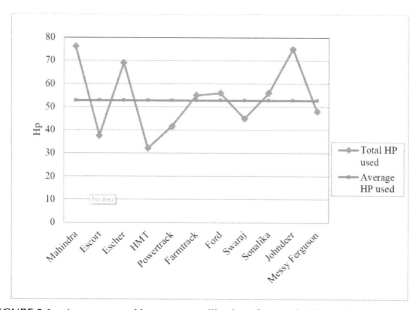

FIGURE 5.1   Average annual horsepower utilization of tractor in Jalgaon Jamod.

Figure 5.1 shows average HP by the selected manufacturers and the average of all the types of tractor used.

5.3.1   *CREDIT FACILITY FOR PURCHASE OF TRACTOR*

It was found that the financial services are available to the farmers for purchasing tractors and implements through a number of nationalized banks, cooperative banks and cooperative societies. However, the rate of interest varies from 15% to 19%. The loan is available only if the land is eligible for mortgage with loan agency. It is clear from the pie chart in Figure 5.2 that the share of cooperative bank and self-hired tractor has more percentage.

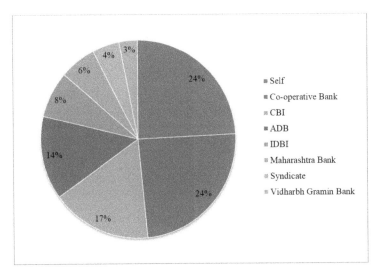

FIGURE 5.2   Percentage-wise contribution of financing organizations.

From the pie chart, it is seen that 24% of the farmers preferred to buy the tractor with their own money, whereas the same number of farmers seek loan from cooperative banks. The percentage of loan from Central Bank of India was 17% compared to 14% share of Agricultural Development Bank. The loan percentage of other banks was 8% by IDBI, 6% by Maharashtra Bank, 4% by Syndicate Bank, and 3% by Vidharbh Gramin Bank.

It was found that the credit for purchasing the tractor mostly comes from the cooperative banks, because the farmers get loan easily from such banks and the procedure of getting loan is not that tedius as the government banks. The most farmers prefer to buy the tractor at their own money: self-credit (Sawkar) because the procedure for getting loan is time consuming and difficult and has a touch of corruption. In the survey, farmers expected more subsidy (up to 50%) for the tractor purchasing and the farmers must get the subsidy at the time of purchasing the tractor or implements because the late subsidy is most of the times the late subsidy is used to pay the interest on the loans.

It can be observed that 36% of farmers possess Mahindra tractor models compared to 14% Escher, 12% Escort, 8% Power track, 6% Farm track, 6% Ford, 6% Sonalika, 5% HMT, 5% Messy Ferguson, 1% John Deer, and 1% Swaraj (Fig. 5.3). These are tractor models mostly found in JJR. The mold board plough, blade harrow, cultivator, V-pass, seed drill, and usually two wheel trolleys are most common agricultural machines among the farmers of the region.

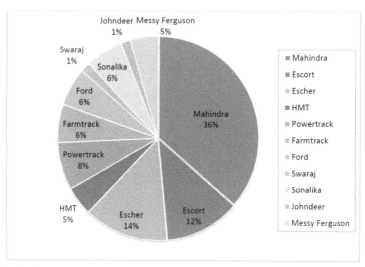

FIGURE 5.3   Percentage-wise tractor model available in the Jalgaon Jamod.

5.3.2 OPINION OF THE FARMERS ABOUT THE TRACTOR— IMPLEMENT SYSTEMS

The most the farmers (about 90%) have expressed satisfaction about the performance of tractor and implements possessed by them. However, they complained: About the high cost of fuel and repairs, maintenance of tractors, and high rate of interest that resisted them from purchasing new tractors.

5.4 CONCLUSIONS

From the study, it is concluded that 76% of tractor sales were on loan that was extended by commercial banks, state land development banks and regional rural banks. Most of farmers (24%) preferred credit from local cooperative banks. Local cooperative banks play a major role for providing the loan. Irrigation facilities reduce reliance on monsoon and allow for adoption of quick yielding varieties of grain. This reduces the cropping cycle to 3–4 months from the traditional cycle of 5–6 months. Reduced cropping cycle requires deep tilling, which translates into higher demand for tractors.

The tractors of 65–75 HP dominate the market. Lately, it is visualized that higher HP sizes has the maximum growth potential. The tractors with high HP will be the future requirement due to the government intention to encourage contract farming through the leasing in and leasing out of farm lands. Most of the farmers have bought the tractors as their status symbol by paying the amount in cash.

KEYWORDS

- Asia
- farm employment
- farming
- horsepower
- India
- mechanization
- operating cost
- tractor
- tractorization

REFERENCES

1. Dixit, V. K.; Baratwaj, J. L. The Impact of Tractorization on Farm Employment in Raipur district, India. *Agric. Situat. India* **1990,** *45*(4), 233–235.
2. Nandel, D. S. Mechanization of Haryana Agriculture: An Overview. *Agric. Situat. India,* **1988,** *XLIII*(1), 3.
3. Pandy, P. H.; Ali, I. Unit Operating Cost of General Purpose Tractors in Raising Wheat Crop. *J. Agric. Eng. I.S.A.E.* **1971,** *VIII*(4), 1–5.
4. Singh, K. N.; Singh, B. Utilization of Different Model of Tractor on Large Mechanized Farms. *Agric. Eng. Abstr.* **1993,** *18*(9), 350–352.
5. Abraham, V. Yield and Employment Effects of Tractor Use in Agriculture. *Indian Dissert. Abs.* **1988,** *XVII*(4), 403.
6. Ward, S. M. Tractor Ownership Costs. *Agric. Mech. Asia* **1990,** *21*(1), 21–33.

CHAPTER 6

PERFORMANCE OF REVERSIBLE MOLD BOARD PLOW

YOGESH V. MAHATALE[1*], DNYANESHWAR V. TATHOD[1], and VISHAL K. CHAVAN[2]

[1]Department of Farm Power & Machinery, College of Agricultural Engineering & Technology, Warwat Road, Jalgaon, District Buldhana, Maharashtra, India. E-mail: yogeshvmahatale@gmail.com, dnyanutathod@gmail.com

[2]AICRP for Dryland Agriculture, Dr. Panjabrao Deshmukh Krishi Vidyapeeth, Krishinagar PO, Akola 444104, Maharashtra, India. E-mail: vchavan2@gmail.com

*Corresponding author.

CONTENTS

Abstract 138
6.1   Introduction 138
6.2   Literature Review 140
6.3   Methods and Material 144
6.4   Results and Dissussion 150
6.5   Conclusions 155
Keywords 156
References 157
Appendix 6.I   Data Sheet for Evaluation of Reversible MB Plow 159
Appendix 6.II   Cost Operation of Reversible MB Plow 159
Appendix 6.III   Data Sheet and Evaluation Data 163

ABSTRACT

The evaluation of reversible MB plow under the actual field conditions is the only way of recognizing the quality and output. It can also help govt. agencies to control the undeserving units, improve the deficiencies in existing implements, standardize critical components, and save energy. During 2014–2015, a locally available three bottom 30-cm reversible MB plow was evaluated at Jalgaon (Jamod) region of Buldhana District. The results show that at average operating speed of 3.8 km/h, the average depth of cut and width of cut were 20 cm and 80 cm, respectively, and actual field capacity, theoretical field capacity, field efficiency, and fuel consumption were 0.26 ha/h, 0.34 ha/h, 76.4% and 4.61 l/h, respectively.

6.1   INTRODUCTION

The tillage implements in India are manufactured by small scale sector. These manufacturing units are mostly located in the states of Madhya Pradesh, Uttar Pradesh, Punjab, Haryana, Bihar, Maharashtra, and Karnataka. About 70% of these units have an investment of less than INR[1] 2×10^5 with only about 4% investment of more than INR 5×10^5, in plant and machinery. With such a small investment, there is no facility for design, development testing, and quality control. These deficiencies result in more energy requirement, frequent breakdowns, less output, and unsatisfactory performance. Although the Bureau of Indian Standards (BIS) has published about 300 standards in the area of farm machinery, yet only 650 product licenses have been granted by BIS as against total number of about 17,000 manufacturers of agricultural machines in India.

The primary purpose of plowing is to turn over the upper layer of the soil, bringing fresh nutrients to the surface, while burying weeds, the remains of previous crops, both crop and weed seeds, allowing them to break down.[12–14] It also provides a seed-free medium for planting an alternate crop. In modern use, a plowed field is typically left to dry out and is then harrowed before planting. Plowing and cultivating a soil homogenizes and modifies the upper 12–25 cm of the soil depth to form a plow layer.

The prime necessity of tillage is to prepare the land or the seedbed, where the plants can easily grow. Using different types of equipment (driven manually or by powered), machines make the soil suitable to place the seeds at a

---

[1]In this chapter, one Indian rupee (INR) = 0.01566 US$.

desirable depth. Tilling the fields hinders or slows the growth of weeds and improves crops' competition against weeds. Moreover, tillage loosens the compacted layers. The history of tillage goes back to 3000 BC in Mesopotamia. People started cultivation in the fertile land close to the river valleys of Nile, Tigris, Euphrates, Indus lands.

Different types of tillage systems have different tillage depths and capacity to change soil physical and chemical properties that affect the crop yield and quality. Time and frequency of tillage also has significant effect on crop production. Important soil physical properties such as bulk density, penetration resistance, water infiltration, hydraulic conductivity, and soil compaction are affected by tillage.

The reversible plow is a unique implement, which is directly mounted on the tractor.[18] This is a mechanically operated basic implement for land preparation.[11] It is very useful for primary tillage in hard and dry trashy land. The moldboard (MB) retains their mirror finish at all-time contributing to well turn furrows. The plow has special wear resistant steel bottoms with bar points for toughest plowing jobs. Bar point bottom ensures longer life as it can be extended or reversed. The MB bottom reversing mechanism is operated by a lever provided on the distributor. When the implement is hitched, plow bottom is free to rotate 180° along the axis of the hollow shaft.

With this plow, plowing can be done without formation of ridges (back furrow) or hallows (dead furrow or valley). Thus, the layout of the field is not disturbed and unidirectional plowing (putting the furrow slice on one direction of the field) can be achieved (Figs. 6.1 and 6.2). The main objectives of quality tillage implements are as follows[20,23,24]:

1. To minimize the mechanical power and labor requirements.
2. To conserve the moisture and reduce soil erosion.
3. To perform the operations necessary to optimize soil tilth.
4. To maximize the field efficiency.

Therefore, testing and evaluation of tillage implements under the actual field conditions is the only way of recognizing the quality and output.[8] This will also help government agencies in controlling the undeserving units, improving the deficiencies in existing implements, standardization of critical components, and saving the energy though adoption of energy efficient implements. For example, a single plowing with soil inversion plow gives better tilt than a four times plowing with country plow, in addition to saving in cost of cultivation, saving in time for seed bed preparation for sowing of the succeeding crop.

This chapter discusses the performance of reversible MB plow.

FIGURE 6.1 Plowing using animal power and tractor power.

FIGURE 6.2 Tractor operated reversible moldboard plow.

6.2 LITERATURE REVIEW

When agriculture was first developed, simple hand-held digging sticks and hoes were used in highly fertile areas, such as the banks of the Nile where the annual flood rejuvenates the soil, to create drills (furrows) to plant seeds in. Digging sticks, hoes, and mattocks were not invented in any one place, and hoe-cultivation must have been common everywhere agriculture was practiced. Hoe-farming is the traditional tillage method in tropical or subtropical regions, which are characterized by stony soils, steep slope gradients, predominant root crops, and coarse grains grown at wide distances apart. While hoe-agriculture is best suited to these regions, it is used in some fashion everywhere. Instead of hoeing, some cultures use pigs to trample the soil and grub the earth.

The most fundamental operations in global agricultural system are tillage and are very vital from crop production point of view. During the past centuries, the soil cultivation is brought about by plowing especially with moldboard plow. Its dual functions of cultivation and soil inversion make it very popular among the farming communities especially to control weeds. The recent economic and environmental concerns have compelled the farming community to reconsider the use of tillage operations and if possible, implement alternative technologies for soil cultivation. Energy consumption and the working efficiency are the basic parameters to assess the performance of the implement.

Investigators have proposed several models to calculate draught force for plowing. These models generally disregard the geometrical characteristics of active surfaces of working parts. For this reason, tests on channel of traction were carried out to check the validity of two models frequently used, namely those of Gorjachkin and Gee Clough. Results showed that for the same farm and similar work conditions, the efforts were different from one model to another. Tests were also carried out on two active forms of surfaces. Draught force was calculated. One of these two models gave the same results for two different surfaces, whereas the values determined on channel were significantly different from one form to another. They proposed a more universal model that connects draught force with soil state and geometrical characteristics of active surfaces. Using the modeling method Buckingham—Vachy, they compared it with Gorjachkin and Gee Clough models, which account for two forms of active surfaces of plows made in Algeria by commercial companies ENPMA (cultural form) and SACRA (cylindrical form). The calculated force using Amara's model was closer to the values recorded on channel, when compared to those calculated with Gorjachkin and Gee Clough.

Adewoyin and Ajav[2] studied that the effects of plowing speed and depth on fuel consumption of tractor. Field experiments were conducted at 5.5, 6.5, and 7.5 km/h plowing speeds for plowing depths of 20, 25, and 30 cm. The soil type in the study site was predominantly sandy loam.[5] Mean fuel consumption for Fiat, MF, and Steyr models were 23.35, 23.58, and 24.55 L/ha, while average of 16.78, 22.02, and 32.67 L/ha of diesel were used to plow 20, 25, and 30 cm soil depths, respectively. Fuel consumption values were increased with plowing depth significantly, indicating 31% increase from 20 to 25 cm and 48% increase from 25 to 30 cm depths. Mean fuel consumption at 5.5, 6.5, and 7.5 km/h plowing speeds were 20.0, 24.25, and 27.23 L/h, respectively. Fuel consumption was increased by 4.25 L (21%) when speed was increased from 5.5 to 6.5 km/h and 2.98 L (12%) when speed was increased from 6.5 to 7.5 km/h. Mean fuel consumption of 23.35, 23.58, 24.55 L/ha for MF, Fiat, and Steyr tractors, respectively, were significantly different at the various speeds and plowing depths ($p < 0.05$, $0.87 \leq r^2 \leq 0.99$).

Penetrometer resistance (PR) was measured in the field 1 month after sowing. Plant density of barley was also counted 1 month after sowing. Significantly higher Ks value was found for shallow tillage at the depth of 15–20 and 25–30 cm. Bulk density was significantly lower for moldboard plowing for the first two investigated depth and it was higher at 35–40 cm but the difference was not statistically significant. Moreover, bulk density was high in both treatments. Significant higher PR value was found for shallow tillage especially at the depth of 5–35 cm but was not enough to reduce the root growth. Plant density and crop yield were significantly higher in shallow tilled treatment than in moldboard plowing. Field water content at 15–20 and 25–30 cm was significantly higher for moldboard plowing. Water retention at 1-m suction was also significantly higher in the treatment with moldboard plowing. However, the difference of the physical parameters due to tillage treatments was sufficient to markedly influence crop performance and yield.

Ahemadali et al.[4] evaluated performance of moldboard and disk plows in central region of Iraq in 2011. With MB plow: best practical productivity was 0.3118 ha/h, volume of disturbed soil was 629.77 m³/h, and field efficiency was 59.85%. The disc plow recorded higher actual plow depth of 21.02 cm and slippage percentage of 9.71%. With increase in forward speeds of the tillage from 1.85 to 3.75 then to 5.62, slippage percentage was increased from 6.39% to 8.26% then to 12.22%, practical productivity was increased from 0.1421 to 0.2845 then to 0.4180 ha/h, and volume of disturbed soil was increased from 299.89 to 592.38 then to 838.24 m³/h, while actual plow depth was decreased from 21.13 to 20.86 then to 20.08 cm, and field efficiency was decreased from 58.98% to 58.21% then to 57.07%. With decrease in soil moisture from 21% to 18%, actual plow depth was decreased from 20.92 to 20.70 cm and slippage percentage was decreased from 10.58% to 7.29%, and there was an increase of practical productivity from 0.2712 to 0.2923 ha/h, volume of disturbed soil from 562.01 to 598.98 m³/h, field efficiency from 55.94% to 60.31%. While with further decrease in soil moisture from 18% to 14%, there was decrease in actual plow depth from 20.70 to 20.46 cm, practical productivity from 0.2923 to 0.2811, volume of disturbed soil from 598.98 to 569.52 m^3/h, field efficiency from 60.31% to 58.01%, and increase in slippage percentage from 7.29% to 9.00%.

Greece engineer determined performance of tillage implements.[25,26] The moldboard plow was operated at two different depths. Two tractors were used for execution of the implement and the wheel slip was measured. The field experiments showed that draught force of moldboard plow was significantly higher than that for disc plow. The tillage efficiency of all implements was similar, with a mean of specific resistance of 58 kN/m^2. In the soil bin, the vertical force preventing penetration occurred for disc plow and disc harrow. The specific resistance of the moldboard plow had the lowest value, while the specific resistance of disc plow was higher by a factor of 1.65. By increasing the wing width of the tines, the draught force was increased but the specific resistance was decreased.

Ahaneku and Ogunjirin[3] evaluated tillage machines at the National Center for Agricultural Mechanization (NCAM). The forward speed was varied from 1.0 to 10.6 km/h and depth of tillage was maintained constant at 20 cm. Results indicate significant differences in soil physical properties under different tractor forward speeds. A forward speed of 7 km/h resulted in appreciable amelioration of soil structure as reflected in improvements in the soil strength properties and maximum reduction in clod mean weight diameter. Soil strength properties generally were decreased with increasing speed but were increased with the depth of tillage. A predictive model depicting the most vital soil physical parameters affected by tractor speed was developed.

Lal and Shukla [21] indicated that the soil compaction is a major concern for agricultural systems and the compaction caused by different implements or trafficking (cattle, tractor tires, harvesting, and tillage equipment) is a great challenge. The greater the compaction, the more adverse effects are observed on seedling establishment, root development, and crop yield. It also affects soil aggregate stability, erosion, and water infiltration.[1] Particle size distribution, organic matter content, moisture content, and bulk density affect soil compaction and strength. Soils with more clay than sand tend to show more soil cohesion and strength.[15–17]

Fuentes [21] concluded that the traffic from different farm implements or animal trampling[10] may cause destruction of soil pores and increase soil resistance to penetration and soil bulk density. The traffic may be restricted to certain field areas. Traffic control is most feasible and economic viable to avoid soil compaction.

Ishaq [22] observed that soil compaction effects can be alleviated by subsoiling, deep plowing, and chiseling, but these methods are temporary because the soil settles back into place.

Researchers at Akola studied the comparative performance by measuring the pulverization index, performance index, and energy requirement for preparation of seed bed with tractor drawn implements and rotavator. It was observed that the combination of plowing + rotavating gave the highest performance index of 85.06%, pulverization index of 3.28 cm, and energy consumption of 2302 MJ/ha. Followed by rotavator operation,[19] they indicated performance index of 76.027, pulverization index of 3.98 cm, and energy consumption of 1462 MJ/ha.

6.3 METHODS AND MATERIAL

6.3.1 MECHANICAL ANALYSIS OF SOIL

Physical test carried on composite soil sample to determine the particle distribution of soil is called mechanical analysis. The process of mechanical analysis involves the separation of all particles from each other, complete dispersion into ultimate particle and measurement of amount of each size group of sample to know composition of soil. For mechanical analysis of soil, following procedure was used to determine moisture content, bulk density, and soil texture.

6.3.1.1 MOISTURE CONTENT

Soil moisture content was computed on dry weight basis. For measurement of soil moisture, core samples of wet soil were taken from at least the different locations randomly selected in the test plot. The sample was weighed. It was placed in hot air oven at 105°C for 24 h. After oven drying, it was cooled for 8 h and dry weight of sample was recorded. The soil moisture content on dry basis was calculated using the following formula:

$$\text{Soil moisture (\% dry wt. basis)} = \frac{\text{Wt. of wet soil sample} - \text{Wt. of dry soil sample}}{\text{Wt. of dry soil sample}} \times 100$$

$$= \frac{\text{Wt. of moisture in soil sample}}{\text{Wt. of dry soil sample}} \times 100 \tag{6.1}$$

6.3.1.2 BULK DENSITY

Bulk density is the oven dry weight of soil per unit volume. The cylinder of core sampler, which has its cutting edge, was driven into the soil and un-compacted soil sample was obtained within the tube. The inner dimension of core cylinder were measured to calculate the volume of core cutter as the volume of the soil core was the same as the inside volume of core cylinder. The samples were carefully trimmed at the both ends of the core cylinders. The initial weight of the soil samples were taken with a weighing balance. These were dried in an oven at 105°C for 24 h until all the moisture was driven off and the samples were weighed again. Bulk density was calculated by using following formula:

$$\text{Bulk density} = \frac{M}{V} = \frac{4M}{\Pi D^2 L} \tag{6.2}$$

where M = mass of core sample (g); V = volume of cylindrical core sample (cm³); D = diameter of cylindrical core sample (cm); and L = length of cylindrical core sample (cm).

6.3.1.3 SOIL TEXTURE

The particle size distribution of soil is often of critical importance in many applications in agricultural engineering.[21,22] A sieve analysis can be performed on any type of nonorganic or organic granular materials. A typical sieve analysis involves a nested column of sieves with mesh cloth (screen). Soil texture is an important soil characteristic that drives crop production and field management practices. Texture refers to the size of particles that make up the soil. Relative proportion of sand, silt, and clay determine textural class.

The texture of soil was determined by wet sieve analysis. The soil samples were collected from the field and stone particles were removed. The soil samples were placed on the top of the sieve unit. After draining the water, soil was collected in moisture box and weighed.

The mechanical analysis of soil sample indicated that it consisted of 40% clay and 45% sand and 5% of silt particles. Then, the soil texture was found using soil texture classification chart.

6.3.2   GENERAL CONSIDERATIONS

6.3.2.1   SELECTION OF TEST SAMPLES

The test sample was taken at random from several locations in the test plot. The sample was thoroughly mixed to have a representative sample.

6.3.2.2   SELECTION OF TEST PLOT

Tillage equipment/machinery gives best performance in rectangular fields. The test plot should be rectangular with side lengths in the ratio of 2:1 as far as possible. If the field is irregular, then a rectangular test plot should be marked for conducting the test. The other portion of the field can be used for initial setup and adjustment of the equipment. As far as possible, the test plot under consideration should not have any previous tillage treatment after the last crop was harvested.

6.3.2.3   FIELD OPERATIONAL PATTERN

Field capacity and field efficiency of an implement are affected by pattern of field operation, which is closely related to the size and shape of the field, the kind, and size of implement. The down time (nonworking time) should be eliminated with adoption of appropriate field operational pattern for the rectangular field.

6.3.2.4   SPEED OF OPERATION

To calculate the speed of operation, two poles 20 m apart were placed approximately in the middle of the test run. On the opposite side, two poles were also placed in a similar position at 20 m apart so that all four poles form corners of a rectangle. The speed was calculated from the time required for the machine to travel a distance of 20 m on the assumed line connecting two poles on opposite sides. The easily visible point of the machine should be selected for measuring the time.

6.3.2.5   WHEEL SLIP

Tractor drive wheels normally slip in all field operations. The distance covered by the tractor is given by number of revolutions of a drive wheel. A number of drive wheel revolutions increases with wheel slip. Therefore, in case of tractor or power tiller operated implement, the wheel slip will affect the speed of operation, and thereby effective field capacity of the implement. The wheel slip is determined as follows:

$$\text{Wheel slip} \left(\%\right) = \frac{\left[N_{\text{L}} - N_{0}\right]}{\left[N_{\text{L}}\right]} \tag{6.3}$$

where N_{L} and N_{0} are total number of revolutions at load and at no load, respectively, for the marked test run.

6.3.2.6   TEST DURATION

The test sample should be operated under different soil and surface conditions for a minimum period of 1 h to establish its performance.

6.3.2.7   FIELD PARAMETERS

Various parameters to define soil characteristics and surface conditions of the test plot (Table 6.1) were observed and recorded.

TABLE 6.1   The Parameters to Define the Test Plot.

Field parameters	
Location of test plot	Jalgaon (Jamod)
Size of test plot (Fig. 6.3)	62 m × 29 m (0.18 ha)
Last crop harvested	Soybean
Details of previous tillage operations, if any (after harvesting of last crop)	No
Topography of field	Level

6.3.3   FIELD EVALUATION

6.3.3.1   RATE OF WORK

6.3.3.1.1   Width of Cut

For determining width of cut, average of three lateral runs was taken (Fig. 6.4). The measurement of composite width was taken at minimum of three equidistant locations along the direction of travel and average working width was determined.

6.3.3.1.2   Effective Field Capacity

The actual output in terms of area covered per hour is expressed as the effective field capacity. For calculating the effective field capacity, the time was recorded on account of real work activity and of other activities such as turning, adjustments, etc. Nonproductive time (h) is the time lost for turning and adjustments, etc. excluding refueling and machine maintenance. Time for refueling was not considered, because usually fuel tank is filled up before starting the test. Effective field capacity was calculated as follows:

$$E_e = \frac{A}{T_p + T_N}$$

(6.4)

where E_e = effective field capacity (ha/h); A = area covered (ha); T_N = nonproductive time (h); and T_p = productive time.

6.3.3.1.3   Field Efficiency

The field efficiency is the ratio of effective field capacity to the theoretical field capacity and is expressed as percentage. For most of the tillage operations, field efficiency ranges from 75% to 90%.

$$\text{Theoretical field capacity} = \frac{\text{Theoritical width of implement (m)} \times \text{speed of opeation (m/h)}}{10,000}$$

$$\text{Field efficiency } (\eta) = \frac{\text{Effective field capacity } (E_e)}{\text{Theoretical field cpacity } (E_t)} \times 100$$

(6.5)

6.3.3.1.4 Depth of Cut

The vertical distance between furrow sole and ground level is referred as depth of cut. To obtain accurate results, the depth was measured at minimum 3 locations and its average was taken (Fig. 6.5).

6.3.4 FUEL CONSUMPTION

The tank was filled to full capacity before and after the test (Fig. 6.6). Amount of refueling after the test was the fuel consumption for the test. While filling up the tank, careful attention was paid to keep the tank horizontal and not to leave empty space in the tank. The fuel consumption gives us an idea of energy requirement by the implement for the operation. The selected field is shown in Figure 6.3.

FIGURE 6.3 Selection of test plot (62 m × 30 m).

FIGURE 6.4 Measuring the width of cut.

FIGURE 6.5   Measuring the depth of cut.

FIGURE 6.6   Measuring the fuel consumption.

6.4   RESULTS AND DISSUSSION

To evaluate the performance of a reversible MB plow, authors determined and evaluated parameters, such as mechanical analysis of soil, soil texture, bulk density, soil moisture, depth of cut by plow, width of plowing, length of strip of plow, area covered, time required for plowing, effective field capacity, theoretical field capacity, field efficiency, speed of operation, number of wheel rotations under loading and unloading, fuel consumption to plow the test area, wheel slip, etc. The performance test was conducted in Jalgaon (Ja) region.

6.4.1 MECHANICAL ANALYSIS OF SOIL

The soil mechanical analysis was carried out at the research farmland and is shown in Table 6.2. The soil is predominantly clay loam, almost neutral and high water retention ability with average moisture content of 13.80% on dry basis and 1403.6 kg/m³ of bulk density.

6.4.1.1 MOISTURE CONTENT

Before conducting the trials, soil moisture was measured at four different locations in the test plot at depth of 6–8 cm, and it was 15.9%, 12.62%, 13.35%, and 13.20% on dry basis. The plot size was 62 m in length and 9.7 m in width with a total area of 0.06 ha for all trials.

6.4.1.2 BULK DENSITY

The bulk density of soil was measured at four different locations in test plot. It was 1406, 1400, 1394, and 1414 kg/m³.

6.4.1.3 SOIL TEXTURE

Based on the mechanical analysis of soil (Table 6.2), soil texture was clay loam.

TABLE 6.2   Soil Analysis.[6,7]

Variables	Units	Soil characteristics
Sand	%	25.58
Silt		32.25
Clay		42.10
Soil texture	–	Clay loam
Soil moisture content (db)	%	13.80
Soil bulk density	kg/m³	1403.6

6.4.3 FIELD EVALUATION

Three bottoms (30 cm each), reversible MB plow was tested in the field, and data are shown in Table 6.3. Field data and sample calculations are shown in Appendices 6.I–6.III. Four test trials of plow were taken at the time of testing. In 42.45 min, the 0.18-ha area was covered. Average fuel consumption rate was 17.36 L/ha (or 4.61 L/h). At the time of operation, an average operating speed was 3.8 km/h. During the test, average depth of cut and width of cut were 20 cm and 80 cm, respectively. Average actual field capacity and theoretical field capacity were 0.26 ha/h and 0.34 ha/h, respectively. Average field efficiency was 76.38%. During the operation, average wheel slip was observed 38.72%. The following parameters were observed, evaluated, and summarized.

TABLE 6.3 Evaluation of Performance Parameters for MB Plow.

Performance parameters	Units	Average value
Width of cut	Cm	27
Depth of cut		19
Actual field capacity	ha/h	0.26
Theoretical field capacity		0.34
Field efficiency	%	76.7
Fuel consumption	Lph	4.58
Wheel slippage	%	38.7

6.4.3.1 WIDTH OF CUT

The actual width of cut was measured at four different locations and it was 27, 28, 29, and 28 cm.

6.4.3.2 DEPTH OF CUT

The actual depth of cut was 20, 19, 22, and 33 cm at four different locations.

6.4.3.3 WHEEL SLIP

The wheel slip varied from 37.03% to 40.74% at different forward speeds. It was observed that the maximum wheel slip was 40.74% at a forward speed of 4.5 km/h; it was minimum of 37.03% at forward speed of 3.5 km/h.

6.4.3.4 THEORETICAL FIELD CAPACITY

The theoretical field capacity varied from 0.29 to 0.40 ha/h at different forward speeds. The maximum actual field capacity was 0.40 ha/h at a forward speed of 4.5 km/h, while minimum value was 0.29 ha/h at forward speed of 3.25 km/h.

6.4.3.5 ACTUAL FIELD CAPACITY

The actual field capacity varied from 0.22 to 0.33 ha/h at different forward speeds (Fig. 6.7). The maximum actual field capacity was 0.33 ha/h at a forward speed of 4.5 km/h, while minimum was 0.22 ha/h at forward speed of 3.0 km/h.

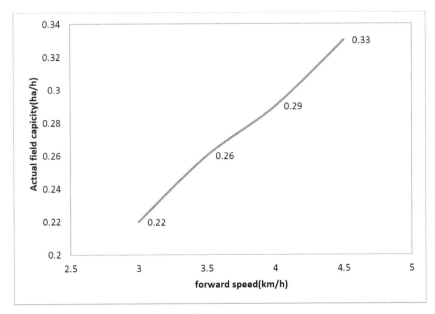

FIGURE 6.7   Actual field capacity (ha/h).

It was observed that the average field capacity was increased with increase in forward speed. In 42.45 min, 0.18-ha area was covered at an average speed of 3.75 km/h. Actual field capacity and theoretical field capacity were 0.24 and 0.30 ha/h, respectively.

These results are close agreement with the findings of other investigators, who concluded that in 44.629 min, 0.1008 ha area was covered. At the time of operation, operating speed was 3.5 km/h. Actual field capacity and theoretical field capacity were 0.1412 and 0.21 ha/h, respectively.

6.4.3.6 FIELD EFFICIENCY

The field efficiency varied from 73.2% to 81% at different forward speeds. The maximum field efficiency was 81% recorded at a forward speed of 3.0 km/h, compared to a minimum value of 73.28% at forward speed of 4.5 km/h. It was observed that the field efficiency was decreased with the increase in forward speed increases as shown in Figure 6.8. Average field efficiency was 76%. Similar findings have been reported by others, who found field efficiency of 67%.

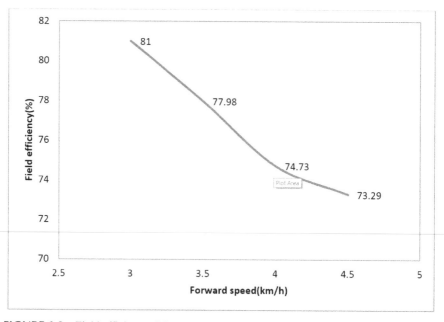

FIGURE 6.8   Field efficiency (%).

6.4.3.7 FUEL CONSUMPTION

The fuel consumption varied from 2.9 (13.11) to 6.3 L/h (21 L/ha) at different forward speeds, as shown in Figure 6.9. It was observed that the maximum fuel consumption was 6.3 L/h (21 L/ha) at a forward speed of 4.5 km/h, compared to a minimum value of 3.1 L/h (13.11 L/ha) at forward speed of 3.25 km/h.

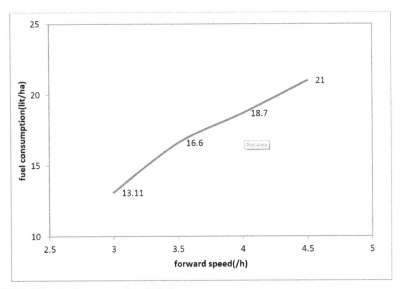

FIGURE 6.9   Fuel consumption (L/ha).

It was observed that the average fuel consumption was increased with an increase in forward speed.[9] An increase of speed from 3.25 to 3.5 km/h resulted in an increase of fuel consumption by 4.5 L (26.7%), etc. The variations in fuel consumption may due to more load on plow as speed was increased. These results are in close agreement with the findings of Adewoyin,[2] who observed an average fuel consumption of 20.00, 24.25, and 27.23 L/ha at a forward speed of 5.5, 6.6, and 7.5 km/h, respectively.

6.5 CONCLUSIONS

The reversible MB plow is a unique implement for preparation of land and can be directly mounted on the tractor. It is very useful for primary tillage in hard and dry trashy land. The MB plow retains its mirror finish at all-time contributing to well turn furrows. Bar point bottom ensures longer life as it can be extended or reversed. The MB bottom reversing mechanism is operated by a lever provided on the distributor. When the implement is hitched, plow bottom is free to rotate 180° along the axis of the hollow shaft. With this plow, plowing can be done without formation of ridges (back furrows) or hallows (dead furrow or valley). Thus, the layout of the field is not disturbed and unidirectional plowing (putting the furrow slice on one direction of the field) can be achieved.

Three bottoms (each 30 cm), reversible MB plow was tested in Ja, district Buldhana, India. At the site soil was clay loam with a bulk density of 1.4 Mg/m^3 and moisture content of 13.80%. Four test trials of reversible MB plow were conducted. In 42.45 min, 0.18 ha of area was plowed.

Average fuel consumption was 17.36 L/ha (4.61 L/h). At the time of operation, an average operating speed was 3.8 km/h. The average depth of cut and width of cut were 20 and 80 cm, respectively. Average actual field capacity and theoretical field capacity were 0.26 ha/h and 0.34 ha/h, respectively. Average field efficiency was 76.38%. During the operation, average wheel slip was 38.72%.

KEYWORDS

- actual field capacity
- animal drawn
- bulk density
- depth of cut
- field efficiency
- fuel consumption
- India
- reversible MB plow
- soil compaction
- theoretical field capacity
- tillage
- tractor
- width of cut

REFERENCES

1. Razzag, A. Comparative Performance of Various Implements in Grassy Land. *Agric. Mech. Asia, Afr. Latin Am.* **1991,** *22,* 30–32.
2. Adewoyin, A. O.; Ajav, E. A. Fuel Consumption of Some Tractor Models for Ploughing Operations in the Sandy-loam Soil of Nigeria at Various Speeds and Ploughing Depth. *Agric. Eng. Int. CIGR J.* **2013,** *3,* 67–74.
3. Ahaneku, I. E.; Ogunjirin, O. A. *Effect of Tractor Forward Speed on Sandy Loam Soil Physical Condition During Tillage.* National Center for Agricultural Mechanization (NCAM), Ilorin, 2005.
4. Ahemadali, A. An Evaluation and Performance Comparison of mould Board Plough and Disc Plough in Soils of Central Iraq, 2011.
5. Al-Suhaibani, S. A.; Ghaly, A. E. Effect of Ploughing Depth of Tillage and Forward Speed on the Performance of a Medium Size Chisel Plough Operating in a Sandy Soil. *Am. J. Agric. Biol. Sci.* **2010,** *3,* 247–255.
6. Blake, G. R. Bulk density. In *Methods of Soil Analysis.* Agronomy Monograph 9 part 1; Black, C. A., Ed.; American Society of Agronomy: Madison WI, 1969; pp 374–390.
7. Brady, N. C. *The Nature and Properties of Soils.* Pearson Publisher: New Delhi, India, 1988.
8. Bukhari, S.; Bhutto, M. A.; Baloch, J. M.; Bhutto, A. B.; Mirani, A. N. Performance of Selected Tillage Implements. *Agric. Mech. Asia, Afr. Latin Am.* **1988,** *19*(4), 9–14.
9. Fathollahzadeh, H.; Mobli. H.; Tabatabaie, S. M. Effect of Ploughing Depth on Average and Instantaneous Tractor Fuel Consumption with Three-share Disc Plough. *Int. Agrophys.* **2009,** *23,* 399–402.
10. Gbadamosi, L.; Magaji, A. S. Field Study on Animal Draught Power for Farmers in Zuguma Village of Niger State, Proceedings of 5th Int. conference and 26th annual General Meeting of Nigeria Institution of Agricultural Engineers (NIAE), 2004, Vol 26; pp 84–85.
11. Grisso, R. D.; Vaughan, D. H.; Roberson, G. T. Fuel Prediction for Specific Tractor Models. *Appl. Eng. Agric.* **2008,** *24,* 423–428.
12. Iraqi, M. E.; Marey, S. A.; Drees, A. M. A Modified Triangle Shaped Chisel Plough (Evaluation and Performance Test). *Misr. J. Agric. Eng.* **2009,** *26,* 644–666.
13. Jain, S. C.; Philip, G. *Farm Machinery an Approach.* Jain Brothers: New Delhi, India, 2002.
14. Kepner, R. A.; Berger, E. L. *Principles of Farm Machinery.* John Wiley and Sons, Inc.: New York, 2003.
15. Mayande, V. M.; Bansal, R. K.; Smith, G. D. A Comparison of Mould Board Ploughs for Dry Season Tillage of Vertisols. *Soil Till. Res.* **1990,** *15,* 349–358.
16. McDonald, J. L.; Daynard, T. B.; Ketcheson, J. W. Results of Soil Tillage Survey, Summer of 1978. University of Guelph: Guelph, ON, 1978.
17. Owende, P. M. O. An Investigation of the Field Performance of Light Mouldboard Ploughs at Slow Speeds. Unpublished Ph.D. Thesis, University College Dublin, Ireland, 1996.
18. Ozarslan, C.; Erdogan, D. Optimization of Tractor Ploughing Performance. *Agric. Mech. Asia, Afr., Latin Am.* **1996,** *27*(3), 9–12.
19. Potekar, J. M.; Tekale, D. D. Comparative Performance of Tractor Drawn Implements Tillage System with Rotavator Tillage System. *Karnataka J. Agric. Sci.* **2004,** *1,* 76–80.

20. Raie, A. E.; Nasr, G. E.; Adawy, W. M. A Study on the Effect of different Systems of Tillage on Physical Properties of Soil. *Misr. J. Agric. Eng.* **2009**, *26*, 664–666.
21. Sahay, J. *Principles of Agricultural Engineering*; Jain Brothers: New Delhi, India, 2005; Vol 1.
22. Sahay, J. *Elements of Agricultural Engineering.* **2010**, pp 224–235.
23. Sayed, A. S.; Ismail, F. S. Effect of different Tillage Techniques on Some Soil Properties & Cotton Yield. *Misr. J. Agric. Eng.* **1994**, *11*, 922–941.
24. Sheikh, G. S.; Ahmad, S. R. *Comparative Performance of Tillage Implements*; AMA: Tokyo, Japan, 1978; Autumn issue.
25. Tarig, O. O.; Moayad, B. Field Performance of Modified Chisel Plough. *Int. J. Natural Sci. Res.* **2014**, *6*, 85–96.
26. Wolf, D.; Hadas, A. Determining Efficiencies of Various Mouldboard Ploughs in Fragmenting and Tilling Air-dry Soils. *Soil Tillage Res.* **1987**, *10*, 181–190.

APPENDIX 6.I   DATA SHEET FOR EVALUATION OF REVERSIBLE MB PLOW

Response no. _____   Date: _____

Evaluator person: _____

Department of Farm Machinery & Power, College of Agricultural Engineering & Technology, Under Dr. Punjabrao Deshmukh Krishi Vidyapeeth [Akola], Jalgaon (Ja), District Buldhana, Maharashtra State, India

GENERAL INFORMATION

1. Name of respondent : Mr. Dilip Bhaurao Patil
2. Village: Wadgaon Patan
3. Taluka: Jalgaon
4. District: Buldhana
5. Age: 52 years
6. Education: Higher
7. Experience in farming: 30 (years)
8. Family details: (a) Male: 2 Female: 2 Children: 1 Total: 5
9. No. of family members in farm work:
10. Male: 1 Female; 1 Children: 0; Total: 2

APPENDIX 6.II   COST OPERATION OF REVERSIBLE MB PLOW

Assumptions made for tractor:

Initial cost of tractor	: Rs. 560,000
Life of tractor	: 10 years
Annual use of tractor	: 1000 h
Salvage value	: 10%
Interest rate	: 10%
Insurance	: 1% of purchase price
Housing	: 1% of purchase price
Repair and maintenance costs	: 10% of purchase price
Price of diesel	: Rs. 59/h
Driver pay	: Rs. 250/day
Depreciation	: Straight-line method

Assumptions made for MB plow

Initial cost of MB plow : Rs. 86,000
Annual use of MB plow : 300 h
Salvage value : 10%
Interest rate : 10%
Housing : 1% of purchase price
Repair and maintenance costs : 10% of purchase price
Depreciation : Straight-line method

A. Initial Cost of operation for tractor

a. Annual fixed cost

(1) Depreciation:

$$D/h = \frac{P - S}{L \times H}$$

where D = depreciation price (Rs./h); P = purchase price (Rs./h); S = salvage value = 10% of purchase price; L = life of the machine in years; and H = number of working hours per year.

$$D = \frac{560,000 - 560,000 \times 0.10}{10 \times 1000} = Rs.\,50.4/h$$

(2) Interest:

$$I/h = \frac{P + S}{2} \times \frac{i}{H}$$

where I = annual interest charge (Rs./h); and i = interest rate (%).

$$I = \frac{560,000 + 560,000 \times 0.10}{2} \times \frac{0.10}{1000} = Rs.\,30.08/h$$

(3) Insurance:

$$Insurance = \frac{1\%\ of\ P}{H}$$

$$I = \frac{0.01 \times 560,000}{1000} = Rs.\,5.6/h$$

(4) Housing rate:

$$HR = \frac{1\% \text{ of } P}{H}$$

$$HR = \frac{0.01 \times 560,000}{1000} = Rs. 5.6/h$$

Total fixed cost = (1) + (2) + (3) + (4) = 50.4 + 30.08 + 5.6 + 5.6 = Rs. 91.68/h.

b.   Operating cost

(5) Repair and maintenance cost are considered as an essential and significant part of machinery ownership. Repairs and maintenance cost was taken 10% of the purchase price of machine per year.

$$\text{Repair and maintenance cost} = \frac{560,000 \times 0.1}{1000} = Rs. 56/h$$

(6) The fuel cost was calculated on the basis of actual fuel consumption in tractor.
Cost of fuel taken Rs. 59
Average fuel required for 1 h = 4.58 L/h
Fuel cost = 59 × 4.58 = Rs. 270/h

(7) Lubricants costs was taken 10% of the fuel cost.
Lubricating cost = Rs. 27/h

(8) The cost of operator was calculated from the actual labor charges paid in rupees per day at the prevailing rates in the study area. Rs. 250/day are paid for tractor operator@8 working hours per day.

$$\text{Driver charge} = \frac{250}{8} = Rs. 31.25/h$$

Total operating cost = (5) + (6) + (7) + (8) = 56 + 270 + 27 + 31.25 = Rs. 384.25 h.
Total cost of tractor = a + b = 91.68 + 384.25 = Rs. 475.93/h

B. Cost of operation of reversible moldboard plow

(i) Depreciation:

$$D = \frac{86{,}000 - 86{,}000 \times 0.10}{10 \times 300} = \text{Rs. } 25.8/\text{h}$$

(ii) Interest:

$$I = \frac{86{,}000 + 86{,}000 \times 0.10}{2} \times \frac{0.10}{300} = \text{Rs. } 15.772/\text{h}$$

(iii) Housing rate:

$$HR = \frac{0.01 \times 86{,}000}{300} = \text{Rs. } 2.87/\text{h}$$

(iv) Repair and maintenance costs are considered as an essential and significant part of machinery ownership. Repairs and maintenance cost was taken 10% of the purchase price of machine per year.

$$\text{Repair and maintenance cost} = \frac{86{,}000}{300} \times 0.10 = \text{Rs. } 28.67/\text{h}$$

Total cost of plow = (i) + (ii) + (iii) + (iv) = 25.8 + 15.77 + 2.87 + 28.67 = Rs. 73.11/h

Total cost of operation = **A** + **B** = 475.93 + 73.11 = Rs. 549 h

From the test results, 0.18 ha area was plowed in 0.7075 h

Therefore, Area covered/h $= \dfrac{0.18}{0.7075} = 0.25\,\text{ha/h}$

Note: Rs. 549 are required to plow 0.25 ha of land

Cost of operation/ha $= \dfrac{549}{0.25} = \text{Rs. } 2196\,\text{ha}$

Therefore, cost for covering 0.18 ha land = 2196 × 0.18 = 395.28.

Total cost of operation= Rs. 395.28.

APPENDIX 6.III   DATA SHEET AND EVALUATION DATA.

Sr. No.	Treatment/ Replication	Forward speed (km/h)	Initial fuel level (L)	Fuel req. to refill tank (mL)	Length of coverage (m)	Time req. (s)	Width of coverage (m)	Depth of plowing (m)	Width of plowing (m)	No. wheel rotation on plowing load (rpm)	No. wheel rotation on no load (rpm)	Initial MC of soil (g)	Final MC of soil (g)	Avg. MC of soil (%)	Area (m2)	Theoretical field capacity (ha/h)	Actual field capacity (ha/h)	Field efficiency (%)	Fuel consumption (L/h) (L/ha)	Wheel slip (%)
1	S1	3.00	Full		62.00	74.00		0.20	0.27	30.00										
		3.05			62.00	73.00		0.19	0.28	25.00										
		3.00			62.00	74.00		0.20	0.27	33.00										
	Avg.	3.00		600	62.00	73.67 / 753	7.38	0.20	0.27	29.33	18.00	80	69	15.9	457.6	0.27	0.22	81.1	2.87 / 13.1	38.6
2	S2	3.33	Full		62.00	67.00		18.00	0.28	25.00										
		3.60			62.00	62.00		19.00	0.26	28.00										
		3.43			62.00	65.00		20.00	0.27	27.00										
	Avg.	3.45		750	62.00	64.66 / 672	7.29	19.00	0.27	27.00	17.00	102	90.57	12.6	452	0.3105	0.24	78	4.02 / 16.6	37
3	S3	3.70	Full		62.00	60.00		18.00	0.26	26.00										
		4.10			62.00	54.00		18.00	0.28	25.00										
		4.21			62.00	53.00		20.00	0.28	26.00										
	Avg.	4.06		840	62.00	55.66 / 591	7.23	18.67	0.27	26.00	16.00	121	106.7	13.5	448.5	0.3654	0.27	74.7	5.12 / 18.7	38.5
4	S4	4.55	Full		62.00	49.00		18.00	0.28	26.00										
		4.46			62.00	50.00		17.00	0.27	28.00										
		4.65			62.00	48.00		19.00	0.29	27.00										
	Avg.	4.55		930	62.00	49.00 / 531	7.14	18.00	0.28	27.00	16.00	120	106	13.2	442.6	0.4095	0.30	73.3	6.31 / 21.1	40.7

PERFORMANCE OF SELF-PROPELLED INTERCULTIVATOR

UDDHAO S. KANKAL[1], M. ANANTACHAR[2], and V. K. CHAVAN[3*]

[1]Department of Farm Machinery and Power, College of Agriculture & Technology, Dr. PDKV, Akola, Maharashtra, India. E-mail: uskankal@gmail.com

[2]College of Agricultural Engineering, University of Agricultural Sciences, Raichur, India

[3]AICRP for Dryland Agriculture, Dr. Panjabrao Deshmukh Krishi Vidyapeeth, Krishinagar PO, Akola 444104, Maharashtra, India. E-mail: vchavan2@gmail.com

*Corresponding author.

CONTENTS

Abstract 166

7.1   Introduction 167

7.2   Literature Review 168

7.3   Materials and Methods 169

7.4   Results and Discussion 186

Keywords 197

References 198

Appendix 7.I   Cost of Operation for Developed Self-Propelled
         Intercultivator 203

ABSTRACT

Weeding operation is one of the important secondary tillage operations which controls unwanted plants between the rows that consume more fertilizers and reduce the crop yield. Controlling weeds is one of the serious problems facing the farmers. At present, in India, weeding is done by hand hoes, which demand costly labor and consume more time. This problem becomes more acute when shortage of labor faced by farmer. In the view the above facts, the research study was conducted to evaluate the performance of self-propelled intercultivator in cotton and red gram fields. Based on field tests, the following conclusions were drawn:

- The effect of approach angle on specific draft and weeding efficiency was significant.
- The specific draft was lower and weeding efficiency was higher for 70° approach angle at all forward speeds and all moisture contents.
- The specific draft was increased as moisture content increased.
- Specific draft was moderate at 13% moisture content.
- Weeding efficiency was lower at higher forward speed but it was higher at 0.42 m/s forward speed compared to other forward speed at all moisture contents and all approach angles. Forward speed 0.42 m/s of weeder was considered optimum.
- The weeding efficiency was decreased as moisture content was increased. It was optimum at 13%.
- The draft of machine for cotton and red gram were 44 and 42.5 kg, respectively.
- The power requirement for cotton and red gram were 0.24 and 0.23 hp, respectively.
- The weeding efficiency for cotton and red gram were 87.19% and 84.19%, respectively.
- Theoretical field capacity for cotton and red gram were 0.071 and 0.072 ha/h, respectively.
- Effective field capacity for cotton and red gram were 0.053 and 0.055 ha/h, respectively.
- Field efficiency for cotton and red gram were 74.64% and 76.36 %, respectively.
- Fuel consumption for cotton and red gram were 0.68 and 0.70 L/h, respectively.
- Performance index for cotton and red gram were 1925.44 and 2013.23, respectively.

- Plants were not damaged during the weeding operation by using prototype because of wide row crop spacing of the cotton and red gram.
- It was observed that operational cost of self-propelled weeder was Rs. 1517.54/ha for cotton compared to Rs. 2000/ha for traditional method of weeding using hand khurpi. The cost saving was 24.12%.
- Operational cost was Rs. 1476.54/ha for red gram compared to Rs. 1980/ha for traditional method of weeding by using hand khurpi. The saving in cost was 25.42%.
- Bending posture adopted in conventional method is eliminated by the use of self-propelled intercultivator. The back pain by the operator is also relieved.

7.1 INTRODUCTION

In India, crops are traditionally weeded manually with traditional hand tools in upright bending posture inducing back pain for majority of labor and require considerable time and labor. Many times, the required number of laborers during peak season of the year is not available. In hand weeding, only the shoots are removed and the root portion remains in ground. Weeds again come out after 3–4 days and the earlier efforts made for removal of weeds becomes useless and cost of cultivation is increased. In single hand weeding, labor input ranges from 300 to 1200 man h/ha.[12] Weeding operation is one of the important intercultural tillage operations, which control unwanted plants between the rows that consume more fertilizers and reduce the crop yield. Controlling weed is one of the serious problems faced by the farmers.[3] The losses (up to 60%) caused by weeds in some of the crops are given in Table 7.1.[52] Weeding methods include physical methods (Hand weeding, burning, flooding, smothering), cultural methods, biological methods (animals and plants like insects, fishes, pathogens, nematodes, etc.), and mechanical methods (pulling, digging, ploughing, harrowing, and mowing). Table 7.2 reveals that about 80% of small and marginal land holdings were below 2 ha.

Most farm operations have been mechanized, whereas intercultivation operation in field crop is not mechanized due to nonavailability of suitable weeders. Indian agriculture is continued to be dependent on human and draught animal power. At present, small capacity power weeders are available in market with a field capacity of 0.07 ha/h,[77] which is greater than bullock operated weeders with a field capacity of 0.058 ha/h.[35] Development of self-propelled weeder can reduce the cost of operation and drudgery of human labor. Recently power tillers are being introduced with rotary tillage equipment having 3.75–5 kW capacity. These machines have not become popular due to clogging of weeds in between tynes and intermediate cleaning is required when used under high moisture content.

TABLE 7.1 Yield Losses in Various Crops Due to Weeds.

Crop	Reduction in yield (%)
Cotton	72.5
Gram	11.6
Groundnut	33.8
Maize	39.8
Millets	29.5
Pea	32.9
Rice	41.6
Soybean	30.5
Wheat	16.0

TABLE 7.2 Land Size Holdings in India and Karnataka.

Size	Land holding (ha)	India (%)	Karnataka (%)
Marginal	1–2	68	63
Small	2–4	18.9	21.83
Medium	4–10	5.4	10.3
Large	More than 10	7.7	4.87

Source: [www.indiastat.com[81]]

Therefore, this chapter presents research efforts to develop a self-propelled low cost drag type weeder for field crops; optimize the operational parameters of weeder in field conditions; evaluate the performance of self-propelled weeder in field conditions; and work out the cost economics of self-propelled weeder for field crops.

7.2 LITERATURE REVIEW

A mechanical weeder has been designed and developed by several investigators: Das,[11] Tajuddin,[67–70] Pradhan et al.,[48] Fielke,[19] Desai,[13] Duraiswamy and Tajuddin,[16] Pullen and Cowell,[49] Victor and Verma,[78] Manian et al.,[35] Muzumdar[39] Olunkunle and Oguntunde,[42] Pullen and Cowell,[51] Rathod,[53] Gobor and Lammers,[21] Dogherty et al.,[15] Yadav and Pund,[79] Tasliman,[71] Goel et al.,[22] Manuwa et al.,[36] Tajuddin,[69] Nkakini et al.,[41] and Rathod et al.[54]

The research studies to optimize the operational parameters of a tiller have been conducted by Thakur,[73] Biswas et al.,[8] Ben and Logue,[5] Yadav et al.,[88] and others. The performance of a mechanical weeder have been evaluated by Ambujam et al.,[2] Guruswamy,[25–27] Das,[11] Fawoll,[18] Oni,[43] Tajuddin et al.,[67–70] Fasina and Anumenechi,[17] Gupta and Pandey,[24] Varma et al.,[74] Rangasamy et al.,[52] Schwazel and Parler,[58] Pradhan et al.,[48] Irla,[30] Varshney et al.,[75] Pullen and Cowell,[50] Sogaard,[64] Piet et al.,[47] Victor and Verma,[78] Dahab and Hamad,[10] Biswas and Yadav,[7] Manian and Kathivel et al.,[35] Muzumdar,[39] Olunkunle and Oguntunde,[42] Remesan et al.,[55] Rathod,[53] Padole,[45] Dixit and Sayed,[14] Manuwa et al.,[36] Veerangauda et al.,[77]and Nkakini et al.[41]

Cost economics of mechanical weeder have been studied and estimated by Ambujam et al.,[1] Rangasamy et al.,[52] Panda,[46] Dahab and Hamad,[10] Manian and Kathirvel et al.,[35] Sthool and Shinde,[66] Muzumdar,[39] Tajuddin,[70] Padole,[45] Dixit and Syed,[14] Goel et *al.*,[22] Mynavathi et al.,[40] and Kathirvel et al.[33] They all agree that manual weeding is one of the costliest operations in crop production, and mechanization can reduce the labor cost.

7.3 MATERIALS AND METHODS

The research study included: design and development of self-propelled weeder with sweep; optimization of operational parameters of the weeder; performance evaluation of developed weeder; and cost economics of weeding operation. Based on optimal design values, the weeding unit of self-propelled weeder was fabricated and evaluated under actual field condition. Present study was conducted in Department of Farm Power and Machinery, College of Agricultural Engineering, University of Agricultural Sciences, Raichur, India.

7.3.1 *AGRONOMIC AND SOIL PARAMETERS*

The self-propelled weeder was designed and developed by considering agronomic and machine parameters. The agronomic parameters like crop, variety, row spacing, and others parameters like weeding interval and physical properties of soil were observed and considered. The row spacing of cotton and red gram varies from 900 to 1050 mm. In the view of this row spacing, overall width of machine was taken as 680 mm. The ground clearance of self-propelled weeder was chosen as 265 mm. Agronomic parameters are listed in Table 7.3.

TABLE 7.3 Agronomic parameters: Crop varieties, row to row spacing, plant to plant spacing and weeding interval of cotton, red gram, and sugarcane.

Crop	Variety	Row to row × plant to plant spacing		Weeding interval
Cotton	Bt-Cotton DHB-105	Irrigated	75 cm × 30 cm	After
	DCH-32		90 cm × 30 cm	30 DAS
	Varalaksmi	Hybrid variety	120 cm × 60 cm	40 DAS
	RAHB-87		90 cm × 60 cm	60 DAS
	Renuka			
	DHC-11			
Red gram	Maruti	90 cm × 30 cm		After 15–20 DAS
	Asha	45 cm × 10 cm		
	Ts-3			
	DWRP-1			
	S-1			
Sugarcane	SNK-044	90 cm × 30 cm		After
	CO-94012			30 DAS
	CBC-671			40 DAS
	CO-86032			60 DAS
	Nayna			

Source: Package of Practice, UAS, Raichur[44]

DAS = days after sowing.

TABLE 7.4 Soil Resistance for Various Soils.

Soil texture	Soil resistance (kg/cm2)
Clay	0.4–0.56
Heavy loam	0.5–0.75
Sandy loam	0.3
Sandy soil	0.2
Silt loam	0.35–0.5

Source: Sharma and Mukesh.[59]

The soil properties relevant to the design were identified as: soil type, moisture, and bulk density. The resistances for different soils are presented in Table 7.4. Soil moisture content affects the draft required for weeding tool of the weeder and slip of cage wheel. Soil having more moisture content gives more slip. Optimum soil moisture is needed at time of weeding to minimize the field losses and energy input. Bulk density of soil is the measure of a compaction of soil condition which influences draft required for weeding.

7.3.2 DESIGN OF SELF-PROPELLED WEEDER

Based on crop and weed parameters, it was proposed to develop a drag type self-propelled weeder for 90-cm row spacing crop. Considering the draft limitations of weeder and ensure good maneuverability, walk-behind type self-propelled weeder was designed and developed.

7.3.2.1 SELECTION OF ENGINE

According to the power requirement, commercial 4 hp, Honda GK-200 petrol start kerosene run engine was selected as the source of power. Engine was mounted in front of the cage wheel axle, so that the engine crankshaft and wheel axle were parallel to each other. The specifications of the engine are presented in Table 7.5.

TABLE 7.5 Specifications of an Engine.

Particulars	Values
Engine (model)	Honda GK-200
Stroke, cylinder	Four strokes 1 cylinder
Fuel used	Petrol start kerosene run
Max. power	4 hp
Engine displacement	197 cm^3
Engine speed	3600 rpm
Weight	18 kg
Ignition system	Point
$L \times W \times H$	330 × 380 × 425 m
Maximum torque	0.9 kg m/2500 rpm

7.3.2.2 POWER TRANSMISSION SYSTEM

The power transmission system was designed to reduce engine output shaft speed from 3600 to 23 rpm on cage wheel shaft. The power reduction was designed in three stages.

In first stage reduction, a chain and sprocket was selected. For the power 4 hp, 10-B-ISO chain number was selected as per IS:2403-1991. Number of teeth on sprocket mounted on engine shaft was assumed as 11 with speed ratio 2.27:1. Number of teeth of sprocket on counter shaft was calculated by using formula from Khurmi and Gupta[34]:

$$\text{Speed ratio} = T_2/T_1 \tag{7.1}$$

where T_1 = no. of teeth on sprocket which mounted on engine shaft; and T_2 = no. of teeth on sprocket which mounted on counter shaft.

Substituting values of T_1 = 11 and speed ratio = 2.27 in Equation (7.1), we get: T_2 = 24.97. Therefore, available sprocket with 25 teeth was selected. The speed of counter shaft in power train of transmission system is calculated as follows[34]:

$$N_1 T_1 = N_2 T_2 \tag{7.2}$$

where N_1 = speed of engine output shaft (rpm): N_2 = speed of counter shaft (rpm); T_1 = No. of teeth on sprocket which mounted on engine shaft; and T_2 = No. of teeth on sprocket which mounted on counter shaft.

Using N_1 = 3600 rpm, T_1 = 11 teeth, and T_2 = 25 teeth in Equation (7.2), we get N_2 = 1584 rpm. Therefore, in first reduction stage, engine speed reduced from 3600 rpm to 1584 rpm.

In second stage, speed reduction was obtained using crown and pinion mechanism. Number of teeth of pinion mounted on counter shaft was assumed as 8 with a speed ratio 9:1. Number of teeth on crown shaft was calculated by using formula Khurmi and Gupta[34] in Equation (7.1), with subscripts 4 and 3 instead of 2 and 1. Therefore, T_3 = number of teeth on pinion shaft = 8, and T_4 = number of teeth on crown shaft, and speed ratio = 9. With these values, we get T_4 = 72. Therefore, 72 teeth on crown gear were selected. The speed of crown shaft (N_4) in power train of transmission system was calculated using Equation (7.2), with subscripts 4 and 3 instead of 2 and 1, Therefore, N_3 = speed of pinion shaft, N_4 = speed crown shaft, T_3 = no. of teeth on pinion shaft, and T_4 = no. of teeth on crown shaft. Using N_3 = 1584 rpm, T_3 = 8 teeth, T_4 = 72 teeth in Equation (7.2), we get: N_4 = [(1584 × 8)/72] = 176 rpm. In second stage, the speed was reduced from 1584 to 176 rpm.

In third stage, speed was reduced using a set of spur gear. We assume no. of teeth of small spur gear mounted on other end of crown shaft = T_5 = 9 teeth and a speed ratio of 7.66:1. No. of teeth on spur gear mounted on cage wheel shaft is calculated using formula.[34]

$$\text{Speed ratio} = [T_6/T_5] \qquad (7.3)$$

$$N_5 T_5 = N_6 T_6 \qquad (7.4)$$

where, T_5 = no. of teeth on spur gear mounted on crown wheel shaft; T_6 = no. of teeth on spur gear mounted on cage wheel shaft; N_5 = Speed of spur gear mounted on crown wheel shaft; and N_6 = Speed of spur gear mounted on cage wheel shaft. Using speed ratio = 7.66, T_5 = 9 in Equation (7.3), we get T_6 = 7.66 × 9 = 68.94 = 69 teeth. Now using $N_4 = N_5$ = 174 rpm, T_5 = 9 teeth, T_6 = 69 teeth in Equation (7.4), we get N_6 = [176 × 9/69] = 22.95 rpm = 23 rpm. In third stage of speed reduction, speed was reduced from 176 to 23 rpm. In power transmission, engine speed reduced from 3600 to 23 rpm for weeding operation.

7.3.2.3 DESIGN OF LUGS

The lugs were welded on the outer circumference of the cage wheel. The soil acceleration force is defined in Equation (7.5).[65]

$$F_{S1} = \frac{P_g}{g} b \times d \times V_0^2 \frac{\sin\theta}{\sin(\theta + \alpha)} \qquad (7.5)$$

$$\alpha = \frac{1}{2}(90 - \varphi) \qquad (7.6)$$

$$B_0 = 3 \times F_{s1} \qquad (7.7)$$

$$M = B_0 \times L \qquad (7.8)$$

$$F_b = \frac{M}{Z_z} \qquad (7.9)$$

$$Z_z = \frac{1}{6} bt^2 \qquad (7.10)$$

$$N = \frac{\pi \times D_g}{S} \tag{7.11}$$

Where, in Equation (7.5): F_{s1} = soil acceleration force (N); b = width of penetration lugs (m); d = depth at penetration of lugs (m); V_0 = forward speed of weeder (m/s); θ = tool lift angle (°); α = angle of forward failure surface (°); ρ = bulk density of soil (kg/m³); g = gravitational force (m/s²); and in Equation (7.6): φ = angle of internal friction; and in Equation (7.7): B_0 = total soil acceleration force on cage wheel (N); and in Equation (7.8): M = maximum bending moment on lug (N mm); L = distance between point of action of soil resistance and top edge of cage wheel (mm); and in Equation (7.9): F_b = stress induced on the material of lugs (N/mm²); Z_z = section modulus of cage wheel (mm³); and in Equation (7.10): t = thickness of lugs (mm); b = width of lugs (mm); and in Equation (7.11): D_g = diameter of cage wheel (mm); S = spacing between lugs; N = numbers of lugs.

The size of each lug on cage wheel was selected as 25 mm width × 10 mm thickness. The projection of lugs is considered from the tip of circumference of cage wheel = 18 mm = depth of lugs penetrated in the soil. Lugs are welded perpendicular to ground wheel with 90° to soil surface. The bulk density of soil was 1500 kg/m³. Internal angle of friction was assumed as 36°. The maximum forward speed was 2.5 km/h. Angle of forward failure surface is calculated using Equation (7.6): α = [1]/[90 − 36] = 27°. Using values of all known parameters in Equation (7.5), we get:

$$F_{S1} = \frac{1500 \times 9.81}{9.81} 0.025 \times 0.018 \times 0.7^2 \frac{\sin 90}{\sin(90 + 27)} = 0.330\,\text{N}$$

Considering three lugs are in contact with soil, total soil acceleration force from Equation (7.7), we get: B_0 = 3 × 0.330 = 0.99 N. Using Equation (7.8), maximum bending moment,[34] M = 0.99 × 9 = 8.91 N mm. The section modules for rectangular section is given by Varshney et al.[76] in Equation (7.10) and is Z_z = [1/6][25 × 5²] = 104.16 mm³. Bending stress produces (0.085 N/mm²) is less than allowable bending stress 70 N/mm². Hence, design is safe. Using spacing between lugs as 67 mm = S, number of lugs wheel is obtained from Equation (7.11) = N = [3.412 × 600/67] = 28.11. Hence, 28 lugs were provided on the cage wheel. The 28 lugs were welded on the outer periphery of the wheel at 67-mm equal interval to facilitate good traction in soil. The lugs 18 mm in height were welded at an angle of 25°–30° with the axis of rotation to reduce the slip.

7.3.2.4   DESIGN OF SWEEP BLADE

Sweep type blade was selected and fixed on self-propelled weeder frame. The performance of sweep blade was better than straight and curved blade with minimum draft force per unit working width and having the highest performance index reported by Biswas and Yadav.[7] Design Equations (7.12)–(7.14) were used to design a sweep blade. While designing the sweep blade, following assumptions were made:

    a.   Row to row spacing = 90 cm
    b.   Depth of the cut (d) = 5 cm[70]
    c.   Crop protection zone = 15 cm
    d.   Angle of internal friction, $\phi_s = 25°$[59]

The cutting width of the sweep tyne[59]

$$S_c = Z_f - Z_p \text{ or} \tag{7.12}$$

$$Z_f = S_c - Z_p \tag{7.13}$$

Effective soil failure zone[59]

$$Z_f = \left[W + 2d \tan \phi_s \right] + 2\left[\left[W_1 + 2d \tan \phi_s \right] \right] \tag{7.14}$$

In Equations (7.12)–(7.14): S_c = row spacing in cm, Z_f = effective soil failure zone, Z_p = crop protection zone (protection zone is multiplied by 2 since protection zone has to be provided on both side of the crop); W = width of full sweep in cm, W_1 = width of half sweep in cm and here $W = 2W_1$. Using $S_c = 90$ and $Z_p = 2 \times 15$ in Equation (7.13), we get $Z_f = 60$ cm. Using $Z_f = 60$ cm, $d = 5$ cm, angle = 25°, and $W = 2W_1$ in Equation (7.14), we get $W = 25.4$ cm = 250 mm (in this chapter) = width of sweep. While designing the sweep, the apex angle, condition for easy undercutting the weeds by the sweep blade should be considered. In practice, approach angle of sweeps ranges from 60° to 90°. Four shape of sweep with different approach angle, namely, 60, 70°, 80°, and 90° were fabricated using 20MN-CR-5 grade steel. The sweeps were attached to the shank with the help of nut and bolt. During the fabrication of sweep, parameters were followed according to the test code (IS: 6451-1972).

7.3.2.5   DESIGN OF SHANK OF THE INTERCULTIVATOR

The square shank was designed to have proper fixing on tool frame of self-propelled intercultivator. Following assumptions were considered while designing shank:

1.   For design of the shank, unit draft of the soil = 0.75 kg/cm²;
2.   Width of the sweep = 25 cm (calculated using Eq. (7.14));
3.   Depth of the soil = 5 cm;
4.   Factor of safety = 2;[59]

$$\text{Draft} = \text{Soil resistance} \times \text{Cross sectional area of cut} \qquad (7.15a)$$

$$= 0.75 \times (1/2) \times 25 \times 5 = 43.75 \text{ kg} = 87.5 \text{ kg for two sweeps.}$$

Power requirement for self-propelled intercultivator[57]
$$= [\text{Draft} \times \text{speed}]/75 \qquad (7.15b)$$

$$= [87.5 \times 0.7]/75 = 0.81 \text{ hp}$$

Therefore, according to the power requirement, commercial 4 hp (Honda GK-200) petrol start kerosene run engine was selected as the source of power.

The maximum draft on sweep type tyne
$$= [\text{Draft} / \text{no. of tynes}] \qquad (7.15c)$$

$$= 87.5 /2 = 43.75 \text{ kgf} = \text{It is a maximum force at the tip of the sweep.}$$

Taking factor of safety 2 and taken 2 times of maximum force for impact loading,

Bending sweep in sweep = 43.75 × 2 × 2 = 175 kgf

Let the height of the shank suitable for the cotton crop be 500 mm.[59]

∴ The maximum-bending moment (M) for the cantilever length of 500 mm = M = 175 kgf × 500 mm = 87,500 kgf mm.

Bending stress is given below[59]: $f_b = MC/I$ or $Z = I/C = M/f_b$ (7.16)

Or $Z = \dfrac{87,500}{30} = 2916.66 \text{ mm}^3 = b^3/6 \text{ or } b = \sqrt[3]{2916.66 \times 6} = 25.9 \text{ mm}$

Therefore, a square shank was made mild of steel material in size 25 × 25 mm. Projected end of shank was fitted to the sweep with nut and bolt and other end to the tool frame of self-propelled intercultivator with the help of clamp.

TABLE 7.6   Components of Self-propelled Intercultivator.

Component name	Material	Section	Dimension (mm)	Quantity
Implement frame	Mild steel	ISA 40406	780 × 220 × 35	1
Shank	Carbon steel	–	25 × 25 × 500	2
Sweep	Spring steel, 20 MNCR-5	V-shape	250 × 4	4
Clamp	Carbon steel	–	140 in length	2
Cage wheel	Mild steel	Round	10 Φ	2
Lugs on cage wheel	Mild steel	Square	25 × 25	28
Spoke	Mild steel	Curved	220 curved length	8
Cage wheel shaft	45-C8	Round	25Φ	2
Sprocket	CI	Round	11 and 25 teeth	2
Roller chain	St-50	–	Std. pitch 15.87	1
Handle	Mild steel	–	25Φ, 560 Curved length	2
Handle grip	Rubber	–	110 length	2
Hitch pin	Carbon steel	–	20Φ	1

7.3.2.6   DEVELOPMENT OF SELF-PROPELLED INTERCULTIVATOR

The prototype was fabricated based on the dimensions obtained from design. The prototype consisted of tool frame, power transmission system, cage wheel, sweep blade, shank, stand, and handle. Components of prototype are presented in Table 7.6. The specification of prototype are presented in Table 7.7. The overall dimensions of prototype were 1830 mm in length, 780 mm in width and 1080 mm in height. The self-propelled weeder was fabricated by using standard production techniques. The prototype is shown in Figure 7.1.

The **cage wheel** provides good traction, in addition to saving in cost compared to pneumatic wheel. The cage wheels were used to get traction under field conditions. Two steel lugged cage wheels of 600-mm diameter were mounted on opposite end of the cage wheel shaft both ends of central shaft connected to the transmission box. The spacing between two wheels can be adjusted based on row spacing of crop. MS rods of 10-mm diameter 8 in numbers were welded as spokes on the central hub of cage wheel. The 50-mm long hub was made to fit on the 25-mm diameter cage wheel shaft.

TABLE 7.7   Specifications of Self-propelled Intercultivator.

Particulars	Details
Name of machine	Engine operated power weeder
Make of machine	CAE, Raichur
Model of machine	MPIC-63001
Overall dimension of machine ($L \times B \times H$)	1830 × 780 × 1080 mm
Weight of machine	56 kg
Power source	4 hp petrol start kerosene run engine
Fuel used	Kerosene
Fuel tank capacity	3.9 L
Engine details	4 stroke, 1 cylinder
Speed at engine	3600 rpm
Displacement	197 cm³
PTO shaft rotation	Counter clockwise from drive end
Weight of engine	18 kg
Ground clearance	265 mm
Gear type	Bevel and spur gear
Chain drive	ISO 10 B bush roller chain
Clutch	Dog clutch
Axle	25 mm in diameter
Cage wheel	600 mm in diameter
Lug	28 no. 25 × 25 mm in size lugs welded at periphery of ground wheel
Details of weeding components	
Frame dimension ($L \times B$) mm	780 × 220 mm
Type of blade	Sweep type
No of blade	6
Distance between blade	Adjustable
Shank	25 mm in dia. and 500 in length

The **trapezoidal tool frame** of size 760 × 680 × 220 mm was fabricated using 35 × 35 × 6 mm MS angle. Front end of tool frame attached to the power transmission box.

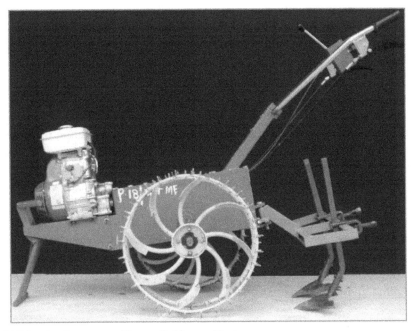

FIGURE 7.1   Prototype self-propelled intercultivator.

Speed control was done through accelerator (engine throttle). It was a knob, which was provided on or near to the handle to facilitate the easy access for the operator. It was connected to the governor. When it was rotated clockwise, more fuel was injected to engine speed increases and vice-versa.

Two handles were made of circular steel pipe 40 mm in diameter was attached to the main frame of the machine. It can be adjusted as to suit the ergonomic working height of the operator. Two grips were provided on the handle because handle yoke affords maneuverability so that it will be affected smoothly and uniformly. Two turning clutches were provided to the both side of the handle.

Stand was made of MS rod and a spring is provided to make the stand in position.

7.3.3   OPTIMIZATION OF OPERATIONAL PARAMETERS OF SELF-PROPELLED INTERCULTIVATOR

The soil of North Karnataka is characterized under 2 and 3 regions as black soil that contains predominantly montmorillonite clay. The soil has good moisture holding capacity, and it swells considerably with moisture content. When dry, soil shrinks and forms cracks. The soil used for testing of sweeps in field has clay texture with 1.84% fine sand, 12.95% coarse sand, 30.51% silt, and 54.70% clay.

Three different shapes of sweeps were tested in field condition to obtain design parameters of sweeps under black soil. It was observed that sweep was most suitable soil working tool under black soil conditions.[8,74] Barnacki et al.[6] recommended that the apex angle should be in the range of 60–90°. Three ranges of approach angles 60°, 70°, and 80° were selected for study. It was decided to test the weeder with different geometry of sweep that had small width but in gang instead of single sweep. A gang width of 300 mm was selected for study. The range of speed was selected from 0.28 to 0.56 m/s, which was ergonomically suited for walking behind implements.[70] The workable range of moisture was from 13% to 18% considering the field capacity and wilting point. The dependent variables were specific draft and weeding efficiency. The levels of variables for evaluation of sweeps are presented in Table 7.8.

TABLE 7.8   Values of Variables for Evaluation of Prototype.

Parameters	Levels	Symbols
	Independent	
Approach angle	3	$\theta_1 = 60°$, $\theta_2 = 70°$, $\theta_3 = 80°$
Forward speed	3	$S_1 = 0.28$m/s, $S_2 = 0.42$m/s, $S_3 = 0.56$m/s
Moisture content	3	$M_1 = 13\%$, $M_2 = 15\%$, $M_3 = 18\%$
	Dependent	
Specific draft	–	S_p
Weeding efficiency	–	E_w

7.3.3.1   FIELD EVALUATION OF PROTOTYPE

The field was leveled and compacted. Water was sprinkled on the soil to maintain desired soil moisture. A hand-held cone penetrometer was used to measure the cone penetration resistance of the soil. The penetrometer had an angle of 30° and base area of 491.1 mm². Readings were taken up to a soil depth of 100 mm. The value of cone penetration resistance was reasonably uniform and ranged from 0.120 to 0.121 N/mm². The layout of experiment is shown in Figure 7.2.

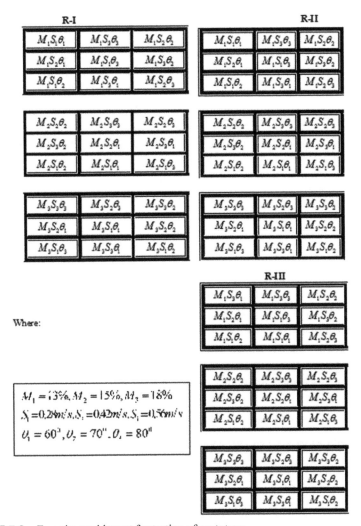

FIGURE 7.2 Experimental layout for testing of prototype.

The experiment was conducted in MARS field. The soil moisture was maintained at nearly 13%. The sweeps of 60° approach angle were mounted on tool frame with gang of 300 mm. The auxiliary mini power tiller was hitched to self-propelled intercultivator with the help of hook joint. A hydraulic dynamometer was connected in between self-propelled prototype and auxiliary power tiller. The no load draft was calculated without fixing the sweeps to tool frame. The load draft was calculated fixing the sweeps to tool frame. Total draft exerted on the wedding tool (sweep) was computed using the formula[4]: $D = D_1 - D_2$, where, D = draft of sweep in kg; D_1 = draft of sweep when sweep in operating condition in kg; D_2 = draft of sweep when sweep not operating condition in kg. The total draft computed (D) was divided by gang width to obtain draft per working width termed as specific draft.

The weeds were allowed to grow in test plots. Depth of operation was 50 mm. The horizontal force was recorded by using hydraulic dynamometer. The prototype was pulled at 0.28 m/s speed. The undisturbed weeds were counted. After weeding, undisturbed weeds (W_2) and disturbed weeds (W_1) were counted from 1-m^2 test plot and were measured. The weeding efficiency was calculated by using the formula:

$$E_W = \frac{W_1}{W_1 + W_2} \times 100 \tag{7.17}$$

Then, the field was maintained at 13% moisture content. The same procedure was repeated with at the speed of 0.42 m/s, and in another field at 0.56 m/s, and so on. The identical test was conducted at three approach angles (60°, 70°, 80°) of tools at 15% and 18% moisture contents. The experiment was replicated thrice for each combination of approach angle, forward speed and moisture content, for specific draft and weeding efficiency. The details of treatments combination are presented in Table 7.9. The data were subjected to statistical analysis.

TABLE 7.9   Values of Parameters Selected for Field Testing.

Moisture content (%)	Speed (m/s)	Approach angle (°)		
M	S	θ		
13	0.28	60	70	80
	0.42	60	70	80
	0.56	60	70	80
15	0.28	60	70	80
	0.42	60	70	80
	0.56	60	70	80
18	0.28	60	70	80
	0.42	60	70	80
	0.56	60	70	80
Replications		Three		
Total number of treatments		3(M) × 3(S) × 3(θ) = 27		

The geometry of weeding tool was selected having a lower draft and higher weeding efficiency at optimum approach angle, speed, and moisture content. The optimum dimensions of weeding tool were computed and used to test the prototype self-propelled intercultivator.

The RNAM test code and test procedure[4,56] were followed for field testing, to cover an area of 450 m². During the field testing, following parameters were recorded:

a. FIELD AND SOIL:
- location
- kind of field
- length
- width
- area
- soil type
- soil moisture
- bulk density

b. CROP:
- crop type and variety
- planting method
- row spacing
- crop height
- crop age
- plant population

c. WEED:
- weed type
- weed population
- weight of weed/m²
- weed height

d. OPERATOR:
- skill
- wages

7.3.3.2 MEASUREMENT OF DIFFERENT PARAMETERS

7.3.3.2.1 Soil Moisture Content

Soil moisture content on dry basis was measured using a procedure by Mohsenin[38] and oven dry method. Five soil samples were collected at random from test plots, weight of each sample was taken using an electronic balance (w_1). Then, these samples were kept in hot air oven at 105°C for 24 h. After that, samples were taken out from the oven and kept in desiccators. The dry weight of sample was recorded by using electronic balance (w_2). The moisture content on dry basis was calculated using following formula:

$$\text{Moisture content (\%) on dry basis} = 100 \times [(w_1 - w_2)/w_2] \qquad (7.18)$$

7.3.3.2.2 Bulk Density

It is the ratio of mass of soil sample to the volume of core cutter. The bulk density of soil was determined by the procedure by Mohsenin.[38] Three soil samples were collected at different locations at random in the test plot using cylindrical core sample. The diameter (D) and length (L) of cylindrical core sampler were measured. The soil samples were kept in hot air oven at 105°C for 24 h. After that, soil sample were taken out and kept in desiccators. The dry weight of soil samples was measured (M). The bulk density of soil (ρ) was calculated by following formula:

$$\rho\left(\text{g/cm}^3\right) = \frac{[\text{soil mass}]}{[\text{volume of core sampler}]} = \frac{[M]}{\left[\pi D^2 L\right]} \qquad (7.19)$$

7.3.3.2.3 Traveling Speed (km/h)

For calculating traveling speed, two poles 20 m apart were placed approximately in middle of the test run. On the opposite side also, two poles were placed in the similar position 20 m apart so that four poles would form corners of a rectangle, parallel to the long side of the plot. The speed was calculated from the time required for machine to travel the distance (20 m) between two poles. Average of such readings was taken to calculate the traveling speed of prototype. The forward speed of operation was calculated by observing the distance traveled and time taken:

$$S(\text{m/s}) = \left[\frac{\text{Distance traveled (m)}}{\text{Time taken (s)}}\right] \qquad (7.20)$$

7.3.3.2.4 Power Requirement

Power requirement was calculated by using the following formula[57]:

$$\text{Power requirement (hp)} = \frac{\text{Draft (kg)} \times \text{Speed (m/s)}}{75} \qquad (7.21)$$

7.3.3.2.5 Weeding Efficiency

The weeding efficiency was defined in Equation (7.17) above in this section.

7.3.3.2.6 Theoretical Field Capacity

Theoretical field capacity (ha/h) was calculated by using following formula[37]:

$$\text{TFC} = \frac{S \times W}{10} \tag{7.22}$$

where W = theoretical width of machine (m) and S = speed of operation (km/h).

7.3.3.2.7 Effective Field Capacity

For calculating effective field capacity (EFC, ha/h), the time consumed for actual work (T_p) and lost for other activities (nonproductive time, T_1) such as turning and cleaning blade when clogged with weeds were taken into consideration. Effective actual field capacity for an area under test (A, ha) was calculated by following formula.[37]

$$\text{EFC} = \frac{A}{T_p + T_1} \tag{7.23}$$

7.3.3.2.8 Field Efficiency

Field efficiency is a ratio of EFC to theoretical field capacity.[37]

$$\text{Field efficiency (\%)} = \frac{\text{EFC}}{\text{TFC}} \times 100 \tag{7.24}$$

7.3.3.2.9 Fuel Consumption (W, m³/h)

Fuel consumption was measured by recording time required (T, s) and the quantity of fuel consumed for specified length of run (V_f, cm³) and the fuel consumption was calculated on hourly basis as follows[37]:

$$W_f = \frac{36 \times V_f \times 10^{-4}}{T} \tag{7.25}$$

7.3.3.2.10 Plant Damage

Plant damage[7] was calculated by counting the number of plants in 10 rows before weeding (P) and number of the plant damaged in 10-m row length after weeding (Q).

$$\text{Plant damage } (\%) = \frac{Q}{P} \times 100 \qquad (7.26)$$

7.3.3.2.11 Performance of Index

Field performance of tools were assessed through the performance index (P) suggested by Gupta et al.[24] It was calculated by using the following formula.

$$P = \frac{a \times q \times e}{p} \qquad (7.27)$$

where a = output (ha/h); q = (100 − % plant damaged); e = weeding index (%); and p = power input.

7.3.4 ECONOMICS OF WEEDING OPERATION BY USING SELF-PROPELLED INTERCULTIVATOR

The cost of operation of the prototype was determined as per the specifications of the Bureau of Indian Standards.[29] The sample calculation of cost economics of weeding with the prototype is presented in Appendix 7.I.

7.4 RESULTS AND DISCUSSION

7.4.1 OPTIMIZATION OF OPERATIONAL PARAMETERS OF SELF-PROPELLED INTERCULTIVATOR

The effects of operational parameters were studied to evaluate the performance of self-propelled intercultivator in terms of specific draft and weeding efficiency. Statistical analysis was carried out to study the effects of operational parameters. A three factor completely randomly block design and factorial variance analysis techniques were used to analyze the effects of approach angles, forward speeds and moisture contents on the performance of prototype. In addition, two and three ways interactive effects were also analyzed and significance levels are presented.

7.4.1.1   EFFECTS OF OPERATIONAL PARAMETERS (APPROACH ANGLE, FORWARD SPEED, AND SOIL MOISTURE CONTENT) ON SPECIFIC DRAFT

Table 7.10 shows the effects of approach angle, forward speed, and soil moisture content on specific draft of prototype. It can be observed that the specific draft at 13% soil moisture content varied from 0.399 to 0.480 N/mm at different approach angles and different forward speeds. A maximum specific draft of 0.480 N/mm was recorded at a forward speed of 0.56 m/s with an approach angle of 80°, while it was minimum (0.399 N/mm) at forward speed of 0.28 m/s and with an approach angle of 70°. At 15% moisture content, the specific draft varied from 0.590 to 0.681 N/mm at different approach angles and different forward speeds. A maximum specific draft (0.681 N/mm) was recorded at forward speed of 0.56 m/s with an approach angle of 80°, whereas it was minimum (0.590 N/mm) at forward speed of 0.28 m/s with an approach angle of 70°. At 18% soil moisture content, the specific draft varied from 0.665 to 0.841 N/mm at different approach angles and different forward speeds. A maximum specific draft (0.841 N/mm) was recorded at forward speed of 0.56 m/s with an approach angle of 80°, while it was minimum (0.665 N/mm) at forward speed of 0.28 m/s with an approach angle of 70°.

TABLE 7.10   Effects of Moisture Content (*M*), Forward Speed (*S*), and Approach Angle (*θ*) for Specific Draft.

Moisture content (M) (%)	Forward speed (S) m/s	Specific draft (N/mm)		
		Approach angle (θ) (°)		
		60	70	80
13	0.28	0.426	0.399	0.455
	0.42	0.440	0.410	0.469
	0.56	0.458	0.431	0.480
15	0.28	0.642	0.590	0.650
	0.42	0.658	0.619	0.669
	0.56	0.670	0.628	0.681
18	0.28	0.770	0.665	0.811
	0.42	0.791	0.678	0.829
	0.56	0.815	0.690	0.841

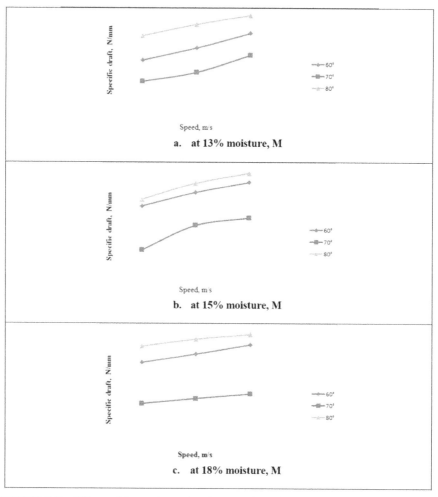

FIGURE 7.3 Effects of forward speed and approach angle on specific draft at 13% (top), 15% (center), and 18% (bottom) moisture content.

From Table 7.10 and Figure 7.3, it is evident that the specific draft was increased as the forward speed was increased for all approach angles of sweep. However, the lowest specific draft was observed at forward speed of 0.28 m/s for an approach angle of 70°. Similar results were observed at all soil moisture contents.

The individual and combined effects of operational parameters on specific draft were analyzed statistically. Table 7.11 reveals that the effects of speed of operation (S), moisture content (M), and approach angle (θ) significantly influenced the specific draft at 1% level of significance and each variable individually influenced the specific draft. The significance was observed in order of moisture content (M) followed by approach angle (θ) and speed of operation (S).

TABLE 7.11 Analysis of Variance for Specific Draft.

Source of variance	DF	SS	MS	F-ratio
Replication (R)	2	0.00163351	0.00081675	21.34**
Treatment	26	1.57094232	0.06042086	1578.44**
Speed of operation (S)	2	0.01293306	0.00646653	168.93**
Moisture content (M)	2	1.29919899	0.64959949	16970.25**
Approach angle (θ)	2	0.16848780	0.08424390	2200.80**
$S \times M$	4	0.00016316	0.00004079	1.07 ns
$S \times (\theta)$	4	0.00012924	0.00003231	<1
$M \times (\theta)$	4	0.08933264	0.02233316	583.44**
$S \times M \times \theta$	8	0.00069743	0.00008718	2.28*
Error	52	0.00199049	0.00003828	
Total	80	1.57456632		

cv = 1.0%; **significant at 1% level; *significant at 5% level; ns = not significant.

The interaction effects of variables ($S \times M$) were not significantly influenced the specific draft. The interaction effect of variables ($M \times \theta$) was significantly influenced the specific draft at 1% level significance. The interaction effect of variables ($S \times \theta$) was not significantly influenced the specific draft. The influence of interaction effect levels on specific draft are in order of ($M \times \theta$) followed by ($S \times M$) and ($S \times \theta$). The combined effect of variables ($M \times S \times \theta$) also significantly influenced the specific draft at 5% level of significance. Since the combination effect of variables on specific draft was significant, optimum operational parameters were considered for design of sweep type weeder.

The correlation matrix between speed of operation (S), moisture content (M), and approach angle (θ) for specific draft is presented in Table 7.12. The effect of forward speed was positively correlated with specific draft. The moisture content was positively correlated with the specific draft at 5% level of significance. The approach angle was positively correlated with specific draft.

TABLE 7.12 Correlation Matrix between Approach Angle (θ), Speed of Operation (S), and Moisture Content (M) for Specific Draft (Y).

Parameters	Specific draft (N/mm)(Y)	Approach angle (θ) (°)	Moisture content (%) (M)	Speed of operation (m/s) (S)
Specific draft (Y)	1.000	0.070	0.865**	0.091 ns
Approach angle (θ) (°)	0.070	1.000	0.000ns	0.000 ns
Moisture content (%) (M)	0.865	0.000	1.000	0.000 ns
Speed of operation (m/s) (S)	0.091	0.000	0.000	1.000

**Significant at 5% level, ns = Non-significant.

The increase in draft with speed is reported by other investigators and by Yadav et al.[80] The lowest draft at 70° approach angle has been recorded due to physiomechanical properties of soil, trihedral wedge theory, theory of rupture, and cutting theory given by Goryachkin[23] and Sineokov.[62] Change in approach angle causes change in flow pattern of the soil along tool surface. The change inflow pattern causes significant variation in draft.[20]

To prevent sticking, the direction of cut can be made normal to the working face of the wedge. For that is necessary, that $\Psi = 90° - \alpha$, where, $\Psi =$ is the angle of the internal friction and $\alpha =$ is cutting angle. The forces acting on the implements are (1) forces causing forward travel of the weeder (P) which acts at right angle to the horizontal; (2) weight of the implement (W); (3) the resistance of the working surface; and (4) the reactions of the supporting surfaces, R. The reaction of the supporting surfaces (R) includes normal and tangential frictional forces. The frictional forces can be eliminated by changing the inclination of the working and supporting planes by angle of friction. The resultant of the all above forces must lie in same line. For the minimum draft, the forces causing forward speed must make an angle with the horizontal equal to the angle of the friction. The draft in present study was the lowest for 70° approach angle, which might be fulfilling this condition. The results are agreement with the findings reported by Biswas et al.[8] and Tewari et al.[72]

It was observed that the relation between specific draft and forward speed is linear. The specific draft was lower at 0.28 m/s. Specific draft was increased with the increase in forward speed. The increase in draft with speed might be explained by change in zone of influence and strain hardening.[61] Also soil strength increases as the rate of the shear is increased. When the sweep tool is operated at higher speed, instead of inverting and throwing the soil, sweep carries soil along with it, which results in bulking and heaving of soil the implement base. Increase in specific draft requirement with increase in speed has been reported by Shrestha et al.[60]

The relationship between specific draft and moisture content was linear. Specific draft to a particular limit, the increase of moisture increases the frictional coefficient. This increase is explained by the growth in the forces of molecular attraction of the soil particles to the surface. With increase in unit pressure on the surface of contact, adhesiveness is increased, which depends on the weight of furrow slice. Therefore, increase in frictional coefficient and adhesiveness might be the reason for higher specific draft at higher soil moisture.

7.4.2.2   EFFECTS OF APPROACH ANGLE, FORWARD SPEED, AND MOISTURE CONTENT ON WEEDING EFFICIENCY OF PROTOTYPE

Effects of approach angle, forward speed and soil moisture content on weeding efficiency are presented in Table 7.13. It can observed that at 13% soil moisture content, the weeding efficiency varied from 75.23% to 85.21% at different approach angles and different forward speeds. A maximum weeding efficiency (85.21%) was recorded at forward speed of 0.42 m/s with an approach angle of 70°, while it was minimum (75.23%) at forward speed of 0.56 m/s with an approach angle of 60°.

TABLE 7.13   Effects of Moisture Content (M), Forward Speed (S), and Approach Angle (θ) on Weeding Efficiency.

Moisture content (M) (%)	Forward speed (S) (m/s)	Weeding efficiency (%)		
		Approach angle (θ) (°)		
		60	70	80
13	0.28	77.00	81.13	79.20
	0.42	79.20	85.21	81.56
	0.56	75.23	81.00	77.5
15	0.28	80.07	85.23	81.37
	0.42	83.76	89.58	85.19
	0.56	76.23	82.00	80.00
18	0.28	75.13	80.00	75.67
	0.42	76.99	83.23	79.38
	0.56	73.57	78.9	76.08

It was also observed that at 15% moisture content, the weeding efficiency varied from 76.23% to 89.58% at different approach angles and different forward speeds. A maximum weeding efficiency (89.58%) was recorded at forward speed of 0.42 m/s with an approach angle of 70°, whereas it was minimum (76.23%) at forward speed of 0.56 m/s with an approach angle of 60°. It was found that at 18% soil moisture content, the weeding efficiency varied from 73.57% to 83.23% at different approach angles and different forward speeds. A maximum weeding efficiency of 83.23% was recorded at forward speed of 0.42 m/s with an approach angle of 70°, while it was minimum (73.57%) at forward speed of 0.56 m/s with an approach angle of 80°.

The effects of approach angle, forward speed on weeding efficiency at 13%, 15%, and 18% moisture content have been presented in Figure 7.4. It is evident that the weeding efficiency was decreased as the forward speeds increased for all approach angles. However, higher weeding efficiency were observed at forward speed of 0.42 m/s for approach angle of 70°. Similar results were observed for all soil moisture contents.

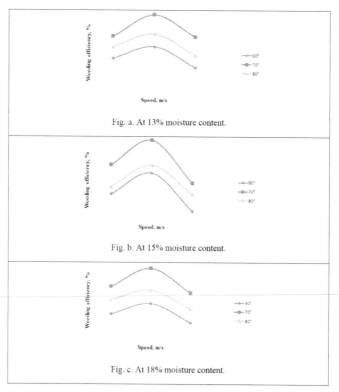

Fig. a. At 13% moisture content.

Fig. b. At 15% moisture content.

Fig. c. At 18% moisture content.

FIGURE 7.4   Effects of forward speed and approach angle on weeding efficiency at 13% (top), 15% (center), and 13% (bottom) moisture content.

The individual and combined effect of operational parameters on weeding efficiency was analyzed statistically and presented in Table 7.14. The effects of speed of operation (S), moisture content (M), and approach angle (θ) significantly influenced the specific draft at 1% level of significance and each variable individually influenced the weeding efficiency. The significant results were observed in the order of: moisture content (M) followed by approach angle (θ) and speed of operation (S).

TABLE 7.14 Analysis of Variance for Weeding Efficiency.

Source of variance	DF	SS	MS	F-ratio
Replication (R)	2	55.401965	27.700983	36.63**
Treatment	26	1193.330180	45.897315	60.69**
Speed of operation (S)	2	251.488225	125.744112	166.27**
Moisture content (M)	2	519.245551	259.622775	343.30**
Approach angle (θ)	2	391.902625	195.951312	259.11**
$S \times M$	4	5.110849	1.277712	1.69 ns
$S \times \theta$	4	14.035331	3.508833	4.64**
$M \times \theta$	4	6.467916	1.616979	2.14 ns
$S \times M \times \theta$	8	5.079684	0.634960	<1
Error	52	39.325501	0.756260	
Total	80	1288.057647		

cv = 1.1%; **significant at 1% level; ns = non-significant.

The interaction effects of ($S \times M$) did not significantly influence the weeding efficiency. The interaction effect of ($S \times \theta$) did significantly influence the specific draft at 1% level significance. The interaction effect of ($M \times \theta$) did not significantly influence the weeding efficiency. The influence of interaction effect levels on specific draft are in order of ($S \times \theta$) followed by ($M \times \theta$) and ($S \times M$). The combined effect of ($M \times S \times \theta$) also did not significantly influence the weeding efficiency. Since the combination effect of variables on weeding efficiency was significant, optimum operational parameters were considered for design of sweep type weeder.

The correlation matrix among speed of operation (S), moisture content (M), and approach angle (θ) for weeding efficiency is presented in Table 7.15. The effect of approach angle was positively correlated with weeding efficiency at 1% level significance. The moisture content was negatively correlated with the weeding efficiency at 1% level of significance. The speed of operation was negatively correlated with weeding efficiency.

TABLE 7.15   Correlation Matrix between Approach Angle (θ), Speed of Operation (S), and Moisture Content (M) for Weeding Efficiency.

Parameters	Weeding efficiency (Y) (%)	Approach angle (θ) (°)	Moisture content (M) (%)	Speed of operation (S) (m/s)
Weeding efficiency (Y) (%)	1.000	0.524**	−0.825**	−0.154 ns
Approach angle (θ) (°)	0.524	1.000	0.000 ns	0.000 ns
Moisture content (M) (%)	−0.825	0.000	1.000	0.000 ns
Speed of operation (S) (m/s)	−0.154	0.000	0.000	1.000

ns = Non-significant; **significant at 1% level; *significant at 5% level.

It is evident from results in this chapter that the weeding efficiency is higher for 70° approach angle sweep for the all forward speeds and all moisture contents. Similar findings reported by Yadav et al.[80] Weeding efficiency was decreased with increase in moisture content. This is due to the fact that moisture content increases the slippage of ground wheel of self-propelled weeder which considerably affects the turning length of weeder. From the field results, weeding efficiency was more at 13% and 15% moisture contents compared to 18% moisture. Weeding efficiency is higher for 300-mm gang width for all moisture contents due to higher gang width, due to more area coverage between the rows. Weeding efficiency was lower in case of higher forward speed but it was higher at 0.42 m/s forward speed compared to other forward speeds at all moisture contents and all approach angles.

7.4.2.3   OPTIMIZATION OF FORWARD SPEED, APPROACH ANGLE, AND MOISTURE CONTENT FOR SWEEP

The operational parameters of sweep were optimized based on lower specific draft and higher weeding efficiency and are presented in Table 7.16. The specific draft was lower and weeding efficiency was higher for 70° approach angle of sweep. Therefore sweep of 70° approach angle of sweep was considered optimum design value. The weeding efficiency was higher at 0.42 m/s speed, since it provides higher field capacity and also gives more weeding efficiency, therefore it was considered optimum. Similarly specific draft and weeding efficiency were moderate at 13% and 15% moisture contents.

However, 15% moisture content was considered optimum as it gives reasonably higher working range. Similar findings reported by Yadav et al.[80]

TABLE 7.16 Optimum Design and Operational Parameters of Self-propelled Weeder.

Particulars	Optimum dimensions
Approach angle (°)	70
Forward speed (m/s)	0.42
Moisture content (%)	15

7.4.3 FIELD EVALUATION OF SELF-PROPELLED INTERCULTIVATOR UNDER OPTIMUM CONDITIONS

The prototype was evaluated for optimum parameters at 15% moisture content, for 70° approach angle and 0.42 m/s forward speed. The prototype was evaluated in a test plot of 30 m × 15 m for actual field capacity, theoretical filed capacity, fuel consumption, specific draft, and weeding efficiency. Parameters for the performance test are indicated in Table 7.17.

TABLE 7.17 Performance Evaluation of Self-propelled Weeder in Cotton and Red Gram.

Particulars	Cotton	Red gram
Actual operating time (min)	49.80	48.3
Time loss in turning (min)	5.48	6
Forward speed (m/s)	0.42	0.42
Area covered (m²)	450	450
Effective width of cut (mm)	476	479
No of runs required in between one rows (run)	2	2
Depth of cut (mm)	54	52
Effective field capacity (ha/h)	0.053	0.055
Theoretical field capacity (ha/h)	0.071	0.072
Field efficiency (%)	74.64	76.36
Draft (kg)	44	42.5
Weeding efficiency (%)	87.19	84.19
Power requirement (hp)	0.24	0.23
Fuel consumption (L/h)	0.68	0.70
Percentage of plant damaged (%)	Nil	Nil
Performance index	1925.44	2013.23

Soil moisture content affects draft of weeder and wheel slip. Soils having more moisture content give more slip and hence increase the draft. Weeder was tested at 15% moisture content. Soil bulk density is a measure of soil compaction. The soil bulk density ranged from 1.52 to 1.55 g/cm³.

Crop variety is an important parameter, which influence the mechanical weeding operation since the growth factor and foliage varies for each variety. In this study, the Bt cotton and Asha variety of red gram were used. The cotton crop was grown at 90-cm row spacing. At the time of testing, crop height of cotton was 151 mm (Avg.) compared to 160 mm for red gram.

The prototype was tested at a speed of 0.42 m/s in cotton and red gram field, and at a depth of operation in the range of 40–58 mm. Similar results were reported by Nkakini et al.,[41] Rangasamy et al.,[52] and Manian et al.[35] Depth of operation was influenced on the weeding efficiency and specific draft. These findings are close agreement with the results reported by Biswas et al.,[8] Rangasamy et al.,[52] Tajuddin et al.,[70] and Yadav et al.[80]

The width of operation of tiller varied from 470 to 480 mm.

Draft of weeder in cotton and red gram fields were 44 and 42.5 kg, respectively. Power requirement of weeder operated in cotton and red gram fields were 0.24 and 0.23 hp, respectively. These findings are in close agreement with the results reported by Yadav et al.[80] and Biswas et al.[8] The weeding efficiency of weeder operated in cotton and red gram fields were 87.19% and 84.19%, respectively. These findings are in close agreement with the results reported by Dixit and Syed[14] and Nkakini et al.[41]

Theoretical field capacity of machine operated in cotton and red gram fields were 0.071 ha/h and 0.072 ha/h, respectively. These findings are in close agreement with the result reported by Nkakini et al.[41] and Kathirvel.[33] EFC operated in cotton and red gram fields were 0.053 and 0.055 ha/h, respectively. These results are in close agreement with the results reported by Yadav et al.,[80] Manian et al.,[35] Veerangauda et al.,[77] Tajuddin,[68] and Rangasamy et al.[52]

Field efficiency of machine for cotton and red gram were 74.64% and 76.36%, respectively. Fuel consumption of weeder for cotton and red gram were observed as 0.68 and 0.70 lph, respectively. These findings are in close agreement with the result reported by Dixit and Syed,[4] Nkakini et al.,[41] Manuwa et al.,[36] Rangasamy et al.,[52] and Tajuddin.[69]

Plants were not damaged during weeding by using self-propelled weeder in cotton and red gram field. Performance index of weeder for cotton and red gram were 1925.44 and 2013.23, respectively. Similar results are reported by Biswas and Yadav,[7] Nkakini et al.,[41] and Rangasamy et al.[52]

7.4.4 ECONOMICS OF WEEDING OPERATION

The cost economics of self-propelled weeder for cotton and red gram and compared with conventional method is presented in Table 7.18. An example of cost estimation of prototype is presented in Appendix 7.I.

TABLE 7.18   Cost Economics of Self-propelled Weeder for Cotton and Red Gram.

Particulars		Weeding cost with the prototype	Conventional method cost
Development cost (Rs.)		42,000	–
Total fixed cost (Rs./h) (A)		27.97	–
Total Variable cost (Rs./h) (B)	Cotton	52.46	–
	Red gram	53.24	–
Total cost (A+B) (Rs./h)	Cotton	80.43	–
	Red gram	81.21	–
Cost of operation (Rs./ha)	Cotton	1516.54	2000
	Red gram	1476.54	1980

It was observed that operational cost of the prototype machine was Rs. 1517.54/ha for cotton, compared to Rs. 2000/ha for traditional method of weeding using *hand khurpi*. The saving in cost was 24.12%. The operational cost was Rs. 1476.54/ha for red gram, compared to Rs. 1980/ha for traditional method of weeding by using *hand khurpi*. This resulted in a cost saving of 25.42%. These findings are in close agreement with the result reported by Tajuddin,[69] Manian et al.,[35] Kathirvel et al.,[33] and Veerangauda et al.[77]

KEYWORDS

- bending posture
- cotton
- draft
- effective field capacity
- field efficiency

- **hand khurpi**
- **intercultivator**
- **operational cost fixed cost**
- **performance index**
- **plant damage**
- **red gram**
- **salvage value**
- **self-propelled intercultivator**
- **theoretical field capacity**
- **total cost**
- **weeder**
- **weeding**
- **weeding efficiency**

REFERENCES

1. Ambujam, N. K. Performance Evaluation of Power Weeder. *AMA* **1993,** *4*(4), 16.
2. Ambujam, N. K.; Arasu, P.; Chandramouli, S. Design and Development of Power Driven Rotary Paddy Weeder. *Unpublished B.E. (Ag.) Project Work*, TNAU: Coimbatore, 1984.
3. Anderson, R. L.; Lyon, D. J.; Tanaka, D. L. Weed Management Strategies for Conservation-till Sunflower. *Weed Technol.* **1996,** *10*, 55–59.
4. Anonymous, *RNAM Test Codes and Procedures for Farm Machinery.* Economic and Social Commission for Asia and Pacific Regional Network for Agriculture Machinery (RNAM), Bangkok, 1995.
5. Ben, Y.; Lougue, M. Optimum Settings for Rotary Tools Used for on-the-row Mechanical Cultivation in Corn. *Trans. ASAE* **1999,** *15*(6), 615–619.
6. Bernacki, H.; Haman, J.; Kanafojski, C. Z. *Agric. Machine Theory Construction*; Reproduced by National Technical Information Service, United State Department of Commerce: Springfield, VA, 1972; Vol I.
7. Biswas, H. S.; Yadav, G. C. Animal Drawn Weeding Tools for Weeding and Intercultural in Black Soil. *Agric. Eng. Today* **2004,** *28*(1–2), 47–53.
8. Biswas, H. S.; Ingle, G. S.; Ojha, T. P. Performance Evaluation and Optimization of Straight Blades for Shallow Tillage and Weeding in Black Soils. *AMA* **1993,** *24*(4), 19–22.
9. Bosai, E. S.; Vernaev, O. V.; Smirnov, I. I.; Sultan Shakh, E. G. *Theory, Construction and Calculation of Agricultural Machines.* Oxonian Press Pvt. Ltd.: New Delhi, 1987.
10. Dahab, M. H.; Hamad, S. F. Comparative of Weeding by Animal-drawn Cultivator and Manual Hoe in En-nohoud Area, Western Sudan. *AMA* **2003,** *34*(3), 27–29.

11. Das, F. S. Development of a Power Weeder for Sugarcane Crop. *Agric. Eng. Today* **1986,** *10*(5), 33–34.

12. De Datta, S. K.; Aragon, K. L.; Malabuge, J. A. Vertical Differences in and Practices for Upland Rice. *Seminar Proceeding: Rice Breeding and Vertical Environment West Africa Rice Development Association,* Monrovia, Liberia, 1974; pp 35–73.

13. Desai, S. R. Development and Evaluation of an Intercultural Tool Bar Attachment for Power Tiller to Suit Row Crops. *Unpublished M. Tech. Thesis,* TNAU: Coimbatore, 1998.

14. Dixit, J.; Sayed, I. Field Evaluation of Power Weeder for Rain Fed Crop in Kashmir Valley. *AMA* **2008,** *39*(1), 53–56.

15. Dogherty, M. J.; Godwin, R. J.; Dedousis, A. P.; Brighton, J. L.; Tillett, N. D. A Mathematical Model of Kinematics of Rotating Disc Inter and Intra Row Hoeing. *Biosyst. Eng.* **2007,** *96*(2),169–179.

16. Duraisamy, V. M.; Tajuddin, A. Rotary Weeder Mechanical Inter-culturing in Sugarcane. *Agro India* **1999,** *3*(1–2), 48.

17. Fashina, A. B.; Anumenchi, G. A Power Operated Inter-row Weeder for the Farmer in Developing Country. *J. Res. Andhra Pradesh Agric. Univ.* **1991,** *19*(1), 17–21.

18. Fawoll Laurece, O. Maintenance of Tree Crop Plantations Using Portable Motorized Rotary Weeder. *AMA* 1988.

19. Fielke, J. M. Interaction of the Cutting Edge of Tillage Implement with Soil. *J. Agric. Eng. Res.* **1996,** *63*(1), 61–67.

20. Girma, G. Dynamic Effect of Speed, Depth and Soil Strength Upon the Forces on Plough Components. *J. Agric. Eng. Res.* **1992,** *51,* 47–66.

21. Gobor, Z.; Lammers, P. S. Prototype of Rotary Hoe for Intra Row Weeding. 12th JFTOMM World Congress, Besancon (Fraxe) June, 2007; pp 18–21.

22. Goel, A. K.; Behra, D.; Behra, B. K.; Mohanty, S. K.; Nanda, S. K. Development and Ergonomic Evaluation of Manually Operated Weeder for Dry Land Crops. *Agric. Eng. Int.: CIGR E-j.* **2008,** *10,* 1–11. *Manuscript PM 08 009.*

23. Goryachkin, V. P. *Collected Works in Three Volumes.* The U.S. Department of Agriculture and National Science Foundation: Washington, DC, 1968; Vol I.

24. Gupta, J. P.; Pandey. K. P. Performance of Spiral and Straight Edge Tynes of Rotary Tiller Under Wet Land Condition. *J. Agric. Eng. Res.* **1991,** *28*(1–4), 211–216.

25. Guruswamy, T. Effect of Blade Shape on Performance of Blade Hoe for Comparative Study. *Agric. Eng. Today* **1985,** *12*(2), 25–27.

26. Guruswamy, T. Field Performance Evaluation of Deep Interculture Implement under Dryland Cultivation. *Agric. Eng. Today* **1985,** *9*(2), 26–34.

27. Guruswamy, T.; Veerangauda, M.; Desai, S. R.; Barker, R. Performance of Triangular Sweep Hoe in Safflower Crop. *Kar. J. Agric. Sci.* **1995,** *8*(3), 296–299.

28. Indian Standards: 1964-1967. *Test Code for Estimating Cost of Farm Machinery.* Government of India, Institute of Indian Standards: New Delhi.

29. Indian Standards: 6451-1972. *Test Code for Specification for Sweep.* Government of India, Institute of Indian Standards: New Delhi.

30. Irla, E. Tending Technique and Mechanical Weed Control in Potatoes. Environmentally Friendly Tending Methods are Successful. *FAT Ber. Switzerland,* **1995,** *462,* 1–7.

31. Kankal, U. S.; Anantachar, M.; Palled, V. Optimization of Operational Parameters and Performance Evaluation of Self-propelled Weeder for Field Crops. *Int. J. Appl. Agric. Hortic. Sci. (Green Farming Int. J.)* **2014,** *5*(1), 56–60.

32. Kankal, Uddhao, Nage, S.; Anantachar, M. *Design, Development and Evaluation of Self-propelled Weeder*. LAMBERT Academic Publishing (LAP), 2014.

33. Kathirvel, K.; Thambidurai, R. D.; Manohar, D. Ergonomics of Self-propelled Power Weeders as Influenced by Forward Speed and Terrain Condition. *AMA* **2009,** *40*(4), 28–32.

34. Khurmi, R.; Gupta, J. *A Text Book of Machine Design.* Eurasia Publishing House (Pvt.) Ltd., Ram Nagar, New Delhi, 2003.

35. Manian, R.; Kathirvel, K.; Reddy, A.; Senthikumar, T. Development and Evaluation of Weeding cum Earthing up Equipment for Cotton. *AMA* **2004,** *35*(2), 21–25.

36. Manuwa, S. I.; Odubanjo, O. O.; Maliumi, B. O.; Olfinkua, S. G. Development and Performance Evaluation of a Row-crop Mechanical Weeder. *J. Eng. Appl. Sci.* **2009,** *4*(4), 236–239.

37. Mehta, M. L.; Verma, S. R.; Mishra, S. R.; Sharma, V. K. *Testing and Evaluation of Agricultural Machinery.* Daya Publishing House: Delhi, 2005.

38. Mohsenin, N. N. *Physical Properties of Plant and Animal Materials.* Gorden and Breach Science Publisher: New York, 1979.

39. Muzumdar, G. Performance Evaluation of Weeders in Cotton Crop. TMC Ann. Rep. 2006-07-TMC-MMI-2.5, 2006, 5 pages.

40. Mynavathi, V. S.; Prabhakaran, N. K.; Chinnusamy, C. Evaluation of Mechanical Weeders in Irrigated Maize. *Indian J. Weed Sci.* **2008,** *40*(3–4), 210–213.

41. Nkakini, S. O.; Akor, A. J.; Anyotamuno, M. J.; Ilkoromari, A.; Efenudu, E. O. Field Performance Evaluation of Manual Operated Petrol Engine Powered Weeder for the Tropics. *AMA* **2010,** *41*(4), 68–73.

42. Olunkunle, O. J.; Oguntunde, P. Design of Row Crop Weeder. In Conference of International Agricultural Research for Development, Dept. of Agricultural Engineering, Federal University of Technology, Ankure, 2006.

43. Oni, K. C. Performance Analysis of Ridge Profile Rotary Weeders. *AMA* **1990,** *21*(1), 43–49.

44. UAS. *Package of Practices.* UAS: Raichur, 2009.

45. Padole, Y. B. Performance Evaluation of Rotary Weeder. *Agric. Eng. Today*, 31, 2007.

46. Panda, B. C. Development and Performance Evaluation of Weeder for Rice. *AMA* **2002,** *33*(3), 21–22.

47. Piet, B.; Rommie, V. D. W.; Dirk, K. Experiences and Experiments with New Intra Row Weeder. In 5th EWRS Workshop on Physical Weed Control, Pisa, Italy, 2002; pp 97–100.

48. Pradhan, S. C.; Behra, B. K.; Mahapatra, M. Development and Evaluation of Power Tiller Power Weeder for Groundnut Crop. *Curr. Agric. Res.* **1995,** *8*(3–4), 128–130.

49. Pullan, D. W. M.; Cowell, P. A. Prediction and Experimental Verification of the Hoe Path of a Rear Mounted Inter Row Weeder. *Trans. ASAE,* **2000,** *77*(2), 137–153.

50. Pullan, D. W. M., and Cowell, P. A. An Evaluation of Performance of Mechanical Weeding Mechanism for Use in High Speed Inter Row Weeding Arable Crops. *J. Agric. Eng. Res.* **1997,** *67*, 27–34.

51. Pullen, D. W. M.; Cowell, P. A. The Effect of Implement Geometry on Hoe Path of Steered Rear Mounted Inter Row Weeder. *Biosyst. Eng.* **2006,** *94*(3), 373–386.

52. Rangasamy, K.; Balasubramanian, M.; Swaminathan, K. R. Evaluation of Power Weeder Performance. *AMA* **1993,** *24*(4), 20–23.

53. Rathod, R. K. Design and Development of Tractor Drawn Inter Row Rotary Weeder, Unpublished M. Tech. Thesis, *MAU*: Parbhani, 2007.

54. Rathod, R. K.; Munde, P. A.; Wandre, R. G. Development of Tractor Drawn Inter Row Rotary Weeder. *Int. J. Agric. Eng.* **2010,** *3*(1), 105–109.

55. Remesan, R.; Roopesh, M. S.; Remaya, N.; Preman, P. S. Wet Land Paddy Weeding—A Comprehensive Comparative Study from South India. *Agric. Eng. Int.: CIGR E-j.* **2007,** *9*, 1–21. *Manuscript PM 07 011.*

56. RNAM. *Test Code and Procedure for Weeder.* 1983.

57. Sahay, J. *Elements of Agricultural Engineering.* Standard Publishers Distributors:1705-B, Nai Sarak, Delhi, 2006.

58. Schwazel, R.; Parler, D. Mechanical Weed in Potatoes Possibilities and Limits. *Reverse Success Agric.* **1993,** *25*(2), 71–74.

59. Sharma, N. D.; Mukesh, S. *A Text Book Farm Machinery Design Principles and Problems.* Jain Brothers: Karol Bagh, New Delhi, 2008.

60. Shrestha, D. S.; Singh, G.; Gebresenbet, G. Optimizing Design Parameters of Mouldboard Plough. *J. Agric. Eng. Res.* **2001,** *78*(4), 377–389.

61. Sial, J. K.; Harison, H. P. Soil Reacting Forces from Field Measurement with Sweeps. *Trans. ASAE* **1978,** *21*(3), 825–829.

62. Sineokov, G. N. *Design of Soil Tilling Machines.* Indian Scientific Documentation Centre: New Delhi. U.S. Department of Agriculture, Agricultural Research Service and National Science Foundation: Washington, DC, 1977.

63. Singh, M.; Kumar, A.; Agrawal, S. Design and Development of Self-propelled Rotary Power Weeder. *IE (I) J.-AG* **2008,** *89*, 14–17.

64. Sogaard, H. T. Automatic Control of Finger Weeder with Respect to Harrowing Intensity at Varying Soil Structure. *J. Agric. Eng. Res.* **1998,** *70*, 157–163.

65. Srivastava, A. K. *Engineering Principles of Agricultural Machines.* ASAE Text Book No. 6. ASAE: St. Joseph, MI, 2003.

66. Sthool, V. A.; Shinde, P. H. Performance Evaluation of Peg Tooth Weeder in Dry Land Agriculture. *J. Maharashtra Agric. Univ.* **2004,** *29*(1), 113–115.

67. Tajuddin, A. Development of an Engine Operated Sweep for Weeding and Interculture. *RNAM Newslett.* **1994,** *51*, 10.

68. Tajuddin, A. Design Development and Testing of Engine Operated Weeder. *Agric. Eng. Today* **2006,** *30*(5–6), 25–29.

69. Tajuddin, A. Development of Power Weeder for Low Land Rice. *IE (1) J.-AG.* **2009,** *90*, 55–57.

70. Tajuddin, A.; Karunanithi, R.; Swaminathan, K. R. Design Development and Testing of Engine Operated Blade Harrow for Weeding. *Indian J. Agric. Eng.* **1991,** *1*(2), 137–140.

71. Tasliman. *Design of Small Multipurpose Power Weeder.* Posted by Pak Tas on September 11, 2008.

72. Tewari, V. K.; Datta, R. K.; Murty, S. R. Field Performance of Weeding Blades of Manually Operated Push Pull Type Weeder. *AMA* **1993,** *55*, 129–141.

73. Thakur, T. C. Design Aspects of Soil Engaging Hand Tools. *Agric. Eng. Today* **1985,** *21*, 15–18.

74. Varma, M. R.; Tiwari, R. C.; Agrawal, A. Adaptation of Field Evaluation of Improved Equipment to Power Tiller for Sugarcane Cultivation. *J. Agric. Eng.* **1991,** *18*(9), 354.

75. Varshney, A. C.; Pacharne, D. T.; Deshmukh, V. D. Feasibility of Use of Power Tiller in Grape Cultivation. *Agric. Eng. Today* **1995,** *19*(3), 16–20.

76. Varshney, R. A.; Tiwari, P. S.; Narang, S.; Mehta, C. R. *Data Book for Agricultural Machinery Design.* Book no. 2004/1, CIAE, Bhopal, 2005.

77. Veerangauda, M.; Anantachar, M.; Sushilendra. Performance Evaluation of Weeders in Cotton. *Karnataka J. Agric. Sci.* **2010,** *23*(5), 732–736.
78. Victor, M.; Verma, A. Design and Development of Power Operated Rotor Weeder for Wetland Paddy. *AMA* **2003,** *14*(4), 27–29.
79. Yadav, R.; Pund, S. Development and Ergonomic Evaluation of Manual Weeder. *Agric. Eng. Int.: CIGR E-j.* **2007,** *9,* 1–9. *Manuscript PM 07 022.*
80. Yadav, S. N.; Pandey, M. M; Saraswat, D. C. Effect of Design and Operating Parameters of Performance of Inter-cultivation Sweep in Vertisols. *AMA* **2007,** *38*(3), 38–44.
81. www.indiastat.com.
82. www.researchgate.net/profile/Uddhao_Kankal/publications.

APPENDIX 7.I   COST OF OPERATION FOR DEVELOPED SELF-PROPELLED INTERCULTIVATOR

Assumptions:

Price to develop self-propelled weeder (C) = Rs. 42,000
Annual use in hours (H) = 250 h.
Average life (L) = 8 years.
Total useful life in hours = 2000 h.
Junk values, S (10% of development cost) = Rs. 4200.

(A) Fixed cost

(i)   Depreciation (D):

Depreciation was calculated by using following formula:

$$\text{Depreciation} = \frac{C-S}{L \times H} = \frac{[42,000 - 4200]}{[8 \times 250]} = 18.9 \, \text{Rs./h}$$

where D = depreciation per hour; C = capital investment; S = salvage value = 10% of capital investment; H = number of working hours per year; L = life of the machine in years.

(ii) Interest (I):
Interest was calculated on the average investment of the machine taking in to consideration the value of the machine in first and last year. It was taken 8% of development cost.

$$I = \frac{C+S}{2} \times \frac{i}{H} \quad \text{or} \quad I = \frac{42000 + 4200}{2} \times \frac{0.08}{250} = \text{Rs. } 7.39/h$$

where I = interest per hour and i = interest rate = 8%

(iii) Housing:
Housing cost was calculated on the basis of the prevailing rate of the locally but roughly housing cost was taken 1% initial cost of the machine per year.

$$= \frac{42,000 \times 0.01}{250} = \text{Rs. } 1.68/h$$

Total fixed cost = (i) + (ii) + (iii) = 18.9 + 7.39 + 1.68 = Rs. 27.97/h.

(B) Operational cost for cotton crop

(i) Fuel cost:
Fuel cost was calculated on the basis of actual fuel consumption for weeding.
Cost of kerosene was taken as Rs. 30.
Fuel required for 1 h = 0.68 lph.
Fuel cost = 0.68 × 30 = Rs. 20.4/h.

(ii) Lubricants cost:
Charge of lubricant was calculated on the actual fuel consumption, but it was taken 30% of the fuel cost.
Lubricating cost = 20.4 × 0.30 = Rs. 6.12/h.

(iii) Labor cost:
Wages cost was calculated on the actual wages of the worker.
Rs. 100/day of is required for skilled labor@8 h of work/day.
Wages cost = 100/8 = Rs. 12.5/h.

(iv) Repair and maintenance:
Repairs and maintenance cost was taken 8% of the initial cost of the machine per year.
Repairs and maintenance cost = [42,000/250] × 0.08 = Rs. 13.44/h.

Total operating cost = 20.4 + 6.12 + 12.5 + 13.44 = Rs. 52.46/h.

Total operational cost of self-propelled weeder per hour for cotton crop.
=Fixed cost + Operational cost (A) + (B) = 27.97 + 52.46 = **Rs. 80.43/h**.

Effective field capacity of self-propelled weeder in cotton crop was 0.053 ha/h.
Operational cost of self-propelled weeder per ha for cotton crop

$$= \frac{80.43}{0.053} = Rs.1517.54/ha$$

(C) Operational cost for red gram crop

(i) Fuel cost:
Fuel cost was calculated on the basis of actual fuel consumption in weeder.
Cost of kerosene taken Rs. 30. Fuel required for 1 h = 0.70 lph.
Fuel cost = 0.70 × 30 = Rs. 21/h.

(ii)   Lubricants cost:
       Charge of lubricant was calculated on the actual fuel consumption, it
       was taken
       30% of the fuel cost.
       Lubricating cost = 21 × 0.30 = Rs. 6.3/h   ix

(iii) Labor cost:
       Wages cost was calculated on the actual wages of the worker.
       Rs. 100/day of is required for skilled labor, 8 h of work for 1 day.
       Wages cost = 100/8 = Rs. 12.5/h.

(iv)   Repair and maintenance:
       Repairs and maintenance cost was taken 8% of the initial cost of the
       machine per year.
       Repairs and maintenance cost = [42,000/250] × 0.08 = Rs. 13.44/h.

       Total operating cost = 21 + 6.3 + 12.5 + 13.44 = Rs.53.24/h.

       Total operation cost of self-propelled weeder per hour = Fixed cost +
       Operating cost = (A) + (C) = 27.97 + 53.24 = Rs. 81.21/h.
       Effective field capacity of self-propelled weeder in red gram crop was
       0.055 ha/h.
       Operational cost of self-propelled weeder per ha for red gram crop

       $$= \frac{81.21}{0.055} = Rs.\,1476.54/ha.$$

PART III
Management of Natural Resources

CHAPTER 8

AGRO-BIODIVERSITY COLLECTION AND CONSERVATION FROM ARID AND SEMI-ARID REGIONS OF INDIA

RAJU R. MEGHWAL and N. K. DWIVEDI[*]

National Bureau of Plant Genetic Resources, Regional Station, CAZRI Campus, Jodhpur 342005, Rajasthan, India. E-mail: drnkdwivedi52@ yahoo.co.in

*Corresponding author.

CONTENTS

Abstract 210
8.1   Introduction 210
8.2   General Features, Status, and Changing Patterns of
      Agroecosystems in Thar Desert of India 212
8.3   Threats to Agro-Ecosystems 215
8.4   Significance of Agro-Biodiversity 217
8.5   Collection and Conservation of Agro-Biodiversity in
      Arid and Semiarid Regions 218
8.6   Diversity in Wild Relatives, Wild Species, and Weedy Forms
      in Arid and Semiarid Regions 221
8.7   Traditional Knowledge 222
8.8   Traditional Management of Farm Pests and Enemies 225
8.9   Conclusions 226
Acknowledgment 226
Keywords 226
References 227

ABSTRACT

Agriculture and allied activities not only provide livelihood to a large section of population but also play a pivotal role in their lifestyle. Agro-biodiversity is an evolutionary divergent, highly interrelated component of biodiversity dealing with agroecosystem and variation in agriculture related to plants, animals, marine life, insects, microbes, avian species, etc. Environmental, biological, sociocultural and economic factors are responsible for the evolution of diverse agroecosystems. Traditional agroecosystems are diverse in that crop husbandry, animal husbandry, and forests constitute complex and interlinked production systems. A total of 17,395 agro-biodiversity collections comprising of cultivated (16,136) and wild (1259) accessions have been made in various crops like cereals and millets (1957), pulses and legumes (5609), oilseeds (1760), fibers (541), vegetables (1551), fruits and minor fruits (991), medicinal and aromatic plants (362), halophytes (21), spices and condiments (198), grasses and fodder plants (291), dye-yielding plants (26), famine food plants (11), wild relatives and weedy forms (433), and miscellaneous (293). A wide range of variability has been recorded in the distribution; plant habit and types; canopy; earliness; bearing; size, shape and color of leaves, flowers, fruits, and seeds; and weight of fruits and seeds. It recognizes that agriculture evolved from bioprospecting, selection, and development of a few species from plants and animal kingdoms, to meet human needs of food, fiber, and fuel. All biotic factors related to agriculture, such as plants, animals, fish, reptiles, insects, birds, and microbes are components of biodiversity. Therefore, the conservation, management, and sustainable use of these genetic resources require specific attention.

8.1   INTRODUCTION

Agro-biodiversity is the result of the interaction between the environment, genetic resources, and the management systems used by people. Thus, agro-biodiversity encompasses the variety and variability of animals, plants and microorganisms that are necessary for sustaining key functions of the agro-ecosystem, including its structure and processes for, and in support of, food production and food security. Local knowledge and culture can therefore be considered as integral parts of agro-biodiversity, because it is the human activity of agriculture that shapes and conserves this biodiversity. Agriculture and allied activities not only provide livelihood to a large section of population but also play a pivotal role in their lifestyle. Agro-biodiversity is an evolutionary divergent, highly interrelated component of biodiversity dealing with agroecosystem and variation in agriculture related to plants, animals, marine life, insects, microbes, avian species, etc. Environmental, biological, sociocultural, and economic factors are responsible for the evolution of diverse agroecosystems. Traditional agroecosystems are diverse, in that crop husbandry, animal husbandry and forests constitute complex, and interlinked production systems.[31] Inaccessibility, environmental heterogeneity, ecological fragility, and marginality have favored the evolution of subsistence production systems sustained with organic matter and nutrients derived from the forests with emphasis on optimizing productivity in the long range.[32,33,36] In any area, huge diversity has been maintained through variety of crop compositions, favored by enormous variations in the edaphic, topographic, and climatic conditions. Increased population pressure from within the region, further exacerbated by external forces from relatively industrialization has resulted in major changes in the agroecosystems and resulted in rapid depletion of agro-biodiversity.[32]

The land use and land cover changes significantly affect a while range of local, regional, and global issues, such as climate change, genetic erosion, biological invasion, land degradation, cropping pattern alternations, and hydrological imbalances.[11,32] The above changes are closely linked with the sustenance and livelihood issues of the traditional societies since these have an effect on resources. Poor scientific understanding of traditional farming systems and related socioeconomic dimensions seriously impedes the identification of sustainable agricultural development solutions in any area. Traditional agro-biodiversity management plays a key role in coping with the uncertainties in the arid and semiarid region in India. There is a strong need to bring desirable changes in the agricultural policy, research and development, land use, breeding approaches, and collection and conservation of biodiversity in reference to particular region. The adoption of appropriate practices and enabling policies are required to provide solutions and to overcome threats from ongoing erosion of genetic resources. Recent concerns about sustainability have brought these issues to the fore and have reawakened interest in the rich traditional agroecosystems. Unfortunately, genetic erosion is happening at a more rapid pace in developing countries because of the somewhat faulty planning to bring about change and increase productivity. Above all, agro-biodiversity which is genetic diversity related to agriculture, is threatened not because of overuse but because it is not used and nurtured as much as it should be.

This chapter attempts to (1) understand the general characteristics, status, and changing scenario; (2) highlight agro-biodiversity and its significance;

(3) assess the rate, casual factors and ecological, social, and policy dimensions and their impacts on agro-biodiversity loss; (4) collection and conservation of biodiversity for future generation; (5) prepare a synthesis and assessment on the maintenance and use of agro-biodiversity by indigenous peoples and rural communities under conditions of climate change; and (6) bring together and make available information on the use of agro-biodiversity by rural and indigenous communities to cope with climate change, and relevant research work on effect of climate change on agriculture and agrobiodiversity.

8.2   GENERAL FEATURES, STATUS, AND CHANGING PATTERNS OF AGROECOSYSTEMS IN THAR DESERT OF INDIA

The Indian subcontinent is one of the 12 mega-centers of biodiversity and is unique in the point of floristics and climate. Two of the 18 globally identified "hotspots" of biological diversity occur in India, one each in Western Ghats and the Northeastern Himalayas. The Indian Thar Desert covers 300,000 km^2 in North Western India, in which Rajasthan cover 61%, Gujarat 20%, Haryana and Punjab 9%. There are 141 endemic genera (37%) belonging to over 47 families of higher plants occur in India. As per botanical survey of India, there are 46,214 plant species are that found (11.9% of global flora) in Indian Thar Desert. Of these, about 17,500 represent flowering plants (7% of the world flora). Of the plant taxa endemic to Thar Desert, 23 are species and 14 subspecies (6.4% of the taxa from this region). It assumes special significance since the biodiversity in this zone survives in a very fragile ecosystem under a highly hostile environment.[35] Most of the area of this desert consists of dry undulating mass of loose sand, forming shifting dunes. The plain is full of sand hills with several low depressions where salt and soda are deposited on dry soil. Extremes of temperature, severe droughts accompanied by high wind velocities, low relative humidity, and scanty and highly erratic rainfall forms the main climatic features of this region. Agriculture is very little developed. The climate and cultures of this region are different,[11] its topography and traditions are unique and its fauna and flora composition is diverse (Fig. 8.1 and Table 8.1).

Desert people have developed lifestyles best suited to the available natural resources for them as well as for their livestock, thereby enriching the biodiversity of this region. Orans and Gochars (village grazing land) were created and preserved under collective control of village Panchayat (the elected body of five persons). Certain trees of the Desert like *Prosopis cineraria* L. (locally known as Khejri) were considered abode of a deity and rituals were conducted to honor the deity which served to preserve the diversity of this species.

FIGURE 8.1   Sand dunes at Western Rajasthan.

TABLE 8.1   Ten Largest Families with Maximum Species Diversity in Indian Desert.

Family	Species
Acanthaceae	22
Asteraceae	44
Convolvulaceae	35
Cucurbitaceae	19
Cyperaceae	36
Euphorbiaceae	23
Fabaceae	65
Malvaceae	28
Poaceae	111
Scrophulariaceae	15

Source: Flora of the Indian Desert.[10]

Oran is derived from the Sanskrit word aranya, meaning forest. Orans are patches of jungle preserved in the name of local deities or saints. There are about 25,000 Orans in Rajasthan covering an area of about 600,000 ha. About 1100 major Orans cover 100,000 ha. Of these, nearly 5370 sq. kilometer Orans in the Thar Desert. The largest Oran—15,000 ha—is situated in Bhadariya district of Jaisalmer. Some may be as small as a few square meters.

Orans are rich in biogenetic diversity and provide a refuge for wildlife. Rajasthan is vertically cut in to two, north to south, by the Aravali hills. To the west of the Aravalis is the Thar Desert. There are about 7.5 million pastoralists and 54.5 million livestock in Rajasthan, all dependent on the Orans.

The prevalence of native diversity in food plants and food habits of the local communities in such areas of subsistence agriculture has much to do with the process of folk domestication. Thus, species with a wide range were domesticated at different places depending on the need of local inhabitants. Overall choice of plants/plant parts, consumed by man in this selective process, has been maintained by society itself since the local inhabitants were aware of the plants possessing special attributes/biological characteristics. Folk selection operated likewise involving local needs/customs and habits and resulted in conservation of biodiversity.

The diversity of landscapes in terms of altitude, topography, climate, forest and desert resources, availability of water for irrigation, socioeconomic and cultural factors play important role in the evolution of a variety of agroecosystem types. The heterogeneity becomes more complex when ecological conditions are superimposed. The traditional agro-biodiversity is complex with strong linkage between crop plants, animals, and the forests.[38] It needs to be emphasized that mountains, in general, are still the safe havens in terms of agro-biodiversity because the modern/commercial agriculture has not yet made significant inroads as in other parts of the country.

In most of the areas traditional farming systems can be categorized into three basic types: livestock farming, mixed livestock and crop farming, and mixed feed crop—livestock farming, resulting in respective ways of life. These lifestyles are nomadism, semi-nomadism and settled agriculture. The introduction of horticultural crops as commercial crops is a recent phenomenon. In recent time, cultivation of off-season vegetables, medicinal plants, and floriculture have also made significant inroads.

The indicators of change, both visible and invisible, relate to the resource base, production flow, and change in resource use and management practices.[29,46] Degradation of the natural resources base has contributed to an increase in frequency and intensity of soil erosion, reduced per capita availability of land and overall decrease in productivity, lack of on-farm employment opportunities resulting in male migration to the parts of India, particularly the cities. Increased population pressure and declined in land area resulting in the reduced shifting agriculture. The plant species belong to wide variety of edible fruits, leafy vegetables, cereals, millets, minor millets, tubers, liquors, medicine, fiber, dyes, and forage are being utilized by local inhabitants. Such extensive use of biodiversity suggests that the tribes and traditional farmers[1,4,11,12,28,41] rely heavily on biodiversity to meet their varied needs.

There is considerable diversity and variability reported in cereals, pseudo-cereals, millet, minor millets, legumes, oilseeds, fruits, and vegetables, etc. in different geographical areas on India (Fig. 8.2).

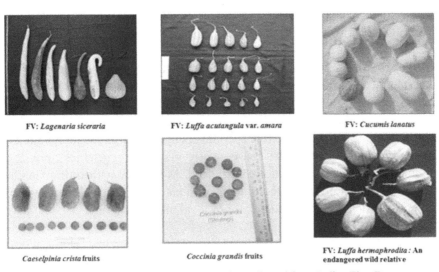

FV: *Lagenaria siceraria*   FV: *Luffa acutangula* var. *amara*   FV: *Cucumis lanatus*

Caeselpinia crista fruits   *Coccinia grandis* fruits   FV: *Luffa hermaphrodita*: An endangered wild relative

FIGURE 8.2   Fruit variability in different species collected from Indian Thar Desert.

8.3 THREATS TO AGRO-ECOSYSTEMS

Agro-biodiversity encompasses an enormous array of biological resources tied with agriculture such as crops, livestock, soil biota, rangelands, etc.[32] There is steady shift from traditional agroecosystems, which maintain and protect biological biodiversity to modern agroecosystems based on intensive cropping (e.g., monocropping, plantations, etc.). These have evolved basically to maximize yield and economic returns with a considerable loss of biological diversity.[31,34,38,42] How much of the genetic base has already been eroded is difficult to say, but state-driven "Green Revolution" impacts, that is supply of high-yielding varieties (HYVs), inorganic fertilizers, pesticides, and irrigation free of cost to farmers, initially in some areas and then highly subsided, have squeezed out the native landraces. Due to above factor, area under variety of traditional crops declined. The traditional crops have been mostly replaced by cash crops. Thus, the reduction in crop diversity is partly because of the introduction of HYVs and partly because of increased emphasis on cultivation of cash crops. Similar information of genetic erosion of crop plants has also been reported by Pratap[44] and Anwar and Bhatti.[5] The rate of erosion of indigenous pearlmillet in Gujarat and Rajasthan; wheat, maize, soybean, and potato in Indian Himalayan region; and loss of rice varieties were particularly high in Baluchistan.[30]

In the days before road were built and pre-independence days (before 1947), green vegetables in the Indian Desert areas were a rarity. With whatever, vegetables available from native plants *Acacia senegal* (seeds), *Capparis decidua* (immature fruits, Fig. 8.3), *Cucumis callosus* (fruits), *Cyamopsis tetrogonoloba*, *P. cineraria* (immature pods, Fig. 8.4), etc., local people were able to adjust their dietary requirement from them. Over the centuries use of native vegetables, which were consumed as emergency foods in the wake of recurring droughts and famines, became part of their stable diet. However, with fast communication facilities people changed their habits to include the fresh vegetables in their diet consequently the dried vegetables drew less attention and to some extent are neglected. Moreover, with Indira Gandhi canal water reaching most Desert areas of India, green vegetables cultivation got boost which also changed the use of local biodiversity.[11]

FIGURE 8.3   Flowering stage of *Capparis decidua.*

FIGURE 8.4 Collecting weedy form of *Cyamopsis tetragonoloba* for fodder purpose.

8.4 SIGNIFICANCE OF AGRO-BIODIVERSITY

Biodiversity plays a significant role in maintaining the long-term stability of traditional agroecosystems in a variety of ways: it helps to minimize crop loss due to drought, high temperature, diseases, insect—pests and nematodes; improve soil fertility by incorporating legumes in the crop mixture; inhibits or suppresses weed growth; increases productivity per unit area; produces a varied diet; conserves soil from erosion; and insures against crop failure.[2]

Furthermore, it is estimated that 20% of current global production of food is met from traditional multi-cropping systems which also help in maintaining ecological equilibrium and sustaining crop genepool admixtures, transgression, other micro-evolutionary processes, etc.[42] The modern approaches of monocropping of HYVs neither recognize the value of mixed cropping for risk avoidance, nor do they realize the difficulty in maintaining a continuous supply of necessary inputs such as fertilizers, improved seeds, pesticides, etc. Once a local traditional crop is displaced, its unique germplasm gets lost forever.[3] In addition to the above benefits, these traditional crops provide a more varied and interesting diets, greater food security, and also possess medicinal properties. It should not be undermined that with the erosion of genetic base, due to loss of landraces and varieties, future breeding options are becoming limited.

8.5   COLLECTION AND CONSERVATION OF AGRO-BIODIVERSITY IN ARID AND SEMIARID REGIONS

The collections of plant genetic resources primarily aims at tapping germplasm variability in different field and horticultural crops, their wild relatives and relate species Based on precautionary principle, in all of these are equally legitimate targets for collecting.

Between 1980 to October, 2012, 97 major explorations comprising of 71 crop/region specific and 29 multi-crops were carried out besides 13 minor ones in the states Gujarat, Haryana, Rajasthan, and adjoining areas of Madhya Pradesh, Punjab, and Uttar Pradesh[13–18,20–26] of India. Of these collection trips, 28 major explorations were jointly carried out with national/international institutes (Central Arid Zone Research Institute, Jodhpur; Central Institute for Arid Horticulture, Bikaner; Central Institute for Jute and Allied Fibers, Barrackpore, WB; CCS Haryana Agricultural University, Hisar and its Agricultural Research Station, Bawali; Indian Institute of Pulses Research, Kanpur; ICRISAT, Hyderabad; Indian Institute of Horticulture Research, Bangalore; Gujarat Agricultural University, Main Rice Research Station, Navagaon, Kheda; Wheat Research Station, Vijapur, Mehsana; National Bureau of Plant Genetic Resources, New Delhi; National Research Centre for Seed Spices, Ajmer and Rajasthan Agriculture University, Agriculture College, Jobner). Six crop-specific explorations were undertaken for the collection of eggplant, okra, and sesame under NBPGR/IPGRI Collaboration Program from parts of Gujarat, Madhya Pradesh, and Rajasthan. The samples were collected from farmers field, threshing yards, farm store, kitchen gardens, weekly village market (cultivated form), waste lands, road side, irrigated and rainfed farmer's fields, river banks, canal areas, annual and ephemeral ponds, terraces, slopes, valley, and hilltops. Biased, random, and bulk sampling procedures were followed for collecting of germplasm. Herbarium sheets were prepared. Passport data were recorded along with important characters. One set of collected seed material was sent to NBPGR Headquarter, New Delhi for conservation in National Gene Bank and assignment of accession numbers. Vegetative propagules were planted at this Regional Station, Jodhpur for establishment in Field Gene Bank.

A total of 17,395 agro-biodiversity collections comprising of cultivated (16,136) and wild (1259) accessions have been made in various crops like cereals and millets (1957), pulses and legumes (5609), oilseeds (1760), fibers (541), vegetables (1551), fruits and minor fruits (991), medicinal and aromatic plants (362), halophytes (21), spices and condiments (198), grasses and fodder plants (291), dye-yielding plants (26), famine food plants (11), wild relatives and weedy forms (433), and miscellaneous (293). Table 8.2 indicates landraces in different crop groups.[14,20,22,25]

TABLE 8.2 The Local Landraces Collected in Different Crop Groups.

	Triticum aestivum	**Safed kanak**	*T. durum*
Wheat	Bodiya Gehun	Sachiya	Arnej Gehun
	Contoli Gehun	Teria	Arnej Katha
	Deshi Gehun	Tukara	Bansi
	Deshi Lal Gehun	Vajya	Bhatiya Deshi
	Dhola Gehun		Daut Khani
	Kotbajiya	*T. durum*	Futel
	Kharchiya Gehun	Jau	Katha
	Lal Gehun	Popatiya	Kathiya
	Moti	Putadiya	Poonagiri, *T. polonocun*
	Nana Gehun	Putragehun	*T. sphaerococcum*
	Pili Kanak	Putrajau	Patoliya
Barley	Desi Jov		
Maize	Desimakki	Hatti Makki	Malan Makki
	Sattimakki		
Rice	Akoli Bhat	Ekngare	Kosure
	Altiya	Futiya	Lakshari Bhuri
	Ankalo	Geton	Lakshari Sona
	Bangalo	Gutniyao	Lal Dangar
	Barihai	Halaki	Markolin
	Batli Bhat	Kauchi	Masuri
	Bhujhavalia Bhat	Hari Bangalo	Nanisal
	Chawara	Hathi	Panijal
	Chocolate	Indrani	Pejje
	Cholli Desi	Jaware	Podhari
	Dabla	Jhadaki	Rata Chawal
	Dabohal	Jhini Sathi	Rodikawod

TABLE 8.2   *(Continued)*

	Triticum aestivum	**Safed kanak**	*T. durum*
	Dabual	Jodohallio	Ratiya Dangar
	Dadrcolon	Kada	Saket, Sathi
	Dalaki	Kajal Hari	Sativa Dangar
	Dang Bhat	Kala Kundiya	Sukhwel
	Dault Puri	Kali Dangar	Sutrasal
	Desi Bath	Kalinga	Taichun
	Desi Dangar	Kansarvav Karojo	Thumbati
	Desi Sathi	Kirali	Tulshya
	Dhauar	Khadasi	Ubhi Dangar
	Dhanpari	Khadsia	Vardala
	Dholi Bhatt	Khasriva Dangar	Vijalpuri
	Dholi Dangar	Kolpi	
Pearlmillet	Mahuda Bajra		
Sorghum	Deshi	Dhorki	Gundri
Sesame	Bhanwari Tilli	Dhili Tilli	Zhumak Tilli
Cotton	Dhumak Kalvan	Mathio	
	Vara	Waged	
Ber	Banarasi	Kaithali	Noki
	Chhuahara	Kalingada	Pathan
	Chonchal	Karawa	Porda
	Dak	Kara	Rashmi
	Daudan	Kathaphal	Sandhura Narnaul Seb
	Gola Kela	Sotai	
	Kakrol	Lasora	Umran
	Ilaichi		
Lasora	Desi Lesua	Lesu	Pemali Lesua
Cucumber	Balam Kakadi	Balam Kheera	
Chillies	Bayana Local	Haripur rati	Raipuri
	Bilara Local	Mathaniya Lambi	Reshampatti
	Ghumari Tikhi	Mehasana Charkhun	Tonkey Local
Fenugreek	Acharmaithidana	Nagouri Maithi	Nainaki Maithi
	Dana Methi		

Source: Dwivedi *et al.*[25]

An effort was made to collect maximum information on samples collected. Local names of landraces, their uses, indigenous knowledge[19,22] and role of women in the conservation of agro-biodiversity by local methods were recorded. A wide range of variability has been recorded in the distribution; plant habit and types; canopy; earliness; bearing; size, shape, and color of leaves, flowers, fruits, and seeds; and weight of fruits and seeds.

8.6 DIVERSITY IN WILD RELATIVES, WILD SPECIES, AND WEEDY FORMS IN ARID AND SEMIARID REGIONS

There are about 166 species of native cultivated plants in Indian gene center. The crops with primary, secondary, and regional centers of diversity represent a part of native and introduced species which account for over 480 species.[36] Diverse agroclimatic and agricultural practices have led to rich diversity of crop species in the form of landraces and cultivars. Besides, the center has over 320 wild relatives.[8,9] The floristic diversity in wild relatives of cultivated/weedy types and related taxa constitutes a useful genepool. The collection of crops species are easy than wild relatives and weedy forms of the taxa. As crop species are accessible for collections in fields, orchards, gardens, markets, and with farmers. On other side, the wild relatives are difficult to locate as they grow in their natural habitats with other wild plants and also difficult to identify without skill in systematic botany.[6,7] In all more than 433 collections of various wild relatives, wild species and weedy forms were collected during this period (Table 8.3). The collections were classified into different crop groups based on their agri-horticultural importance.[9] The distribution pattern and collection of important diversity in wild relatives of crop plants from Himalaya, Rajasthan Desert, Gujarat, Odisha, and West Bengal regions of India, respectively have studied.[37,40,43,45,47] It recognizes that agriculture evolved from bioprospecting, selection, and development of a few species from plants and animal kingdoms, to meet human needs of food, fiber, and fuel. All biotic factors related to agriculture, such as plants, animals, fish, reptiles, insects, birds, and microbes are components of biodiversity. Therefore, the conservation, management and sustainable use of these genetic resources require specific attention.[39]

TABLE 8.3 Wild Relatives, Wild Species, and Weedy forms of Crop Groups Collected from Arid and Semiarid Regions of Indian Thar Desert.

Crops groups	Species diversity
Cereals and millets	*Avena fatua, Echinochloa colona, Panicum antidotale, Sorghum halepense*
Legumes	*Cyamopsis tetragonoloba, Lathyrus aphac, mucuna bracteata, M. capitata, M. pruriens, Rhynchosia minima, Vicia sativa, Vigna aconitifolia, V. mungo* var. *silvestris, V. radiate* var. *Sublobata, V. trilobata* var. *trilobata, V. umbelleta*
Oilseeds	*Ricinus communis, Sesamum indicum* ssp. *Anamalayensis* and *S. radiatum*
Fibers	*Corchorus aestuans, C. capsularis, C. depressus, C. pseudo-olitorius, C. tridens, C. trilocularis, Crotalaria alata, C. burhia, C. medicaginea, Gossypium arboreum, Hibiscus cannabinus, H. radiatus, Linum perenne* ssp. *perenne*
Vegetables	*Abelmoschus ficulneus, A. manihot, A. tetraphyllus, A. tuberculatus, Amaranthus blitum, A. caudatus, A. spinosus, A. viridis, Carissa congesta, Chenopodium album, Citrullus colocynthis, Cordia myxa, C. rothii, Cucumis callosus, C. melo* var. *agrestis, C. prophetarum, C. sativus* var. *hardwickii, C. sativus* var. *hardwickii, Lufa acutangula* var. *amara, L. hermaphrodita, Momordica balsamina* (Fig. 8.7), *M. charantia* var. *muricata, M. cochinchinensis, M. denudata, M. dioica, M. tuberosa, Solanum ferox, S. giganteum, S. indicum, S. nigrum, S. surattense*
Tubers	*Colocasia esculenta, Dioscorea alata, D. esculenta*
Fruits	*Carissa congesta, C. spinarllm, Emblica officinalis, Morus alba, M. serrata, Phoenix rupicola, Vitis parvifolia, Ziziphus mauritiana, Z. nummularia, Z. rugosa, Z. xylopyrus*
Medicinal and aromatic plants	*Achyranthes aspera, Asparagus racemosus, Cannabis sativa, Carthamus lanatus, C. oxyacanthus, Chlorophytum tuberosum, Citrullus colocynthis, Mentha longifolia, Rauvoljia serpentina, Tagetes minuta, Withania somnifera*

8.7 TRADITIONAL KNOWLEDGE

Local cultivars were preferred by subsistence farmers due to their desirable traits and adaptability, nutritive value, better taste, and higher fodder yield, for example, *Triticum durum, T. polonicum* (grown in Madhya Pradesh), *T. dicoccum* and *T. sphaeroccum* (grown in Gujarat). Cucurbits, namely, *Citrullus vulgaris* var. *fistulosus, C. callosus, C. melo* var. *agrestis, C. melo* var. *momordica, C. hardwickii, C. sativus* and *Luffa hermaphrodita* are used as vegetables in various forms. Leaves of *Aloe vera* (IC 333202) sweet type are used for curing many disease like diabetes, blood pressure, cholesterol, arthritis/rheumatism, liver disorder, colds, indigestion, constipation, flatulence, urine problems, skin disorder, piles, bleeding, healing, burns, cuts and wounds as antiseptic, killing intestinal worms, etc. Root extract of *C. prophetarum* for cure of earache and fruits to cows and buffaloes for conception. Taramira (*Eruca sativa*) makes soil poor and has allelopathic effects on subsequent crops. Its oil is edible and seed cakes are given to cattle for higher milk production.

Dried flower of *Diospyros melanoxylon* are used to cure urinary and skin troubles, its fruits are edible and leaves are used in making "Bidi" (slim local smoking pipe). Seeds or seed cakes of *Hibiscus cannabinus* are mixed with guar, bajra, etc. and feed to buffaloes for higher milk production and good quality ghee. Wild relatives and wild species of crop plant (*Corchorus* spp.; *Diospyros montana*; *Solanum* spp.; *Vigna radiata* var. *sublobata*, *V. trilobata*, etc.) have important role and value in the lives of village and tribal people.[19,22,27] Ash of cow dung, neem, and nirgundi leaves or dried leaves of neem and nirgundi are in general used for storage besides keeping seeds intact (chick pea pods/maize cobs) or in straw (*Triticum* spp.) in air-tight traditional vessels/containers locally known as Kinara, Pitara, etc. (Figs. 8.5 and 8.6). Figures 8.8 and 8.9 show the extent of loss of agro-diversity.

FIGURE 8.5   Kinara—made from *Clerodendron phloidis* for traditional seed storage.

FIGURE 8.6   Seed treatment with ash for pest free storage of leguminous seeds.

FIGURE 8.7   Fruit of wild *Momordica balsamina.*

FIGURE 8.8   Loss of agro-biodiversity due to over grazing.

FIGURE 8.9 Loss of agro-biodiversity, where single species is dominated at Jaisalmer district of western Rajasthan.

8.8 TRADITIONAL MANAGEMENT OF FARM PESTS AND ENEMIES

Various studies have been undertaken on farm pest management by Bhil, Garasia, Kathodia, Shahriya, and Damor tribes of southeast Rajasthan. While sleeping on machans (raised platforms on branches of trees), use of scarecrows and use of neem, akda, and tobacco leaves, and cow urine as insecticides is known among farmers from various parts of Rajasthan. Tribals have unique systems of management of pests and farm enemies. Porcupines are smoked out from their burrows and killed. Gerbils and rats are also forced to come out of burrows by flooding their burrows, and then they are killed. Crows are killed by leaving a poisonous gruel of maize mixed with crushed roots of Vajkand (*Dioscorea* spp.) for them to feed on. Vegetable crops such as cucurbits are especially vulnerable to attack by various insects. Smoke created by burning semi-dry wood and neem leaves is used to drive away insects in the field. For larger insects such as moths and fruit flies guggal gum (resin of *Commiphora whitti* tree) and cow dung cakes are burnt.

In total seeds of 36,652 accessions of different crop groups and 752 live plants are conserved in Medium Term Storage Facility and Field Gene Bank, respectively, at this Regional Station of the NBPGR in Jodhpur. These 36,652 accessions of different crop groups includes the collections/selection/breeding material (11,010 acc.) from different ICAR Institutes, SAU, T. University, NGO and progressive farmers of these region.

8.9 CONCLUSIONS

There are substantial amounts of work on maintenance of local crop cultivars and of local genetic diversity in production systems in many countries. This work is demonstrating that diversity continues to be maintained and managed by farmers throughout the world as an integral part of their coping strategies. It is essential that development strategies integrate this knowledge effectively and that the importance of crop genetic diversity to agricultural resources management and sustainable production. In spite of the many virtues of traditional agroecosystems (crop, livestock, soil biota), we are still losing the precious genetic diversity, the rivets of ecosystems stability. If a serious view of the existing situation is not taken into account, this region will lose ecological and economic security. Due this situation the arid and semiarid region may lose traditional knowledge of cultivation and uses of these crops and livestocks forever, and many also lose the chance of being a diverse and nutritive food producing region. In-situ conservation of traditional crops, cultivars, and livestock can succeed when these are strongly linked with the economic development. Poor scientific knowledge of traditional agroecosystems and the socioeconomic dimensions of desert farming society is serious impediment in identifying sustainable agricultural development solutions in desert.

ACKNOWLEDGMENT

Authors are very thankful to Director, NBPGR for providing all necessary facility for this research work.

KEYWORDS

- agro-biodiversity
- arid
- collection
- conservation
- India
- semiarid
- weedy
- wild

REFERENCES

1. Abraham, Z.; Latha, M.; Biju, S.; Lakshmi Narayanan, S. Changing Patterns of Plant Biodiversity in Southern Western Ghat. In *Plant Genetic Resources Management*; Dhillon, B. S., Tyagi, R. K., Lal, A., Saxena, S., Eds.; Narosa Publishing House: New Delhi, 2004; pp 113–135.
2. Altieri, M. A. Traditional Farming in Latin America. *The Ecologist* **1991**, *21*, 93–96.
3. Anonymous. *Under-exploited Tropical Plants with Promising Economic Value*. National Research Council—National Academy of Sciences: Washington, DC, 1975; p 189.
4. Anonymous. Shifting Cultivation. *FAO For. News Asia Pac.* **1978**, *2*, 1–26.
5. Anwar, R.; Bhatti, M. S. *Status of Genetic Resources Potential of the Mountains of Pakistan*. MFS Discussion Papers Series No. 21. International Centre for Integrated Mountain Development: Kathmandu, Nepal, 1990.
6. Arora, R. K.; Chandel, K. P. S. Botanical Source: Areas of Wild Herbage Legumes in India. *Trop. Grassland* **1972**, *6*(3), 213–221.
7. Arora, R. K. Biodiversity in Crop Plants and their Wild Relatives. In Proc. First National Agricultural Sci. Cong., National Academy of Agricultural Sciences, New Delhi, India, 1993; pp 160–180.
8. Arora, R. K. Wild Relatives of Cultivated Plants. In *Flora of India. Introductory Volume. Part II*; Botanical Survey of India: Calcutta, India, 2000; pp 218–234.
9. Arora, R. K.; Nayar, E. R. *Wild Relatives of Crop Plants in India*. Sci. Monograph 7. National Bureau of Plant Genetic Resources: New Delhi, India, 1984.
10. Bhandari, M. M. *Flora of the Indian Desert*. Scientific Publishers: Jodhpur, Rajasthan, 1990.
11. Bhandari, M. M.; Dwivedi, N. K. Changing Patterns of Plant Biodiversity in Western Rajasthan. In *Plant Genetic Resources Management*; Dhillon, B. S., Tyagi, R. K., Lal, A., Saxena, S., Eds.; Indian Society of Plant Genetic Resources & Narosa Publishing House Pvt. Ltd.: New Delhi, India. 2004; pp 108–112.
12. Brookfield, H.; Padoch, C. Appreciating Biodiversity: A Look at the Dynamism and Diversity of Indigenous Farming Practices. *Environment* **1994**, *36*, 6–11, 37–45.
13. Chandel, K. P. S.; Bhandari, D. C. Collection of Germplasm Resources in North-eastern Rajasthan. *Indian J. Plant Genet. Resour.* **1989**, *2*(2), 150–156.
14. Dwivedi, N. K.; Bhandari, D. C. Collecting Cultivated and Wild Species of *Corchorus* in Gujarat, India. *Indian J. Pl. Genet. Resour.* **1996**, *9*(9), 323–326.
15. Dwivedi, N. K. Collection of Khejri (*Prosopis cineraria*) and Other Minor Fruits from Parts of Rajasthan (May–June, 1990*). Germplasm Collect. Rep.* **1991**, *1*(2), 32–43.
16. Dwivedi, N. K. Multicrop Collection Specially Jute and Cotton Germplasm from Gujarat State (January, 1990). *Germplasm Collect. Rep.* **1991**, *1*(2), 44–57.
17. Dwivedi, N. K. Exploration for the Collection of Cultivated and Wild Species of Okra and Eggplant from Madhya Pradesh and Rajasthan. *Germplasm Collect. Rep.* **1991**, *1*(3), 50–63.
18. Dwivedi, N. K. Collection and Conservation of Genetic Diversity in Legume Crops from Arid and Semi-arid Regions. *J. Arid Legumes* **2007**, *4*(1), 35–39.
19. Dwivedi, N. K. Indigenous Herbal Therapy for Controlling Diabetes in Parts of Arid and Semi-arid Regions of Rajasthan, India. *Res. Link*—60 **2009**, *VIII*(1), 18–19.
20. Dwivedi, N. K.; Bhandari, D. C.; Bhatnagar, N. Collecting Wild Species of Crop Plants from Aravali Hills, Rajasthan, India. *Indian J. Pl. Genet. Res.* **1998**, *11*(2), 41–48.

21. Dwivedi, N. K.; Bhandari, D. C.; Verma, S. K. Collecting *Prosopis cineraria* (L.) Mac. Bride Germplasm in Western India. *IPGRI Newslett. Asia, Pac. Ocecinea* **1997**, *22*, 19.

22. Dwivedi, N. K.; Bhandari, D. C.; Gopalakrishnan, S.; Ram Meghwal, R. Plants Used for Curing Piles in Arid and Semi-arid Regions of Western Rajasthan, India. *J. Econ.— Taxon. Bot.* **2009**, *33*(4), 787–802.

23. Dwivedi, N. K.; Bhandari, D. C.; Aghora, T. A. Collecting Pea Germplasm from Rajasthan, India. *IBPGR Newslett. Asia, Pac. Ocecinea* **1994**, 15:11.

24. Dwivedi, N. K.; Kumar, D.; Patel, S. S. Collecting *Triticum* species from Salt and Heat Affected Areas of Gujarat and Rajasthan. *Indian J. Plant Genet. Resour.* **1999**, *12*(2), 263–267.

25. Dwivedi, N. K.; Bhatnagar, N.; Bhandari, D. C. Collection of Plant Genetic Resources from Parts of Arid and Semi-Arid Regions in India. *Indian J. Plant Genet. Resour.* **2001**, *14*(2), 268–272.

26. Dwivedi, N. K.; Bhatnagar, N.; Gopalakrishnan, S. Collecting and Characterising Cowpea (*Vigna unguiculata* (L.) Walp.) Germplasm in Arid Region. *J. Arid Legumes* **2005**, *2*(2), 328–329.

27. Dwivedi, N. K.; Dhariwal, O. P.; Gopala Krishnan, S.; Bhandari, D. C. Distribution and Extent of Diversity in *Cucumis* species in the Aravalli Ranges of India. *Gen. Res. Crop Evol.* 2009.

28. Hore, D. K.; Sharma, B. D. Changing Patterns of Plant Biodiversity in North- eastern Regional India. In *Plant Genetic Resources Management*. Dhillon, B. S.; Tyagi, R. K.; Lal, A.; Saxena, S., Eds.; Narosa Publishing House, New Delhi, 2004; pp 103–107.

29. Jodha, N. S. *Rural Common Property Resources: The Missing Dimension of Development Strategies*. World Bank: Washington, DC, 1992.

30. Kothari, A. Eco-Regeneration, Hope for future. *Survey of the Environment*; The Hindu Publication: Madras, India, 1994; pp. 173–177.

31. Maikhuri, R. K.; Rao, K. S.; Saxena, K. G. Traditional Crop Diversity for Sustainable Development of Central Himalayan Agroecosystem. *Int. J. Sustain. Dev. World Ecol.* **1996**, *2*, 1–24.

32. Maikhuri, R. K.; Rao, K. S.; Palni, L. M. S. Agro-biodiversity Conservation and Management in the Indian Himalayan Region in the Wake of Changing Patterns of Biodiversity. In *Plant Genetic Resources Management*. Dhillon, B. S.; Tyagi, R. K., Lal, A., Saxena, S., Eds.; Narosa Publishing House: New Delhi, 2004; pp 91–102.

33. Maikhuri, R. K., Rao, K. S.; Semwal, R. L. Changing Scenario of Himalayan Agro-Ecosystems: Loss of Agro-biodiversity, an Indicator of Environmental Change in Central Himalaya, India. *Environmentalist* **2001**, *21*, 23–39.

34. Maikhuri, R. K.; Semwal, R. L.; Rao, K. S.; Saxena, K. G. Eroding Traditional Crop Diversity Imperials the Sustainability of Agriculture Systems in Central Himalaya. *Curr. Sci.* **1997**, *73*, 777–782.

35. Nayar, M. P. Hot-spots of Plant Diversity in India—strategies. In *Conservation and Economic Evaluation of Biodiversity*; Pushpangadan, et al. Eds.; Oxford and IPH Publishing House: New Delhi, India, 1997; pp 59–80.

36. Nayar, E. R.; Pandey, A.; Venkateswaran, K.; Gupta, R.; Dhillon, B. S. *Crop Plants of India: A Check-list of Scientific Names*. National Bureau of Plant Genetic Resources: New Delhi, India, 2003.

37. Negi, K. S.; Pant, K. C.; Koppar, M. N.; Thomas, T. A. Wild Relatives of Genus *Allium* L. in Himalaya. Indian 1. Pl. *Genetic Res.* **1991**, *4*(1), 73–77.

38. Palni, L. M. S.; Maikhuri, R. K.; Rao, K. S. *Conservation of Himalayan Agro-ecosystems: Issues and Priorities.* Technical Paper III. Himalayan Eco-regional Cooperation Meeting, Kathmandu, Nepal, 1998.
39. Pandey, A.; Bhandari, D. C.; Bhatt, K. C.; Pareek, S. K.; Tomar, A. K.; Dhillon, B. S. *Wild Relatives of Crop Plants in India: Collection and Conservation*; National Bureau of Plant Genetic Resources: New Delhi, India, 2005; p 73.
40. Pandey, R. P.; Padhye, P. M. Wild Relatives and Related Species of Cultivated Crop Plants and their Diversity in Gujarat, India. *J. Econ. Taxon. Bot.* **2000,** *24*(2), 339–348.
41. Pandravada, S.; Rao, N.; Shivraj; Veraprasad, K. S. The Changing Patterns of Plant Biodiversity in the Eastern Ghats. In *Plant Genetic Resources Management*; Dhillon, B. S., Tyagi, R. K., Lal, A., Saxena, S., Eds.; Narosa Publishing House: New Delhi, 2004; pp 136–152.
42. Paroda, R. S. *Emerging Concern for Agro-biodiversity in National Context: An Introduction.* Keynote address workshop on National Concern for Agro-biodiversity Conservation, Management and Use: Shimla, India, 1997.
43. Patra, B. C.; Saha, R. K.; Patnaik, S. S. C.; Marandi, B. C.; Nayak, P. K.; Dhua, S. R. Wild Rice and Related Species in Orissa and West Bengal. In Proc. Plant Resource Utilization for Backward Areas Development; Sahu, S., et al. Eds.; Regional Research Laboratory, Bhubaneshwar, 28–29 Dec. 2002, Allied Publ. Ltd.: New Delhi, India, 2002; p 87.
44. Pratap, T. Biological Diversity and Genetic Resources: Some Issues for Sustainable Mountain Agriculture. Paper Presented in the International Symposium on Strategies for the Sustainable Mountain Agriculture, ICIMOD: Kathmandu, Nepal, 1990.
45. Singh, V.; Pandey, R. P. An Assessment of Wild Relatives of Cultivated Plants in Indian Desert and their Conservation. In *Scientific Horticulture*; Singh, S. P., Ed.; Scientific Publishers: Jodhpur, India, 1996; pp 155–162.
46. Shrestha, S. *Mountain Agriculture*: Indicators of Unsustainability and Options for Reversal. MFS (Mountain Farming Systems) Discussion Paper 32. International Centre for Integrated Mountain Development: Kathmandu, 1992.
47. Subudhi, H. N.; Saha, D.; Choudhury, B. P. Collection of *Desmodium* Des. and Wild relatives. *Orissa J. Econ. Taxon. Bot.* **2000,** *24*(3), 695–699.

CHAPTER 9

WATERSHED DEVELOPMENT AND MANAGEMENT

R. K. SIVANAPPAN

College of Agricultural Engineering & Technology, Tamil Nadu Agricultural University (TNAU), Coimbatore, Tamil Nadu, India. E-mail: sivanappanrk@hotmail.com

CONTENTS

Abstract 232

9.1   Introduction..232

9.2   Watershed..233

9.3   Concept of Watershed Development and Management..................234

9.4   Objectives of Watershed Management 234

9.5   Components in the Watershed...235

9.6   The Experiences and Challenges 243

9.7   Participation of People..244

9.8   Constraints 244

Keywords 245

References...246

ABSTRACT

The watershed development project can be successful only if the participation of people is emphasized in all phases of planning, implementation, and maintenance in order to get positive results in the program. We must also carefully analyze three aspects of cost, return, and risks because of their importance in every project. In order to achieve the goals and to fulfill the aspiration of the people, we should have interdisciplinary watershed management agency in each region/state/district. This agency will be responsible to formulate the projects incorporating various land development/rural uplifting activities, exercise, and coordinate development works. This will help in achieving the watershed development objectives in a speedy and efficient manner.

9.1   INTRODUCTION

Watershed development and management is an integration of technology within the natural boundary of a drainage area for optimum development of land, water, and plant resources to meet the basic minimum needs of the people in a sustained manner. The poor persons in the rural areas, who are struggling for survival, cannot be expected to pay heed to the conservation strategy unless their daily needs of food, fiber, shelter, and fuel are met with. A still more urgent need is for assured and full employment for all.

Integrated watershed development and management is not only the most effective solution to many of the problems mentioned above, but also the effective solution to many other common problems like drought, floods, climate change, etc. It includes the integration of many scattered programs of soil conservation, afforestation, minor irrigation, crop production, tree planting, fodder development, and other development activities into a well prepared micro-watershed project based on study of climate, land, water, and crop resources on one hand and man, animal resources on the other [1–3]. This integration offers hope for bringing about sustainable development of natural resources. It also provides solutions to many environmental problems like soil erosion, siltation, improper land use, soil salinity, lowering ground water table, etc. The overall productivity, income of the family and employment opportunity can be improved, once such problems are adequately addressed and solved.

This chapter presents basics of development and management of watershed.

9.2 WATERSHED

The United States Environmental Protection Agency (USEPA) defines watershed as "the area of land where all of the water that is under it or drains off of it goes into the same place. According to geographer John Wesley Powell, a watershed is that area of land, a bounded hydrologic system, within which all living things are inextricably linked by their common water course and where, as humans settled, simple logic demanded that they become part of a community. Watersheds come in all shapes and sizes. They cross county, state, and national boundaries. In the continental US, there are 2110 watersheds; including Hawaii, Alaska, and Puerto Rico, there are 2267 watersheds," <http://water.epa.gov/type/watersheds/whatis.cfm>. The rain water after absorbed by the soil, flows as runoff in small gullies, rivulets and joins the stream and form the river system. This represents a natural drainage system. The river basin at macro level and watershed/sub-watershed at micro-level represent the *Natural Drainage System.* A watershed is an area from which runoff, resulting from precipitation flows past a single point into a large stream, river, lake, or an ocean. In other words, a watershed is that area in which all the precipitation converges and drains past a particular point. The term watershed, catchment area, or drainage basin can be used interchangeably. A watershed may be only a few hectares as in the case of small ponds, or hundreds of square kilometers as in the case of rivers or big reservoirs. For convenience, watersheds are classified in terms of size into basins, catchments, sub-catchments, watershed, sub-watershed, mini- and micro-watersheds. Each watershed is an independent hydrological unit and any modification of the land use in the watershed will be reflected on the water as well as in the sediment yield of the watershed.

The watershed can be demarcated from the topo-sheet. But for a small (micro) watershed, a detailed topographical survey has to be made and a contour may have to be prepared. The ridge points are marked and the area below the ridge line is known as the watershed area. This contour map can be imposed on the village map. In case of small watersheds, it can be demarcated by walking over the ridge point.

Watershed has become an acceptable unit of planning for optimum use and conservation of soil and water resources. A watershed is hydrological unit which produce water as an end product by interaction of rainfall and watershed factor. The quality and quantity of water produced by a watershed is an index of amount and intensity of rainfall and the nature of watershed management. In some watersheds, the aim may be to produce maximum quantity of water distributed throughout the year. In another watershed, the priority may be to reduce the peak rates as runoff for minimizing the floods. Yet in another, the concern may be to increase infiltration so as to improve the soil moisture regime for maximum production of crops. Therefore, different measures are to be introduced in order to achieve the objectives of different watersheds. The size of the watershed may be of any size, but for proper planning and implementation in a year or two, the size of the viable watershed may be about 1000–5000 ha.

9.3   CONCEPT OF WATERSHED DEVELOPMENT AND MANAGEMENT

The concept of integrated watershed development involves development and management of the resources in the watershed so as to achieve higher production that can be sustained without causing any deterioration in the resource base or causing no ecological imbalances. This calls for formulation and implementation of a package of programs for action for optimum resource use in the watershed without adversely affecting the soil and water base or life-supporting system. The concept assumes more importance in the context of planning for sustained development. The watershed development aims at preventing watershed degradation that results from the interaction of physiographic features, eliminate unscientific land use, in appropriate cropping pattern, soil erosion thereby improving and sustaining productivity of resources leading to higher income and living standard for the inhabitants in the watershed area. Therefore, it involves restoration of the ecosystem and protecting and utilizing the locally available resources within a watershed to achieve sustainable development.

9.4   OBJECTIVES OF WATERSHED MANAGEMENT

a.   Utilizing the available land to its maximum productivity by adopting various/suitable measures according to the land capability and without any environmental degradation.

b.   Maximizing productivity per unit area, per unit time, and per unit water to meet the food, fodder, and fuel requirements of the people living in the watershed.

c.   Conserving as much rain water as possible in the place where it falls and also increasing the ground water level to get water throughout the year and maintaining it for sustainability.

d. Preventing soil erosion by means of suitable soil and water conservation measures.
e. Draining the excess water safely and avoiding gully formation and
flooding the areas.
f. Maximizing the water storage capacity in the watershed, both in the
soil and storage structures.
g. Improving the infrastructural facilities in the watershed.
h. Increasing the level of income and status of the people living in the
watershed.

9.5 COMPONENTS IN THE WATERSHED

The following are some of the ways and means by which the watershed
should be treated in order to achieve the objectives of the watershed development. The treatment may be different for different soil conditions, topography, and rainfall pattern. The following factors should be considered while
developing a watershed:

- Size and shape of watershed.
- Topography.
- Soil characteristics.
- Precipitation—its quantity and pattern of distribution.
- Land use.
- Vegetative cover including trees.
- Animal population.
- Human/community activities.
- Social systems.

Based on the above factors, watershed development plan can be prepared
and implemented with the cooperation and involvement of people/farmers in
the watershed. The following are the activities and strategies (Figs. 9.1–9.15):

Land development	Erosion control devices
Contour ploughing	Contour stone wells
Land leveling	Contour and staggered trenching
Contour bunding/graded bunding/ Compartmental bunding	Gully control structures/sunken pits
	Construction of silt detention tanks
Terracing in steep slopes	Providing vegetative and stone barriers
	Providing retaining walls
	Preventing stream bank erosion

Water resources development	Agricultural development
Construction of earthen embankments	Improved dry land (rainfed) farming techniques
Construction of check dams	
Temporary and permanent	Suitable crop selection
Subsurface dams	Agroforestry techniques
Construction of farm ponds.	Introducing horticultural and plantation crops
Construction of percolation tanks	
Providing drainage to remove excess rain water	
Conveyance of water by pipe lines	
Providing protective irrigation	

In situ moisture conservation techniques

(i) **For agricultural crops**	(ii) **For tree crops**
Micro-catchment	Saucer basin/crescent bund
Basin/furrow farming	"V" ditch forming
Broad bed and furrows	Tie ridging
Tie ridging/random tie ridges	Catch pits
Water spreading	Deep pitting

The coordination with development departments/non-government agencies (NGOs) is necessary to implement it effectively. The various departments involved are forest/agricultural/horticultural/agricultural engineering, water resources organizations (both surface and ground water), animal husbandry/fisheries highways and rural works/rural development department, social welfare/education/cooperation. Apart from these departments, NGO's working in the area can also be involved.

The various activities may overlap, but these are complementary to one other. If these works are taken up on watershed basis, then all the unproductive/waste/fallow lands can be brought under agriculture production and the productivity of the existing low-yield areas can be increased substantially [4–6].

To implement the project, it requires expertise in a variety of fields like agriculture; horticulture; forestry, soil, and water conservation; hydrology; socioeconomic; and community development, etc. Further, the approach should be in a holistic manner to achieve the goal. Therefore, adequate training of personnel in these subjects will have to be provided to ensure the success of this new and complex activity in the watershed development.

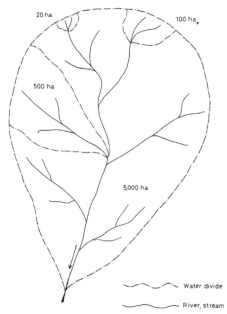

FIGURE 9.1 A typical watershed.

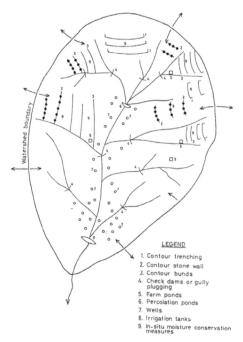

FIGURE 9.2 Components (soil and water conservation measures) of a typical watershed area.

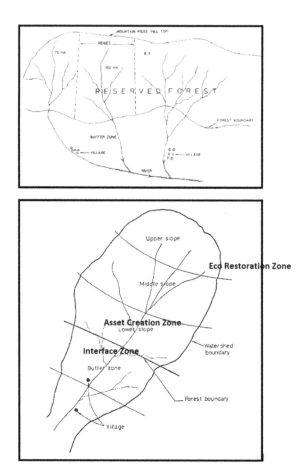

FIGURE 9.3   Forest watershed showing four zones.

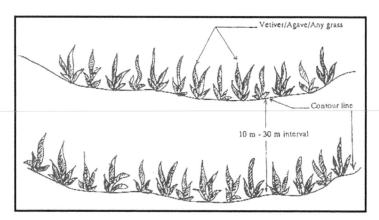

FIGURE 9.4   Vegetative barrier on contour.

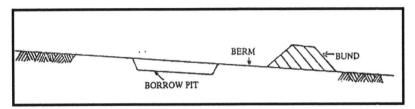

FIGURE 9.5   Cross section of a contour bund.

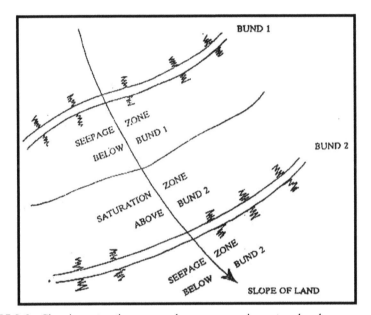

FIGURE 9.6   Showing saturation zone and seepage zone in contour bund area.

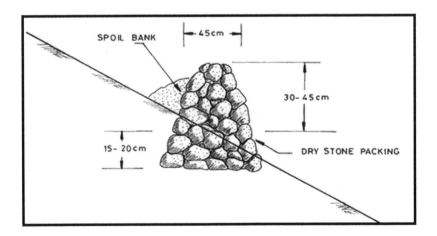

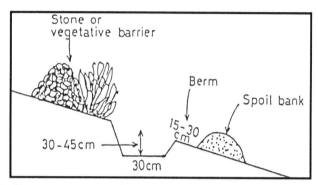

FIGURE 9.7   Top: contour stone wall; bottom: cross section of a contour trench.

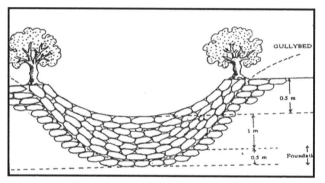

FIGURE 9.8   Front view of loose stone check dam.

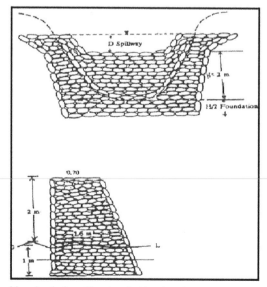

FIGURE 9.9   Boulder check dam (front and side views).

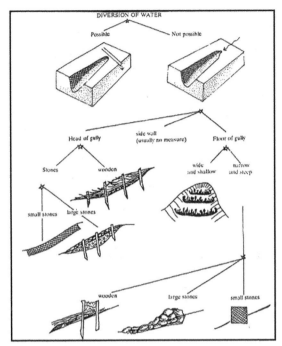

FIGURE 9.10   Gully control structures.

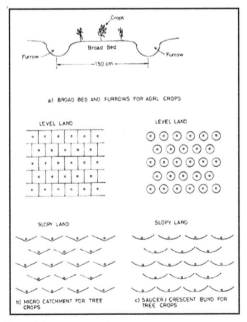

FIGURE 9.11   In-situ moisture conservation measures.

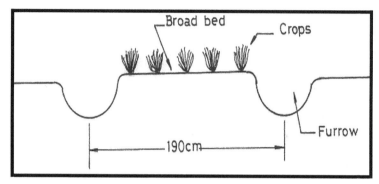

FIGURE 9.12   Broad bed and furrows with crops.

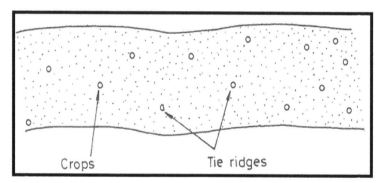

FIGURE 9.13   Tie ridging with crops.

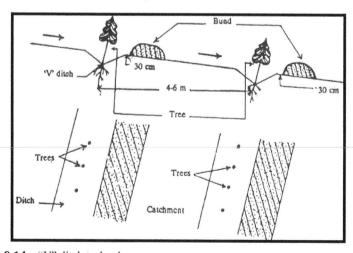

FIGURE 9.14   "V" ditch technology.

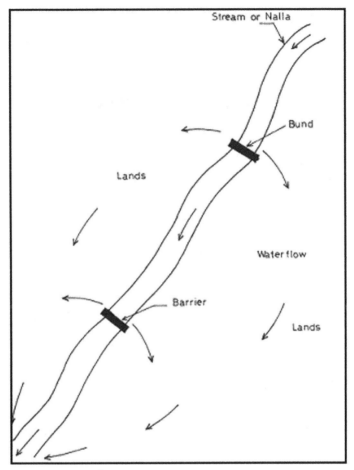

FIGURE 9.15   Water spreading.

9.6   THE EXPERIENCES AND CHALLENGES

The watershed development and management programs are being implemented throughout India by Government agencies and NGOs. There is ample need and necessity of watershed development and management programs to be implemented with involvement of the people living in the watershed, as indicated by the experiences in Ralegoan Siddhi in Maharashtra, Arvari watershed in Alwar district of Rajasthan, Kabilanalla watershed in Karnataka, numerous watersheds in Jhabuva district in Madhya Pradesh.

In Tamil Nadu and Kerala also, there are a few model watersheds, which have changed the livelihood of the people in the area by proper implementation of the watershed development management program. Due to the implementation of these programs in a systematic and scientific manner, there is little water scarcity problems and the living condition of the people in the watershed area have improved.

As the rainfall is erratic, not evenly distributed and falls within a short period (3–4 months), it is necessary to take up the comprehensive watershed development and management programs in a systematic and scientific manner to have water available throughout the year. There are many schemes with different names with the same objective but these are being implemented by different agencies without the cooperation and involvement of the people. Hence, the programs are not giving the expected results, despite inversion of large amounts that are being spent every year for this purpose. It is really a challenge for the people to take up the work as has been done in Ralegoan Siddhi and Arvari watersheds.

9.7   PARTICIPATION OF PEOPLE

All the above activities could not be done without the active support of the farmers and people in the area. There are NGO and VO in many places and their services can be utilized profitably. The people should be educated through various means to understand the seriousness of the problems and the remedial measures. Many pilot projects/demonstration plots can be introduced at village/block/taluk level to convince the need and advantage of this technology for their sustained livelihood. Voluntary and NGOs can play a vital role in this regard.

9.8   CONSTRAINTS

To achieve the above goals, there are many constraints. These constraints can be classified as follows:

(i)   **Sociocultural and economic constraints**
    a. Caste, community.
    b. Religious institutions.
    c. Illiteracy.
    d. Poor economic status of the majority of the people.
    e. Farm size/fragmentation.
    f. Land ownership pattern/tenants.

(ii) **Institutional cum political constraints**
   a. Policy instruments.
   b. Credit instruments—banks.
   c. Marketing institutions (regulated markets).
   d. Research and educational institutions.
   e. Appropriate technology.
   f. Rainfed technology.
   g. Extension agency needed for popularization of technology.
   h. Role of VO/NGOs.
   i. Policy and decision-making level (government/donor agency).

There may be numerous obstacles and constraints for the dissemination and implementation of technologies and practices, which have been proved successful. Factors preventing and factors to promote and spreading of successfulness area should be identified and acted upon for the future development and management activities in the watershed area.

KEYWORDS

- conservation
- constraints
- embankments
- gully control
- micro-catchment
- people's participation
- productivity
- saturation zone
- strategy
- subsurface dams
- tie ridges
- vegetative barriers
- water spreading
- watershed

REFERENCES

1. National Rainfed Area Authority. *Common Guidelines for Watershed Development Projects*, National Rainfed Area Authority: New Delhi, 2008.
2. CRIDLA. *Field Manual on Watershed Management*. Central Research Institute for Dry Land Agriculture (CRIDLA), Hyderabad, India, 1990.
3. Ramasamy, K. *Dryland Development and Maximizing Crop Productivity*. Agricultural Engineering College and Research Institute, Tamil Nadu Agricultural University, Coimbatore, India, 2003.
4. Sivanappan, R. K. *Technology for Profitable Agriculture in Dry Lands and Fallow Lands in Tamil Nadu*. Tamil Nadu Agricultural University, Coimbatore, 2013.
5. Sivanappan, R. K. Watershed Development and Management Practices, Strategies and Constraints—Role of Agricultural Engineers. Tamil Nadu Agricultural University, Coimbatore, 2005.
6. Sivanappan, R. K. Soil and Water Conservation and Water Harvesting. *Tamil Nadu Social Forestry Project, Indo-Swedish Forestry Coordination Project, Madras, 1992 and 2002*, 2002.

CHAPTER 10

WATER RESOURCES OF INDIA: PROBLEMS, SOLUTIONS, AND PROSPECTS

R. K. SIVANAPPAN

College of Agricultural Engineering & Technology, Tamil Nadu Agricultural University (TNAU), Coimbatore, Tamil Nadu, India. E-mail: sivanappanrk@hotmail.com

CONTENTS

Abstract 248

10.1 Introduction 248

10.2 National Water Development Agency 248

10.3 Water Resources of India 249

10.4 Water Crisis 253

10.5 Interlinking of Indian Rivers: Need and Importance 254

10.6 Recommendations 267

Acknowledgments 268

Keywords 269

Refrences 269

Appendix 10.I   Chronological Developments to Interlink
        Indian Rivers [Based on Data in this Chapter]. 271

Appendix 10.II   Proposal by Captain Dastur to Interlink
        Indian Rivers 272

Appendix 10.III   Investment Needs and Potential Economic
        Impact Due to Interlinking of Indian Rivers 273

Appendix 10.IV   Water Resources of India 274

ABSTRACT

Authors have discussed the need to establish *National Center for Water Resources, Research Development & Management*, with objectives of taking up research on water/water-related subjects and coordinating water-related issues/activities like developments and managements. Future water scarcity is inevitable. The necessity of interlinking of rivers in India, particularly interlinking of peninsular rivers is also discussed.

10.1 INTRODUCTION

The land and water resources are always constant and these resources are abundant in India. But these resources are not equidistributed to all states/parts of the country. The demand for more land area, for the food production to the escalated population, can be met by increasing the intensity of cultivation from the present 120% to about 200–300%. However for this, water is necessary. Hence, water will be the constraints in development and to increase the agricultural production in the future.

Water is a scarce commodity and it is valuable natural resources not only to human beings but also for the entire ecosystem for their sustenance and development. With the population explosion, increased demand of food grains and industrial growth, the demand of water will be many fold, but the availability is more or less fixed quantity. To understand the vulnerability, it is necessary to analyze critically the rainfall of the country on regional basis, the average annual surface flows in rivers, the replenishable ground water resources, the water stored in the tanks and ponds, etc. Further, it is necessary to calculate the water needs for various sectors namely: irrigation, industries, house-hold purposes, animals, maintaining ecology, and environment for the living population.

10.2 NATIONAL WATER DEVELOPMENT AGENCY

Inter-basin water transfer can be a powerful tool to minimize the scarcity of water and regional imbalance in the supply and demand for water, though it may raise number of constitutional issues. Such *inter-basin transfer* (IBT) will help in carrying the surplus water from one river (basin) to the water scarce area of another basin. Many suggestions have been made from time to time for interlinking the Indian rivers. In 1980, after carefully examining all such suggestions for transferring water for irrigation and other purposes from well-placed/surplus areas to deficit areas in the country for the optimum development of water resources, the Ministry of Irrigation of GOI (now water resources) and the *Central Water Commission* (CWC) formulated a National Perspective Plan (NPP) for water development keeping existing uses undisturbed and tribunal awards in view and providing for the reasonable needs of the basin States for the foreseeable future. The NPP comprises two components: Himalayan Rivers Development and Peninsular Rivers Development. The National Water Development Agency (NWDA) was set up in July 1982 by the Government of India (GOI) to investigate into the matter further and give a concrete shape to the Peninsular Rivers Development Component of the NPP to start with.

In this chapter, author has used following units:

1 acre—ft = 43,560 cu ft = 1233.65 m³.

1 cubic foot = 28.32 L = 7.48 US gallon = 0.037 cu yard = 0.0283 cu meter.

cumec = cubic meter per second.

cusec = vubic feet per second.

1 cubic meter = 1000 L = 264 US gallon = 35.31 cu ft.

1 ha = 107,600 sq. ft = 10,000 sq. m = 2.47 acre.

1 ha-m = 10,000 m³ = 8.10 acre ft.

1 km = 0.621 mile.

M = Million

1 MCM = Million cubic meter/million M³ = 35.31 M. cft.

M. ha = Million hectare.

1 MHM = Million hectare meter = M (10,000 sq m) meter = 10,000 Million cubic meter = 10,000 Million m³ = M (107,600 sq. ft.)3.281 ft. = 353,035.6 M. cu. ft. = 353 TMC = 8.1 M. acre ft.

MT = Million tons

1 sq. km = 1000,000 sq. m = 100 ha = 0.386 sq. mile.

1 sq. mile = 2590,000 sq. m = 640 acre = 259 ha = 2.59 sq. km.

TMC = thousand million cubic feet

10.3 WATER RESOURCES OF INDIA

The average annual rainfall for India is about 1150 mm which comes from south west and north east monsoons. The total rain works out to about 400 Million hectare-meter (MHM). The assessment of water resources of the Country in each of the 20 basins have been attempted by various bodies from time to time. The latest being 186.937 MHM by the CWC in the year 1993.[3] After reviewing the studies and water availability studies for the various basins by the NWDA, the National commission for *integrated water resources development plan* has reassessed the country's water resources at 195.29 MHM (Table 10.1). The total storage capacities built in various river basins, the projects under construction, and identified would account about 38 MHM of the total flow. After accounting the minor storage, the total storage capacitor in the country would be about 41–42 MHM.

TABLE 10.1   Mean Flow, Utilizable Surface, and Ground Water Resources: Basin-wise.

River basin in India	Mean flow	Utilizable flow	Replenishment	Utilizable***
	Surface water		**Ground water**	
	BCM (10 BCM = 1 MHM)			
Indus	73.31	46.0	26.50	24.3
Ganga	525.02	250.0	171.00	156.8
Brahmaputra	629.05*	24.0	26.55	24.4
Barak	48.36	–	8.52	7.8
Godavari	110.54	76.3	40.64	37.2
Krishna	69.81**	58.0	26.40	24.2
Cauvery	21.36	19.0	12.30	11.3
Submarekha	12.37	6.8	1.82	1.7
Brahmani-Bartarni	28.48	18.3	4.05	3.7
Mahanadi	66.88	50.0	16.50	15.1
Pennar	6.32	6.9	4.93	4.5
Mani	11.02	3.1	7.20	6.6
Sabarmati	3.81	1.9	–	–
Narmada	45.64	34.5	10.80	9.9
Tapi	14.88	14.5	8.27	7.6
West flowing rivers between Tapi and Tadri	87.41	11.9	7.70	16.2
West flowing rivers between Tadri and Kanyakumari	113.53	24.3	–	–
East flowing rivers between Mahanadi and Pennar	22.52	13.1	11.22	10.3
East flowing rivers between Pennar and K. Kumari	16.46	16.7	18.80	17.2

TABLE 10.1   *(Continued)*

River basin in India	Mean flow	Utilizable flow	Replenishment	Utilizable***
	Surface water		Ground water	
	BCM (10 BCM = 1 MHM)			
West flowing rivers of Kutch, Saurashtra, and Luni	15.10	15.0	0	0
Area of inland drainage in Rajasthan	0	–	–	–
Minor rivers draining into Bangladesh and Myanmar	31.00	–	18.12	16.8
Total	**1952.87**	**690.3**	**431.32**	**395.6**

Source: CWC, Publication 6/93—Reassessment of Water Resources Potential of India. Ground Water Resources of India CGWB-1995.[3]

*Includes additional contribution of 19.81 BCM being flow of nine tributaries joining Brahmaputra. **CWC assessment is based on mean flow of the yield series accepted by KWDT award. The figure of the CWC assessed from run-off data at Vijayawada is 78.12 BCM. ***Computed on proportionate basis from annual replenishment.

According to the CWC estimates,[4] the total utilizable water through conventional schemes in all the 20 basins is about 69.03 MHM, which is about 35% (24,840 TMC) as shown in Table 10.1. This implies that about 65% of the river flow (45,360 TMC) is allowed to go to the sea without using it, though the demand to divert water for the deficit areas by interlinking of rivers is there in the last 30–40 years. Table 10.2 shows annual yield of west flowing rivers in Karnataka state of India.

The latest assessment of replenishable ground water resources has been made as 43.20 MHM in the year 1994–1995 by the Central Ground Water Board (CGWB) through large volume of hydrologic and related data.[1] The utilizable ground water resource has been assessed as 39.56 MHM (7 MHM for domestic and industrial uses and 32.56 MHM for irrigation). The basin wise details of various water resources and their utilizable components are given in Tables 10.1 and 10.2. The assessed gross available water and utilizable water resources of the country based on conventional development technology are therefore 238.5 MHM and 108.6 MHM, respectively. Based on the population, the total available water and utilizable water resources per capita per year for the years 1991, 2001, 2011, 2025, and 2050 are calculated (Table 10.3).

TABLE 10.2   Annual Yield of West Flowing Rivers in Karnataka State.

Sub-basin	Catchment area	Average yield
	Sq. km	MCM
Aghanashini	1330	3028
Barapole	560	1274
Bedthi	3574	5040
Chakra River	336	991
Independent catchment between Bedthi and Aghanashini	401	906
Independent catchment between Netravathy and Barapole	1320	4474
Independent catchment between Sharavathi and Chakra River	1042	3086
Independent catchment between Varahi and Netravathy	3067	9457
Kalinadi	412	934
Mahadavi	412	934
Netravathy	3222	9939
Shravathi	3592	8816
Varahi	759	2263
Total		**57,489 MCM or 2000 TMC**

Source: Water resources development organizations, Government of Karnataka, Bangalore.

TABLE 10.3   The Available Water and Utilizable Water per Capita per Year.

Year	Population in India	Available water per capita per year (238.5 MHM)	Utilizable water per capita per year (108.60 MHM)
	million	M3/p/y	M3/p/y
1991	851	2831	1288
2001	1030	2316	1054
2011	1210	1971	908
2025	1350–1400 (estimated)	1700	780
2050	1650 (estimated)	1445	660

M^3 = cubic meter, MHM = million hectare meter.
Source: Estimated by R. K. Sivanappan.

10.4   WATER CRISIS

Tables 10.1–10.3 indicate that the utilizable water resource per capita per year vary from 3020 M^3 in Narmada Basin and about 180 M^3 and less in Sabarmati basin. During 1991, out of 20 basins, 4 basins had more than 1700 M^3/p/y (M^3/person/year) utilizable water resources, while 9 basins had between 1000 and 1700 M^3/p/y, 5 basins had between 500 and 1000 M^3/p/y, and 2 basins had less than 500 M^3/p/y, when the population of India was 851 million.

The population in 2050 is expected to be 1650 million and the food grain requirements may be around 550–600 million tons, including for losses in storage and transportation, seed requirements, and carryover for years of monsoon failures (an allowance of 15%).

The total storage buildup in various river basins through major and medium projects up to 1995 was about 17.37 MHM. The major and medium projects under construction and identified would account for 7.54 MHM and 13.23 MHM, respectively. The total is about 38.15 MHM. After taking into account the minor storage structures including tanks/ponds (about 4 MHM), the total storage capacity would be about 42 MHM. This accounts for the population of 1210 million in 2011, the storage capacity created in the country per person comes to about 350 M^3, compared to 5961 M^3 in USA, 2486 M^3 in China,[2] etc.

In this connection, it is not out of place to mention that there are about 45,000 large dams in the World of which 46% are in China, 14% are in USA, 9% are in India, and Japan/Spain are having 6% and 3%, respectively, as reported by the committee on large dams. The above two facts indicate that India's water storage capacity and dames constructed are very meager compared to various countries in the world taking into consideration of the population.

The data in Table 10.3 reveal that the utilizable water of the country (108.60 MHM) per capita per year in 2011 is only 908 m^3, which is less than 1000 m^3 indicating the water supply begins hamper to health, economic development, and human well-being and nations are considered water scarce. Even if the total water available in the country (238.50 MHM) is taken into consideration, the country will come under water stress as per the world bank and UN bodies after 2025, that is, after 9 years (Table 10.3), the GOI and CWC has not taken seriously to solve the water stress/scarcity of India to utilize the water, which is allowed to go to the sea every year. Further, there are demands from the state governments to divert the water from the surplus basins to the deficit basins. The National Water Policy (2002) clearly states that water should be made available to water short areas by transferring from other areas including transfers from own river basin to another after taking into account the accounts of the basin.

In Tamil Nadu (TN), the per capita availability of water is as low as 383 M^3 in some east flowing river basins. It is estimated that about 1/3 of the area is drought prone in India: Karnataka 79%, TN 64%, Andhra Pradesh 45%, Maharashtra 40%. Although the average availability of water for India at present is about 2000 $M^3/p/y$, yet it is less than 650 $M^3/p/y$ in TN, and it will be further reduced in the years to come as the population is increasing at the rate of about 2% per annum. As the demand of water is increasing day by day for drinking, industrial uses, the availability/allotment of water for irrigation will be reduced from the present 85–70% in another 15–20 years. However, the food production has to be increased for the growing population, that is, from 240 MT in 2010 to 500 MT in 2050. As the availability of water is only 650 $M^3/p/y$, TN is already a water scarcity state according to UN/World Bank norms. Therefore, there is a need to find out sufficient water for South India in general and TN in particular. This is possible in view of the fact that out of 195 MHM of surface/river flow in a year in India only about 35% is useable quantity.

Though there are many institutions/departments/boards in states and at the center, the water problem is not viewed seriously. Even the research institutions like Water Technology Centers, Institute of Water Studies, etc. in various states in the country are taking only on water management and augmenting rain water by harvesting and conservation. The NWDA[16] was created to assess the surplus water available in river basins in order to divert the same into deficit basins, but this is not being implemented since the states having surplus water are not willing to divert the water to the deficit basins/ states during the last 20 years.

10.5 INTERLINKING OF INDIAN RIVERS: NEED AND IMPORTANCE[20-22]

10.5.1 PENINSULAR RIVER DEVELOPMENT (PRD)

The NWDA conducted feasibility studies and found that it was possible to connect the six rivers[16] namely—the Mahanadhi, the Godavari, the Krishna, the Pennar, Cauvery, and Vaigai to provide a long-term solutions to the problem of water scarcity being faced by the farmers in South India, namely, Andhra Pradesh, Karnataka, TN, and Pondicherry states.

The Mahanadhi and the Godavari were found to have surplus water to an extent of 811 TMC (280 + 531) or 2.25 MHM. Instead of allowing the excess water to flow into the Bay of Bengal, it was possible to transport the water through canals providing for irrigation of farmlands. According to the NWDA[16] as conceived by Dr. K. L. Rao,[18,19] the then Minister of Water Resources Govt. of India, the Mahanadhi, the Godavari, the Krishna, the Pennar, and the Cauvery, Vaigai, and Gundar are to be connected by a canal having a length of 3716 km and with an estimated cost of Rs. 350,000 million (Rs. 500,000 million or more at present value) (see Fig. 10.1).

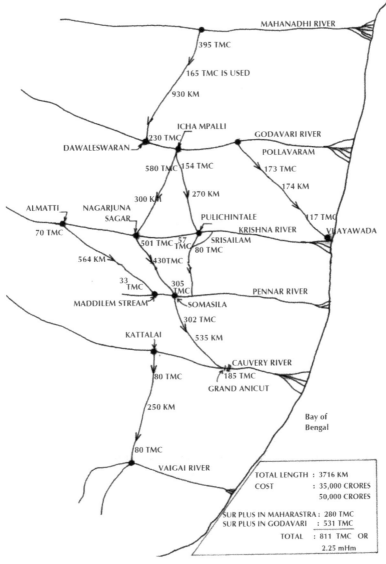

FIGURE 10.1   Proposal submitted to the Union Ministry of Water Resources, GOI.[13]

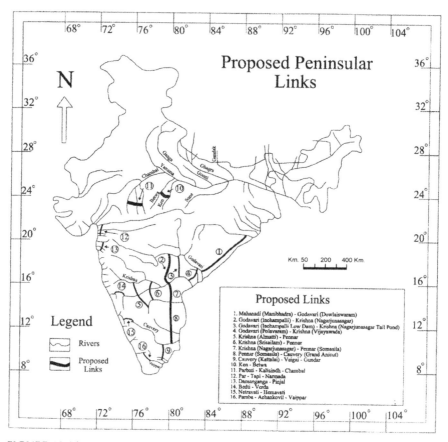

FIGURE 10.2A   Proposed Peninsular links [NWDA].

The NWDA has also prepared blue prints to connect the west flowing rivers (Figs. 10.2a and 10.2b) in the Western coast of India,[16] according to which the Pamba and Achankovil rivers carry about 250 TMC in Kerala, will be diverted to Vaippar river in TN to the extent of 22 TMC to be used in the drought prone Tirunelveli, Tuticorin, and Virudhunagar Districts to irrigate about 2.26×10^5 acres of dry lands at an estimated cost of Rs. 14,000 million and could be implemented in 8 years period in the year 1999 [The Hindu of 6th April 1999[8]]. The author has collected data and worked out the surplus water in Kerala state, which is now going as a waste every year to the Arabian Sea. The details are as follows:

Water resources	MHM	TMC
Total run off in all rivers in Kerala sate	7.80	2810
Total utilizable water	5.75	2070
Water requirement for irrigation, drinking, industry, salinity control, etc.	4.50	1620
Surplus available for transfer	1.25 or	450

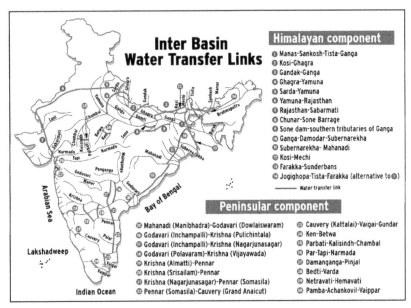

FIGURE 10.2B   Inter-basin water transfer links proposed by NWDA.

In fact, the NWDA has estimated the surplus water in Kerala which is going to the sea every year is more than 1000 TMC.[16] This quantity of water can be diverted for the benefit of drought prone areas of TN to grow food crops, which can feed the growing population not only in TN but also in Kerala and other states [National perspective for water resources development, New Delhi, January 1999,[12,14]].

There is another project, namely—Pandiar and Ponnumpuzha river project: The origin of the river is in TN running toward west and joins to the Cholaiar river in Kerala and falls in Arabian sea. The catchment area of the river is in TN and about 14–15 TMC of water from this catchment can be diverted to east, that is, to TN to irrigate about 2×10^5 acres. Although negotiations were going on between Kerala and TN since 1972, yet there has been no agreement and the project was not taken up even after 43 years.

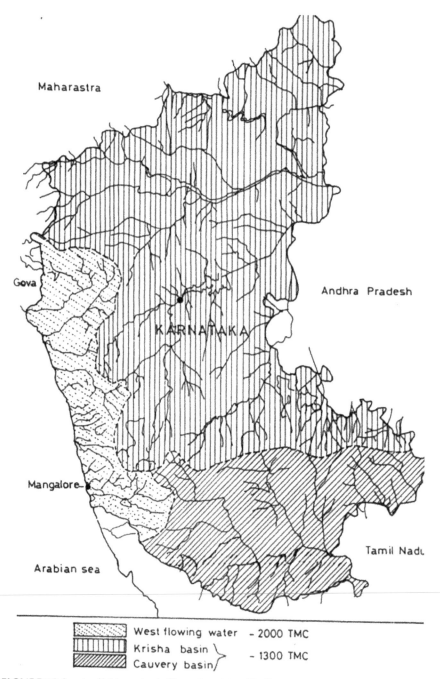

FIGURE 10.3   Available water in Karnatka state of India.

In Karnataka, the Western Ghats which is about 13% of the geographical area of the state having 60% of the state's water resources in terms of quantity due to high intensity of rainfall and every drop of it is running as waste to the sea. The balance 87% of the area of the state mostly comprising Krishna and Cauvery basins have only 40% of the waters for which Karnataka has water disputes with TN and Andhra Pradesh fighting in tribunals. The West flowing rivers in Uttara Kannada and Dakshin Kannada Districts of Karnataka state like Nethravathi, Kumardhara, Varahi, Aghanashini, etc. have in all about 2000 TMC annually (Table 10.2) as against Krishna and Cauvery put together of 1300 TMC (Fig. 10.3).

We can easily and economically without disturbing the environment and ecology of the forests and without displacement of people, divert the West flowing water to the East in Karnataka across the Ghats through pump storage schemes, utilizing the wasted existing thermal power in the nighttime, during monsoons for removing shortage of supplies for irrigation, Industry and drinking water. By this, it is possible to share the excess water and power with TN and Andhra Pradesh.

Till today, no serious thoughts have been given or plans made to divert this western flow to the eastern water deficit plains of Krishna and Cauvery basins through financially and economically feasible engineering projects. If pump storage schemes are thought off and planned for implementation, the problem of power requirements, for diverting the water can be overcome. Basically a pump storage scheme with reversible turbines, consists of two small pools of water, one situated at a high elevation and another at the lower level connected by penstocks with reversible machines in a power house. Water is alternated between the two pools by operating the power house, in dual modes.

In one mode, water is let down from the upper pool to the lower pool through the pen stocks, to generate hydroelectric power. In the other mode, water is lifted up by the reversing the operations of the machines working as a pump turbines into the upper pool. The operation in the generating and pumping modes is done dually in 24 h of the day, the length of each mode depending on the requirement of a particular mode. To minimize environmental and ecological damage, the power house and penstocks could be underground in tunnels with only small pools seen in the surface. The above-mentioned diversions, if implemented in the next 10–15 years, it will not only solve the disputes between the southern states, but also the water demand/requirement of these states.

The NPP has the distinctive features to eventually work on gravity flows expect in small reaches where low lifts not exceeding 120 M may be involved. The plan was *prima facia* to be technically feasible and economically viable when compared with the present cost of development of irrigation facilities. The NWDA has done an excellent job and identified 17 links under the peninsular river development plan and has also prepared the pre-feasibility reports for all the 17 links. The feasibility plan prepared by the experts of NWDA was submitted to the *Union Ministry of Water Resources*[13,16] for its approval and follow up action including a central legislation.

Mr. Suresh Prabhu (Former Chairman Taskforce—Interlinking of rivers) has prepared a detailed report and has submitted the report to the then PM of India during 2003–2004 mainly on Peninsular Rivers development. In April 17, 2003, Mr. Prabhu said that the priority is for the peninsular river project and all the southern states have agreed this proposal for the peninsular rivers development project.

10.5.2 HIMALAYAN RIVERS DEVELOPMENT PROJECT

Under the Himalayan Rivers Development[7] proposed interlinking canal system will be provided to transfer surplus flows of the Kosi, the Gandak, and the Ghagra to the west. In addition, the Brahmaputra–Ganga link[6,7] will be constructed for augmenting dry weather flows of the Ganga and the surplus water available in the Brahmaputra and its tributaries can be transferred to South by extending Brahmaputra and its tributaries with Ganga[6,7] and Ganga link up to Mahanahi. The Himalayan component would provide additional irrigation of about 22 Mha and generation of about 30 M kW of hydropower, besides providing substantial flood control in the Ganga and Brahmaputra basins (Figs. 10.4 and 10.5).

The Japanese scientists also suggested the Brahmaputra–Ganga water diversion and water transmission to the west and the south of India.[9] If these proposals are realized, the present severe flood disasters can be reduced in the North and the people of India shall be blessed in the long run by brilliant industrial and Agricultural developments. Though the cost of the project to implement may be very high but this is a challenge to consider in the next 25–30 years. The Scientists in the University of Texas, USA have also detailed the IBT of water in Ganga and Brahmaputra basin in cooperation and collaboration with Nepal and Bangladesh and a detailed report was prepared and published in 1993.[6]

By the execution of proposed water links (transfers), it is possible to bring an additional area of about 35 Mha (25 Mha from surface and 10 Mha from increased groundwater storage) besides augmenting drinking water to the people of India. The cost may be high but not at prohibitory level.[1] However, the interlinking projects can be taken in a Phased manner, to start in the peninsular rivers development including diverting west flowing rivers to the east. The GOI should have a will and commitment for achieving the goal in the near future without postponement.

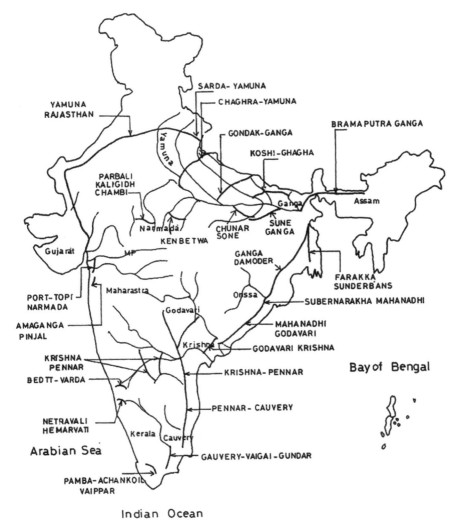

FIGURE 10.4   Proposed inter-basin water transfer links [NWDA].

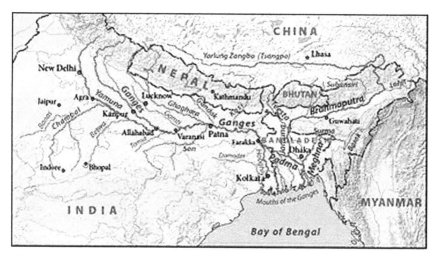

FIGURE 10.5   Map of the Ganges, Brahmaputra, and Meghna rivers [NWDA].

10.5.3   SALIENT FEATURES TO BE CONSIDERED TO IMPLEMENT THE PROJECT ON INTERLINKING OF RIVERS

The proposal of interlinking of rivers in India is there for more than 50 years[1] and the Union Minister for water resources, GOI, Honorable Dr. K. L. Rao[18,19] started and D. J. Dastur[5] gave an alternate proposal in the 1970s. The highest Judiciary of Indian Supreme Court took up the issue twice in 2002 and again in 2012 and directed the GOI to implement the projects without wasting time, since it may cost more if it is postponed. In fact GOI constituted an independent organization in 1982 namely *NWDA* to investigate the matter and to give a concrete shape to the peninsular rivers development component. The agency prepared the feasibility reports and gave a concrete shape to the peninsular rivers development in January 1999.[12,14] Subsequently, a task force with a chairman was created by the GOI in the beginning of the 21st century and the Task force submitted a detailed proposal to the GOI in 2003–2004. But no action has been taken on this subject till today.

The farmers, water experts, and the development-oriented persons of India are insisting the State and Central governments to take up the interlinking of rivers work early but some environmentalists are resisting giving some reasons or others. In order to feed the population of 1650 million in 2050, India needs about 500 MT of food grain. To achieve this, it is necessary to bring the area under irrigation from 100 Mha at present to 150 Mha between 2025/2050 apart from providing water for drinking, industrial use, and other purposes.

Mr. Suresh P. Prabhu, Chairman of the Task force on interlinking of rivers New Delhi submitted a detailed report to the then PM of India in 2004 with details about interlinking of rivers in peninsular India. He stated, "I have been arguing time and again that politics should not be involved in crucial issues, including water. I have been quite consistent on my stand. In fact, I strongly believe there is a need for a political consensus on such issues to address the national challenges. I am happy that the Supreme Court had referred to the report submitted by the task force. The task force had submitted its report after conducting over 5000 meetings with all stakeholders, sectors, IITs, IIMs, business and trade organizations, political parties, chief ministers, leaders of opposition and editors. Several aspects of the inter-linking of rivers were discussed, including the good examples and bad examples at global level. Inter-basin water transfer projects are implemented quite effectively in North America, Australia, Africa. The experiences there can come quite handy while carrying out implementation in India. The project cost was estimated at Rs 5.60×10^6 million at 2002 prices. The project aimed to deliver 173 billion cubic meters of water through a 12,500 km of canals to irrigate 34 million hectares and to supply drinking water to 101 districts and five metro cities. The task force had gone into details of the challenges of water, energy, agriculture and also those pertaining to rehabilitation." <http://www.publishyourarticles.net/eng/articles2/inter-linking-of-rivers-essay/2438/>

In Peninsular India (South India), according to the data by GOI and many studies show large quantity of water is wasted /allowed to go to the sea, that is, Andhra Pradesh 4000 TMC [AP Govt. Report appeared in the Hindu dt.19.11.2011[8]]. Karnataka, 2000 TMC [The data by WR Development organization, Govt. of Karnataka, Bangalore], Kerala more than 1000 TMC [Assessed and reported by the NWDA, New Delhi 1999[16]]. All these 7000 TMC waste can be used to solve the water and energy problems of southern states using the latest technology without affecting the environment with less energy through pump storage schemes to solve the water and energy problems of southern states.

The following salient points must be considered to take up the projects of interlinking of rivers[20–23]:

1. The surfaces water resource of the country is about 195 MHM and the replenishable ground water resource is 43 MHM as reported by the National Commission for integrated water resources and development plan, Govt. of India, New Delhi in January 1999 totaling

about 238 MHM. The total utilizable water is noted only as 69 MHM out of 195 MHM of surface water (i.e., 35% of the water). This statement is there for the last 40–50 years [National Commission of Agriculture and national Commission of Irrigation—1972/1976]. These data (utilizable water) do not seem to be correct, and it should be revised/updated and accurately assessed.

2. The available water for an Indian citizen is =195 + 43 = 238 MHM divided by population (2011) of 1210 million = 2000 $M^3/p/y$. At the same time, the available water for a TN citizen is =2.42 SW + 2.24 GW = 4.70 MHM divided by the present population of 72 million = 4.70 MHM/72 M = 650 $M^3/p/y$, whereas the required water is higher than 1700 M^3. If available water is 1000 M^3, then we can observe the water scarcity. Therefore, we cannot imagine what will be the fate of our population that will increase to 1650 M in 2050.

3. The water storage capacity created per person in different countries in the world are as follows [The Hindu, August 26, 2010[8]]:

Country	M^3
USA	5961
Australia	4717
Brazil	3388
China	2486
India	200

4. There are about 45,000 large dams in the world, of which 46% is in China,[2] 14% is in USA, only 9% is in India, Japan 5%, Spain 3%, etc. [Report of the committee on large dams]. Indian environmentalists are against construction of reservoirs (dams) to store the flood water going as waste giving some reasons.

5. The storage capacity created in USA is about 65 MHM compared to 18 MHM in India, and it may go to about 36 MHM after all works under construction and contemplated are completed. The fact is that both countries have the same quantity of water from the rain, and USA population is only 300 M compared to 1210 M of India.

6. The Aswan Dam in Egypt can store 2 years of rainwater from the catchment area. The Boulder dam in USA can store all the runoff water and no water is overflowing from the dam after its construction[17].

7. China's ambitious $80 billion projects to divert waters of southern river (Yantze) to the arid north (Yellow river) is nearing completion.[2] There is another plan to divert Indian Brahmaputra's waters to Northern China [The Hindu dated February 7, 2012[8]], which will affect India's water resources. There are many reports and proposals of NWDA prepared from the year 1982 up to 2012, implying 30 years' work with more than 200 engineers on feasibility reports and cost estimates for some diversions especially for Peninsular rivers and west flowing rivers to east in Kerala State.

8. In the historic pronouncement on October 31, 2002, the Supreme Court asked center to setup a high level taskforce to work out the modalities for *Interlinking of rivers in India* within 10 years.

9. The recent judgment of the Supreme Court (February 2, 2012) directed the GOI to implement interlinking of rivers in a time bound manner and constitute a panel of ministers, experts, and activists to short out and to execute the project as it has been already delayed resulting the increase in cost.

10.5.4   OBSTACLES

• Unfortunately, the GOI has made little use of the powers vested in it vide Entry 56 of List I. **Constitutional Provisions**: Subject "water" is placed in the Constitution in Entry 17 of List II (State List) of Schedule VII. However, the caveat is Entry 56 of List I (Union List), which says, "Regulations and development of inter-state rivers and river valleys to the extent to which such regulation and development under the control of the Union is declared by Parliament by law to be expedient in the public interest."

• The result is that by virtue of Article 246 read with Entry 17, List II, states have exclusive jurisdiction over waters that are located within their territories, including inter-state rivers and river valleys. This provision prevents Union and Judiciary to settle the issue. It has also stopped the GOI from clearly defined water rights among states and end the long drawn legal battles.

• It is arguably this status of water in the Constitution that constrains the highest in the executive and the judiciary, despite their pronouncements on and commitment to resolving the problem.

- It has also stopped the GOI from establishing allocation rules and clearly defined water rights among states that have unending disputes over the sharing of inter-state water resources.
- The latest example is the second Krishna Water Disputes Tribunal, which has turned into a warzone, with a battery of lawyers, technical staff, and irrigation department officials from Maharashtra, Karnataka, and Andhra Pradesh fighting to win the maximum allocation of the Krishna River water for their respective state.
- Indian supporters of environment and socialists are arguing that ecology will be disturbed, etc.
- Countries Bangladesh, China, and Nepal will have to be consulted to deal with International water rights.

10.5.5 INTERNATIONAL EXAMPLES OF INTERLINKING OF RIVERS[7,10]

1. **Rhine–Main–Danube Canal**—completed in 1992, and also called the Europa Canal, it interlinks the Main river to the Danube River, thus connecting North Sea and Atlantic Ocean to the Black Sea. It provides a navigable artery between the Rhine Delta at Rotterdam in the Netherlands to the Danube Delta in eastern Romania. It is 171-km long, has the summit altitude (between the Hilpoltstein and Bachhausen locks) of 406 m above sea level, the highest point on Earth reachable by ships from the sea. In 2010, the interlink provided navigation for 5.2 million tons of goods, mostly food, agriculture, ores, and fertilizers, reducing the need for 250,000 truck trips per year. The canal is also a source for irrigation, industrial water, and power generation plants.

2. **Illinois Waterway** system consists of 541 km of interlink that connects a system of rivers, lakes, and canals to provide a shipping connection from the Great Lakes to the Gulf of Mexico via the Mississippi River. It provides a navigation route. Primary cargos are coal to power plants, chemicals and petroleum upstream, and agriculture produce downstream primarily for export. The Illinois water way is the principal source of industrial and municipal services water needs along its way. It serves the petroleum refining, pulp and paper processing, metal works, fermentation and distillation, and agricultural products industries.

3. **Tennessee–Tombigbee Waterway** is a 377-km man-made waterway that interlinks the Tennessee River to the Black Warrior–Tombigbee River in the United States. The Tennessee–Tombigbee Waterway links major coal producing regions to coal consuming regions and serves as commercial navigation for coal and timber products. Industries that utilize these natural resources have found the Waterway to be their most cost-efficient mode of transportation. The water from this Waterway is a major source of industrial water supply, public drinking water supply, and irrigation along its way.

4. **Gulf Intracoastal Waterway** completed in 1949, interlinks 8 rivers and is located along the Gulf Coast of the United States. It is a navigable inland waterway running approximately 1700 km from Florida to Texas. It is the third busiest waterway in the United States, handling 70 million tons of cargo per year, and a major low cost, ecologically friendly, and low carbon footprint way to import, export, and transport raw materials and products for industrial, chemical, and petrochemical industries in the United States. It has also become a significant source for fishing industry as well as for harvesting and shipping shellfish along the coast line of the United States.

5. Other completed rivers interlinking projects include the Marne–Rhine Canal in France, and the All-American Canal and California State Water Project in the United States.

10.6   RECOMMENDATIONS

Even if all the available water (238.50 MHM) is brought under-utilization without allowing to go into the sea, the water problem/scarcity will be there in the country after 2025 as can be observed in Table 10.2. We have to take action to utilize all the water available, but it is not considered by GOI seriously. Water is considered as a liquid gold in the South India. The quarrel between the states for sharing of water and the increase demand of water in all sectors warrant that more precise research on water availability, storage development, management and drainage, etc. are essential to solve the crisis and to guide the administrators and politicians.

There is an immediate need to increase the productivity of almost all crops especially rice (e.g., from 3 t/ha to 9–10 t/ha as obtained in Egypt, Korea, etc.). Rice crop consumes 45% of the total irrigated water in the country and the water management for rice can bring water saving and increasing productivity substantially. Similarly, there is great potential to introduce drip

(27 Mha) and sprinkler irrigation (42.5 Mha) systems as suggested by the Task force on micro-irrigation (27 + 42.5 = 69.50 Mha). Another area of more research includes water quality, environmental problems, and land and water pollution, which has to be given top priority.

An International social group foresees India's water demand exceeding availability by a factor of two by 2030.[15] Time is now for India to take on the daunting task of formulating a unifying National water policy for which the *National Institute or Center for Water Resources, Research, Development & Management* is the need of the hour. All problems envisaged above are very important and urgent to unite the country from the water disputes and to solve water scarcity problems, which will be very serious in the coming years. Under these circumstances, it is high time to wake up and give top priority on research on water, water-related subjects, and also to coordinate water-related issues of different states in the country.

In order to tackle the problem and find solutions, it is suggested than a *National Center for Water Resources, Research Development & Management* may be established immediately so that this institution or center can take up research on water/water-related subjects and also can coordinate water-related issues/activities like developments and managements. A task force can be formed with experts to work out the details by the GOI. This problem is very serious and has to be solved/tackled, immediately without wasting time.

It is necessary and a must to take up the interlinking of rivers in India, particularly the interlinking of peninsular rivers without wasting time and to implement the supreme court orders/direction as early as possible. The projects on rivers interlinking will help India's water crisis by conserving the abundant monsoon water bounty, storing it in reservoirs, and delivering this water—using rivers interlinking project—to areas and over times when water becomes scarce. Beyond water security, the projects are also seen to offer potential benefits to transport infrastructure through navigation, as well as to broadening income sources in rural areas through fish farming and more employment opportunities.

ACKNOWLEDGMENTS

Author expresses his sincere thanks to Dr. N. Mahalingam, for a source of inspiration and encouragement; Dr. S. K. Mishra (Editor of Journal of Indian Water Resources Society) for his encouragement on the theme of inter-linking of Indian rivers; Dr. S. R. Sreerangasamy (Former Dean at Tamil

Nadu Agricultural University) to offer useful suggestions and comments; Thiru O. Arumughaswamy (Chairman, Vijalakshmi Charitable Trust, Coimbatore) for printing and publishing booklet on "Tamil Nadu water vision and interlinking of Indian rivers – need and importance." The part of this chapter includes information from this booklet.

KEYWORDS

- available water
- BCM
- CGWB
- CWC
- GOI
- ground water resources
- India
- interlinking of rivers
- million hectare meter
- river basin
- Tamil Nadu
- tributaries
- utilizable water
- water resources of India
- water scarcity
- water stress

REFRENCES

1. Carg, S. K. *Necessary and Prospects of Inter Basin Transfers of Water in India, and River Water Disputes in India.* Laxmi Publications (P) Ltd.: New Delhi, 1999.
2. China Water Vision. *Meeting the Water Challenge in Rapid Transition.* China Water Vision: The Hague, The Netherlands, March 2000.
3. CWC. *Reassessment of Water Resources Potential of India.* Publication 6/93 by Ground Water Resources of India, CGWB-1995, 1993.
4. CWC. *Water and Related Statistics.* Central Water Commission: New Delhi, 2000.

5. Dastur, D. J. *The Garland Canal Project—Answer to India's Flood, Food and Unemployment Problems*. Forum of Free Enterprise: Bombay, 2004.

6. Eaton, David. J. The Ganges–Brahmaputra Basin—Water Resources Cooperation between Nepal, India and Bangladesh. University of Texas: Austin, 1992.

7. Gourdji, S.; Knowlton, C.; Platt, K. *Indian Inter-linking of Rivers: A Preliminary Evaluation*. Master of Science Thesis for Natural Resources and Environment Department at the University of Michigan, 2005. <http://rivers.snre.umich.edu/ganga/India/RiverLinkingFinal.pdf>.

8. Hindu. *Interlinking of Indian Rivers*. HINDU, the Indian express newspapers: New Delhi, 1974–2004..

9. Hori, H. *Macro Engineering Super Scale Water Resources Development in Tropical Continents—Brahmaputra–Ganges—Water Diversion and Water Transmissions to the West and South India*, Government of India: New Delhi, 2004.

10. https://en.wikipedia.org/wiki/Indian_Rivers_Inter-link. Indian Rivers Inter-link, May 20 of 2015.

11. IWRS. *Water Resources Day*. Theme Paper on Water—Vision 2050, New Delhi, 1999.

12. Mahalingam, N. *Reporting India for Faster Economic Development through Interlinking River Basins*. Kisan World, April, 1999.

13. Ministry of Water Resources, *Water Resources Development Plan of India—Policy and Issue*. Government of India, Ministry of Water Resources: New Delhi, January, 1999.

14. Mohan, S. *Interlinking of Indian Rivers—Operational and Legal Issues and Options*. Indian Institute of Technology, Chennai.

15. Narashiman, *The Hindu. January 25. Daily Indian Newspaper*, 2010.

16. NWDA, *National Perspective for Water Resources Development*. National Water Development Agency: New Delhi, January, 1999.

17. Postel, Sandra, *Pillars of Sand*. W. W. Norton and New York: New York, 1999.

18. Rao, K. L. *Inter Basin Transfer of Water and its Importance for Development*. River Development Division, State Government of Tamil Nadu: Tamil Nadu, 2004.

19. Ministry of Agriculture, Govt. of India. *Report of the National Commission on Agriculture-Resource Development Part V*. Ministry of Agriculture, Govt. of India: New Delhi, 1976.

20. Sivanapan, R. K. Inter Basin Transfer of Water to Solve Water and Energy Problem of South India. In Paper Presented at Seminar, Water Technology Center, TNAU: Coimbatore, TN, 2004; pp. 30.

21. Sivanappan, R. K. Key note address—*National Conference on Interlinking of Indian Rivers—Problems and Perspectives*. Mahendra Engineering College: Tiruchangodu, Oct, 2003.

22. Sivanappan, R. K. *Linking of Peninsular Rivers—Need and Importance*. Papers Presented in Many Seminars from 1990 to 2010 in colleges and Universities, 2010.

23. Sivanappan, R. K. *Tamil Nadu Water Vision and Interlinking of Indian Rivers—Need and Importance*, 2012.

APPENDIX 10.I   CHRONOLOGICAL DEVELOPMENTS TO INTERLINK INDIAN RIVERS [BASED ON DATA IN THIS CHAPTER].

—   Pre-independence: During the British raj, Engineer **Sir Arthur Cotton** had sought to link the Ganga and the Cauvery to improve connectivity for navigation purposes.

1972   Ganga–Cauvery link proposed by Dr. K. L. Rao.[18]

1974   "Garland canal" proposal by Captain Dastur.[5]

1980   The NWDA was set up to carry out feasibility studies.

1999   A National Commission for Integrated Water Resources Development Plan (NCIWRDP) was set up.

2002   August 15: President Abdul Kalam mentions the need for river linking in his Independence Day speech.

2002   October: Supreme Court of India recommends that the GOI formulate a plan to link the major Indian rivers by the year 2012.

2002   December: GOI appointed a task force on interlinking of 37 rivers led by Former Minister Suresh Prabhu. The deadline was revised to 2016.

2004   Mr. Suresh P. Prabhu, Chairman of the Task force on interlinking of rivers New Delhi submitted a detailed report to the then PM of India in 2004 with details about interlinking of rivers in peninsular India.

2005   Ken-Betwa river link: MoU was signed among Union Water Ministry, CMs of MP and UP. It is the only project for which the detailed project report has been prepared. Approximately, 8650 ha of forestland in Madhya Pradesh is likely to be submerged for the project and part of that forestland is a part of the Panna National Park.

2012   February: Supreme Court gave its go-ahead to the interlinking of rivers and asked the government to ensure that the project is implemented expeditiously.

APPENDIX 10.II   PROPOSAL BY CAPTAIN DASTUR TO INTERLINK INDIAN RIVERS.[5]

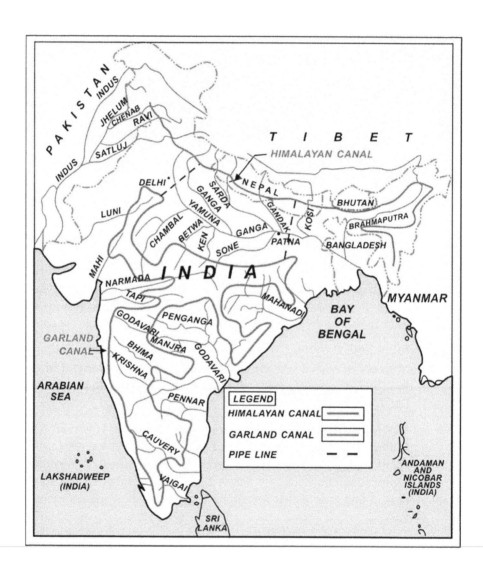

APPENDIX 10.III   INVESTMENT NEEDS AND POTENTIAL ECONOMIC IMPACT DUE TO INTERLINKING OF INDIAN RIVERS.[10,11]

Interlink project	Length (km)	Estimated cost in the year2003 or earlier# (Rs.)	New irrigation capacity added (ha)	Potential electricity generation capacity (MW)	Drinking & industrial water added (Mm3)
Cauvery–Vaigai–Gundar Link	255.6	Rs. 2673 crore (US$420 million)	337,717	–	185
Damanganga Pinjal Link	42.5	Rs. 1278 crore (US$200 million)	–	–	44
Godavari–Krishna Link	299.3	Rs. 26,289 crore (US$4.2 billion)	287,305	70	237
Ken Betwa Link	231.5	Rs.1988.74 crore (US$320 million)	47,000	72	2225
Krishna–Pennar Link	587.2	Rs. 6599.80 crore (US$1.0 billion)	258,334	42.5	56
Mahanadi Godavari Link	827.7	Rs. 17,540.54 crore (US$2.8 billion)	363,959	70	802
Nagarjunasagar Somasila Link	393	Rs. 6320.54 crore (US$1.0 billion)	168,017	90	124
Pamba Achankovil Vaippar Link	50.7	Rs. 1397.91 crore (US$220 million)	91,400	500	150
Par Tapi Narmada Link	395	Rs. 6016 crore (US$950 million)	169,000	93	91
Parbati Kalisindh Chambal	243.7	Rs. 6114.5 crore (US$970 million)	225,992	17	89
Polavaram–Vijayawada Link	174	Rs. 1483.91 crore (US$240 million)	314,718	72	664
Srisailam Pennar Link	203.6	Rs. 1580 crore (US$250 million)	187,372	17	49

#The cost conversion in US $ is at latest conversion price on the historical cost estimates in Indian rupees; one crore = 10 million.

- Cost of the project was estimated at Rs. 5600,000 million [NWDA]; the true cost can be known only when the detailed project reports of the 30 river link projects are drawn up; and so far only Ken–Betwa project is under survey.

Benefits:

- Irrigating 35 million hectares; enabling full use of existing irrigation projects; and generating power to the tune of 34,000 MW with added benefits, including flood control.

APPENDIX 10.IV　WATER RESOURCES OF INDIA.[3,4]

Sl. No.	States / Union Territories	Annual Replenishable Ground Water Resource					Natural Discharge during non-monsoon season	Net Annual Ground Water Availability	Annual Ground Water Draft			Projected Demand for Domestic and Industrial uses up to	Ground Water Availability for future irrigation	Stage of Ground Water Development (%)
		Monsoon Season		Non-monsoon Season		Total			Irrigation	Domestic and industrial uses	Total			
		Recharge from rainfall	Recharge from other sources	Recharge from rainfall	Recharge from other sources									
	States													
1	Andhra Pradesh	16.04	8.93	4.20	7.33	36.50	3.55	32.95	13.88	1.02	14.90	2.67	17.65	45
2	Arunachal Pradesh	1.57	0.00009	0.98	0.0002	2.56	0.26	2.30	0.0008	0	0.0008	0.009	2.29	0.04
3	Assam	23.65	1.99	1.05	0.54	27.23	2.34	24.89	4.85	0.59	5.44	0.98	19.06	22
4	Bihar	19.45	3.96	3.42	2.36	29.19	1.77	27.42	9.39	1.37	10.77	2.14	16.01	39
5	Chhattisgarh	12.07	0.43	1.30	1.13	14.93	1.25	13.68	2.31	0.48	2.80	0.70	10.67	20
6	Delhi	0.13	0.06	0.02	0.09	0.30	0.02	0.28	0.20	0.28	0.48	0.57	0.00	170
7	Goa	0.22	0.01	0.01	0.04	0.29	0.02	0.27	0.04	0.03	0.07	0.04	0.19	27
8	Gujarat	10.59	2.08	0.00	3.15	15.81	0.79	15.02	10.49	0.99	11.49	1.48	3.05	76
9	Haryana	3.52	2.15	0.92	2.72	9.31	0.68	8.63	9.10	0.35	9.45	0.60	-1.07	109
10	Himachal Pradesh	0.33	0.01	0.08	0.02	0.43	0.04	0.39	0.09	0.03	0.12	0.04	0.25	30
11	Jammu & Kashmir	0.61	0.77	1.00	0.32	2.70	0.27	2.43	0.10	0.24	0.33	0.42	1.92	14
12	Jharkhand	4.26	0.14	1.00	0.18	5.58	0.33	5.25	0.70	0.38	1.06	0.56	3.99	20
13	Karnataka	8.17	4.01	1.50	2.25	15.93	0.63	15.30	9.75	0.97	10.71	1.41	6.48	70
14	Kerala	3.79	0.01	1.93	1.11	6.84	0.61	6.23	1.82	1.10	2.92	1.40	3.07	47
15	Madhya Pradesh	30.59	0.96	0.05	5.59	37.19	1.86	35.33	16.08	1.04	17.12	1.74	17.51	48
16	Maharashtra	20.15	2.51	1.94	8.36	32.96	1.75	31.21	14.24	0.85	15.09	1.51	15.10	48
17	Manipur	0.20	0.005	0.16	0.01	0.38	0.04	0.34	0.002	0.000 5	0.002	0.02	0.31	0.65
18	Meghalaya	0.79	0.03	0.33	0.00	1.15	0.12	1.04	0.00	0.002	0.002	0.10	0.94	0.18
19	Mizoram	0.03	0.00	0.02	0.00	0.04	0.004	0.04	0.00	0.000 4	0.0004	0.0008	0.04	0.90
20	Nagaland	0.28	0.00	0.08	0.00	0.36	0.04	0.32	0.00	0.009	0.009	0.03	0.30	3
21	Orissa	12.81	3.56	3.58	3.14	23.09	2.08	21.01	3.01	0.84	3.85	1.22	16.78	18
22	Punjab	5.98	10.91	1.36	5.54	23.78	2.33	21.44	30.34	0.83	31.16	1.00	-9.89	145
23	Rajasthan	8.76	0.62	0.26	1.92	11.56	1.18	10.38	11.60	1.39	12.99	2.72	-3.94	125
24	Sikkim	-	-	-	-	0.08	0.00	0.08	0.00	0.01	0.01	0.02	0.05	16
25	Tamil Nadu	4.91	11.96	4.53	1.67	23.07	2.31	20.76	16.77	0.88	17.65	0.91	3.08	85
26	Tripura	1.10	0.00	0.92	0.17	2.19	0.22	1.97	0.08	0.09	0.17	0.20	1.69	9
27	Uttar Pradesh	38.63	11.95	5.64	20.14	76.35	6.17	70.18	45.36	3.42	48.78	5.30	19.52	70
28	Uttaranchal	1.37	0.27	0.12	0.51	2.27	0.17	2.10	1.34	0.05	1.39	0.06	0.68	66
29	West Bengal	17.87	2.19	5.44	4.86	30.36	2.90	27.46	10.83	0.81	11.65	1.24	15.33	42
	Total States	247.87	69.51	41.84	73.15	432.43	33.73	398.70	212.37	18.05	230.41	29.09	161.06	58
	Union Territories													
1	Andaman & Nicobar	-	-	-	-	0.330	0.005	0.320	0.000	0.010	0.010	0.008	0.303	4
2	Chandigarh	0.016	0.001	0.005	0.001	0.023	0.002	0.020	0.000	0.000	0.000	0.000	0.020	0
3	Dadra & Nagar Haveli	0.059	0.005			0.063	0.003	0.060	0.001	0.008	0.009	0.008	0.051	14
4	Daman & Diu	0.006	0.002	0.000	0.001	0.009	0.0004	0.008	0.007	0.002	0.009	0.003	-0.002	107
5	Lakshadweep	-	-	-	-	0.012	0.009	0.004	0.000	0.002	0.002	-	-	63
6	Pondicherry	0.057	0.067	0.007	0.029	0.160	0.016	0.144	0.121	0.030	0.151	0.031	-0.008	105
	Total Union Territories	0.138	0.075	0.012	0.031	0.597	0.036	0.556	0.129	0.052	0.181	0.050	0.365	33
	Grand Total	248.01	69.59	41.85	73.18	433.02	33.77	399.25	212.50	18.10	230.59	29.14	161.43	58

CHAPTER 11

PRESENT STATUS OF WATER RESOURCES IN TAMIL NADU[*]

R.K. SIVANAPPAN

College of Agricultural Engineering & Technology, Tamil Nadu Agricultural University (TNAU), Coimbatore, Tamil Nadu, India. E-mail: sivanappanrk@hotmail.com

CONTENTS

Abstract 276

11.1 Introduction 277

11.2 Rainfall in Tamil Nadu 278

11.3 Water Resources in Tamil Nadu 282

11.4 Land, Water, and Food Production: Current Situation 285

11.5 Water Policy: Tamil Nadu 294

11.6 Methods to Increase the Area Under Irrigation and
    Food Production 295

11.7 Conclusions 299

Acknowledgments 299

Keywords 300

References 300

Appendix 11.I   Water Used and Area Irrigated in Tamil Nadu:
    Sample Calculation. 302

Appendix 11.II   Illustrations [Photo Courtesy: Author] 303

---

[*]This chapter is an edited and modified version of Sivanappan, R. K., 2012. Tamil Nadu water vision and interlinking of Indian rivers: need and importance, Part I. Special report prepared for GOI and TNSG. Pages 50. Vijalakshmi Charitable Trust, Coimbatore, Tamil Nadu, India.

ABSTRACT

Water resources are more and more important to sustain the agricultural growth and production to meet the increasing population and its desires in Tamil Nadu (TN). Rivers; streams; surface runoff; ground water; and stored water in dams, reservoirs, and tanks are the main sources of water in TN. The water resources of TN are only about 2% of India with a population share of about 6.0%. TN is a water scarcity state, as the available water is only 650 m^3/P/Y compared to 2000 m^3/P/Y for India. This causes critical concerns, anxiety, and insecurity in food production in the coming years in the state. The quantum of rain is fluctuating imposing water scarcity in many districts of TN. The ground water potential is rather critical as the water table has gone beyond 700–1000 feet in many districts. The overuse of water in agricultural production is to be discouraged. Competing and ever increasing water demand from industries and community and municipal needs is alarming. These activities add to the huge volume of effluents and sewage water with injurious quality and toxicity to the health and environment. Thus, there is an immediate need to initiate multipronged steps to redefine and revitalize the water storage ways and water use management practices leading to an increase in agricultural production by bringing more area under irrigation and to produce more per unit quantity of water and per unit of land.

Saving of water and reuse of waste water after reclamation are advocated for agricultural, industrial, and community-based requirements. To be able to manage the expectations of the food requirements of 20/30 million tons by 2025/2050 in TN for the escalating population, the farming community, government departments, community, society, and educators need to join together and unite, and reorient the protocols and practices in water use and water management strategies thereby saving large quantities of available water and also developing new water resources and applying to irrigate additional area of about 2.0 Mha to make TN self-sufficient. This is achievable. The chapter on TN water vision defines the road map leading to the achievements of the goals.

In this context, it is strongly advocated the diversion of surplus flood waters from the west-flowing rivers of Kerala and Karnataka, and east-flowing rivers in Orissa and Andhra Pradesh to the southern states. This can provide an additional irrigation area of 10 Mha and also solve the water problems of the southern state, especially TN.

11.1 INTRODUCTION

Fresh water availability in a territory (country or state) helps in defining the water scarcity. A country/state will suffer water scarcity only occasionally, if renewable fresh water availability on an annual per capita basis exceeds 1700 m³. When the annual per capita fresh water availability falls below 1000 m³, countries will experience chronic water scarcity, and lack of water begins to hamper economic development, human health, and well-being. When renewable fresh water supplies fall below 500 m³/P/Y, countries will experience absolute water scarcity. The 1000 m³/P/Y has been accepted as a general indicator of water scarcity by the World Bank and other analysts.

The state of Tamil Nadu (TN) receives about 925-mm rainfall annually, which is less than the average rainfall of 1150 mm in India. Meeting the water demand is a challenge for TN, as it envisions producing more food in the coming years using almost the same quantity of water available. To achieve this, TN has to initiate newer and improved measures notably on water resources development, water harvesting, storage and delivery systems, and increasing the water use efficiency and community mediated and community participated approaches so that with existing water resources, TN should be able to produce about 20/30 million-tons by 2025/2050.

Table 11.1 indicates the amount of fresh water available (in million hectare meter [MHM]) and the quantity available per person per year for India and TN. According to the norm fixed by the World Bank and experts, India as a country has no water scarcity problem till its population reaches 1500 million, whereas the state of TN is already under **chronic water scarcity**, as is evident from the data in Table 11.1.

TABLE 11.1 Water Availability in India and in Tamil Nadu.

Description	India	Tamil Nadu
Surface water (MHM)	195 (187)	2.42
Ground water (MHM)	43	2.24/2.37
Total (MHM)	**238**	**4.66/4.79**
Population in million (2011)	1210	72
Quantity available per person/year as per 2011 census (m³)	2000	650

Source: Water Related Statistics, CWC, GOI, 2000. *Economic Appraisal*. Govt. of Tamil Nadu, Chennai, 2008–2009.[9]

Under these circumstances, it is important to understand the factors involved to solve the water crisis in TN. This will lead to an increase the food production, bring more area under irrigation, provide water for households, industries, etc. Therefore, author has made earnest efforts to stress problems of water scarcity and presents solutions to tackle water crisis in TN.

11.2 RAINFALL IN TAMIL NADU

TN receives water only from rainfall. Being in a monsoon region, rain is received during the two monsoons (North East and South West), which is erratic and uneven. Rainfall fails once in 3–4 years. The normal rainfall from 1950–1951 to 2003–2004 was 945–964 mm, and after 2004–2005, onward it is 911.6 mm (Table 11.2). The actual rainfall was 700–1222 mm. Taking the average annual rainfall as 940 mm, the total rainfall amounts to about 12 MHM in TN in a year. Water potential due to rivers and groundwater is shown in Tables 11.3 and 11.4.

TABLE 11.2 Normal and Actual Rainfall (mm) in Tamil Nadu from 1950–51 to 2008–2009.

Year	Normal rainfall	Actual rainfall	Year	Normal rainfall	Actual rainfall
50–51	975.7	781.6	80–81	942.8	669.3
51–52	975.7	762.3	81–82	942.8	952.7
52–53	975.7	686.2	82–83	942.8	662.6
53–54	975.7	1016.0	83–84	942.8	1222.5
54–55	975.7	969.0	84–85	942.8	791.4
55–56	975.7	824.7	85–86	942.8	950.9
56–57	975.7	979.0	86–87	942.8	700.9
57–58	975.7	909.9	87–88	945.0	982.8
58–59	942.8	747.4	88–89	945.0	708.8
59–60	942.8	826.9	89–80	945.0	910.7
60–61	942.8	978.0	90–91	945.0	714.6
61–62	942.8	867.0	91–92	925.2	898.9
62–63	942.8	931.4	92–93	925.2	862.0
63–64	942.8	907.6	93–94	925.2	1171.9
64–65	942.8	859.1	94–95	925.2	933.8

TABLE 11.2 *(Continued)*

Year	Normal rainfall	Actual rainfall	Year	Normal rainfall	Actual rainfall
65–66	942.8	870.9	95–96	923.1	668.3
66–67	942.8	1152.8	96–97	976.6	1121.2
67–68	942.8	958.8	97–98	981.3	1132.2
68–69	942.8	682.0	98–99	981.3	1080.4
69–70	942.8	1036.7	99–2000	977.4	896.8
70–71	942.8	918.8	2000–01	979.3	785.4
71–72	942.8	968.8	2001–02	974.5	795.2
72–73	942.8	990.7	2002–03	964.2	730.8
73–74	942.8	839	2003–04	958.4	1034.6
74–75	942.8	643.9	2004–05	911.6	1078.9
75–76	942.8	857.2	2005–06	911.6	1304.1
76–77	942.8	941.4	2006–07	911.6	859.7
77–78	942.8	1123.7	2007–08	911.6	1164.8
78–79	942.8	949.8	2008–09	911.6	1023.1
79–80	942.8	1091.3	–	–	–

Source: Department of Statistics, Chennai.[6]

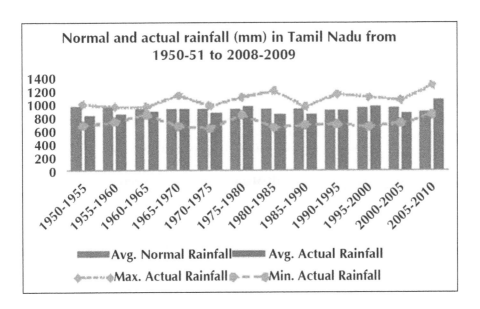

TABLE 11.3 Surface Water Potential of River Basins in Tamil Nadu.

River basins	Surface water potential at 75% dependability	
	MCM	TMC
Chennai	784	28
From Krishna water as per agreement	340	12
Palar	1758	62
Varahanadhi	412	15
Ponniyar	1310	46
Paravanar	144	5
Vellar	963	34
Agniar	1084	38
Pambar and Kottakaraiyar	653	23
Vaigai	1579	56
Gundar	568	20
Vaippar	611	22
Kallar	125	4
Thambaraparani	1375	49
Nambiar	204	7
Kodaiyar	925	33
Parambikulam and Aliyar as per agreement	864	30
Cauvery in Tamil Nadu area	4655	164
From Karnataka as per interim Tribunal order	5805	205
Total	**24,159 or 2.42 MHM**	853 TMC

TABLE 11.4 Status of Ground Water Potential in Tamil Nadu, January 1998: District-wise Data Based on *Groundwater Resources Estimation Committee Norms* (1997).

Districts	Annual GW recharge (MCM)	Net GW recharge (MCM)	Gross GW draft (MCM)	Balance GW available (MCM)	Probable number of wells feasible	Category as on January 1998
	MCM				Nos.	–
Tiruvallur	1230.71	1107.64	913.41	131.66	7132	Semi-critical
Kancheepuram	1401.81	1261.63	675.29	551.23	29,858	Safe
Cuddalore	1629.48	1466.53	995.28	433.19	23,465	Semi-critical
Villupuram	2239.43	2015.49	1797.55	162.22	8787	Critical
Vellore	1394.58	1255.12	1326.17	−144.67	0	Over exploited
Tiruvannamalai	2060.54	1854.49	1490.15	325.22	17,616	Semi-critical

TABLE 11.4 *(Continued)*

Districts	Annual GW recharge (MCM)	Net GW recharge (MCM)	Gross GW draft (MCM)	Balance GW available (MCM)	Probable number of wells feasible	Category as on January 1998
			MCM		Nos.	–
Dharmapuri	1383.41	1245.07	1086.08	108.37	5870	Critical
Salem	1183.59	1065.23	1220.51	−193.44	0	Over exploited
Erode	1384.36	1245.92	642.16	564.01	30,551	Safe
Coimbatore	1159.52	1043.57	772.22	235.20	12,740	Semi-critical
Namakkal	694.96	625.46	485.97	117.08	6342	Semi-critical
Nilgiris	49.91	44.92	1.14	31.03	1681	Safe
Trichy	1029.64	926.68	456.47	440.86	23,880	Safe
Perambalur	810.66	729.59	507.31	198.07	10,729	Semi-critical
Karur	504.52	454.07	199.54	239.50	12,973	Safe
Pudukottai	1123.27	1010.94	179.66	805.27	43,619	Safe
Madurai	919.82	827.84	389.92	411.47	22,288	Safe
Dindigul	919.84	827.86	692.04	105.41	5706	Semi-critical
Theni	599.06	539.15	441.2136	82.36	4461	Semi-critical
Sivagangai	810.99	729.89	104.21	608.39	32,955	Safe
Ramanatha-puram	308.68	277.81	34.75	221.68	12,008	Safe
Virudhunagar	707.57	636.81	281.56	334.39	18,113	Safe
Tirunelveli	1254.41	1128.97	379.93	713.38	38,641	Safe
Tuticorin	311.04	278.94	159.47	101.70	5509	Safe
Kanyakumari	376.72	339.05	17.21	292.75	15,857	Safe
Thanjavur	850.34	765.31	–	–	–	Safe
Tiruvarur						
Nagapattinam						
Total	**26,338.86**	**23,704.98 or 2.37 MHM**	**15,249.36 or 1.52 MHM**	**6906.33**	**390,782**	

Notes:

1. The assessments given above (Tables 11.3 and 11.4) were made by Chief Engineer (SG & SWRDC). They are only tentative, subject to change after reconciliation with Central Groundwater Board, Ministry of Water Resources, GOI, which has assessed the annual groundwater recharge as 22,800 MCM. In this report, 22,380 MCM is taken as the ground water potential.[10]

2. **Safe**: Annual Extraction to rechargeable groundwater below 70% in 1977 and 85% at 2000.

3. **Semi-critical**: Annual extraction to rechargeable groundwater below 70% and 90%

4. **Critical:** Annual extraction to rechargeable groundwater below 90% and 100%.

11.3 WATER RESOURCES IN TAMIL NADU

The average rainfall in India is about 1150 mm compared to the world's average rainfall of about 840 mm. The average annual rainfall in TN is about 925–940 mm. The rainfall, water resources (surface water and groundwater), cultivated area, water available per person/year, and the available land/person are given in Table 11.5 for India and TN.

TABLE 11.5 Rainfall, Water Available, Cultivated and Irrigated Lands in India and Tamil Nadu, 2010.

Item	Units	India	Tamil Nadu
Geographical area	Mha	329	13
Rainfall	Mm	1150	925–940
Total rain	MHM	400	12
Surface water		195	2.42
Groundwater, GW		43	2.37/2.24
Tanks	10^5	5	0.39
Well open/bore well	10^6	17	1.80
Water availability	Persons/year	2000	650
Cultivated land	Mha	142	5.00–5.40
Land availability	Persons/ha	0.15	0.09
Irrigated area, 2010	Mha	100	3.40–3.50
Irrigation potential	Mha	140	5.50
Water storage capacity	MHM	36	1.67

The population of TN is 6% of India's population, whereas the availability of water resources of TN is only 2% and the land area is 4%. Hence, the water availability per person in TN is only 1/3rd of the available water for an Indian. Therefore, TN is under severe water scarcity even today. In this connection, it is not out of place to compare with Israel, where the rainfall is 300–400 mm and the availability of water per individual is about 400–425 m³/person/year, and they are managing the water crisis by implementing various water augmentation and management programs. Under these situations, how to solve the water crisis, which is threatening the TN state especially when the neighboring states of Kerala, Karnataka and Andhra Pradesh are not favorable to share their excess water available with TN, but allow it to go to the sea, becomes more important and urgent. There is lack of patriotism.

In India, the water storage capacity created is only about 18 MHM. It may reach 36 MHM after all the works under construction and those planned are completed. Even then, it will be only about 20% of the surface water. The storage capacity created in the USA is about 65 MHM and the fact is that both countries receive the same quantity of rain, but the population of USA is only 300 million. In TN, the capacity of 39,000 tanks and 79 reservoirs put together comes to 1.674 MHM, which is equivalent to the total surface water 1.68 MHM obtained from TN area (Tables 11.6 and 11.7). Hence there will not be any problem for storing water if the storage structures are maintained properly, and to use in the drought years and during scarcity times. The usage of surface and ground water are almost exhausted in TN. Action is being taken to harvest and store the flood water.

About 10–12% of the available water is used for drinking (households use), industrial, and other purposes. About 80–85% of this water goes as sewage/effluent (polluted water) and left in the rivers/*nallas* and over land surface, polluting the surface and ground water. If this polluted water is reclaimed, there would be about 120 TMC, which can easily irrigate about 0.56 Mha.

In this connection, it is not out of place to mention that in Israel, no fresh water is provided for irrigation and the farmers use only reclaimed sewage/effluent water. Similarly, in Europe and other developed countries the waste water/effluent water is reclaimed and again the same is reused in their industries, and only the actual water consumed is taken daily for the industries. Thus, both the water problem and the pollution problem are tackled in these countries.

The *Action Plan* about the measures to be adopted is detailed in this section under augmentation and water management including bringing large areas under drip and sprinkler irrigation, recycling sewage/effluent water after reclamation, using salt water by selecting salt tolerant crops and sea water after desalination by which it is possible to increase the ultimate irrigation potential to about 5.00 to 5.50 Mha in the next 10–15 years, that is, by 2025 from the available water resources.

TABLE 11.6   Total Water Potential in Tamil Nadu.

Details of water potential	MCM	TMC
	Surface water potential (A)	
1. Within the state	16,769	592
2. From neighboring states	7391	261
Total	24,160	853
	Ground water potential (B)	
	22,380	790
Total water = A + B	46,540	1643

Source: State Planning Commission Office, Chennai, 2005.

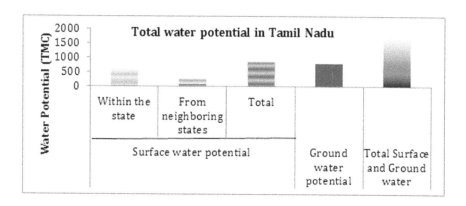

TABLE 11.7   Basin-wise Reservoirs and Tanks Capacity in MCM.

River basins, Tamil Nadu	Tanks		Reservoirs		Total capacity
	Nos.	Capacity	Nos.	Capacity	
Chennai	1519	1373	4	320	1693
Palar	661	355	2	8	363
Varahanadhi	1421	276	1	17	293
Ponniyar	1133	240	7	311	551
Vellar	457	70	5	115	185
Agniar	3975	560	–	–	560
Pambar and Kottakaraiyar	1042	154	–	–	154
Vaigai	1497	410	5	659	1069
Gundar	649	331	–	–	331

TABLE 11.7 *(Continued)*

River basins, Tamil Nadu	Tanks		Reservoirs		Total capacity
	Nos.	Capacity	Nos.	Capacity	
Vaippar	862	559	6	66	625
Kallar	199	43	1	4	47
Thambaraparani	880	196	7	367	563
Nambiar	597	95	2	6	101
Kodaiyar	2922	268	9	404	672
Parambikulam and Aliyar as per Agreement	0		9	32	32
Cauvery	21,186	4910	21	4586	9496
Total	**39,000**	**9840**	**79**	**6895**	**16,735 or 1.67 MHM**

11.4 LAND, WATER, AND FOOD PRODUCTION: CURRENT SITUATION

11.4.1 LAND

The geographical area of TN is 13 Mha. The land-use pattern has changed significantly in the decade 2000–2010, compared to the last century from 1950 to 2000. Forest area has increased from 1.8 Mha in 1950s to 2.1 Mha in 2000–2010. Fallow and waste land increased from 1.7 Mha to 2.5 Mha in the last 60 years. The net and gross cultivated areas were 5.6–6.0 Mha and 6.5–7.5 Mha, respectively, in 1950–2000. This has reduced to 5.0–5.2 Mha and 5.8–6.0 Mha, respectively, during 2000–2010 (Tables 11.8 and 11.9). This indicates that the area of cultivation has reduced. The cropping intensity has also reduced from 123 to 114% in the same period. This is a warning to increase the productivity and production for the escalating population. It is not possible to increase land area for cultivation but it is possible to increase the cropping intensity up to 300% provided soil moisture/water is available to take 2–4 crops in the same field in a year. Again, water is the constraint already and the question is how to increase the availability of water not only for agriculture but also for other uses. In addition, about 2.6 Mha of land are under waste land category and we cannot afford to leave it when more land is required in the coming years for the escalating population.

TABLE 11.8 Land-use Pattern in Tamil Nadu, During 1950–2010 (Area in 10^5 ha).

Classification	1950s	1960s	1970s	1980s	1990s	2000–2010
1. Forest	18.14	19.06	20.05	20.76	21.44	21.00
2. Barren and un culturable land	9.73	8.85	7.05	5.57	4.95	5.00
3. Land put to nonagricultural uses	12.70	13.57	16.00	17.95	19.07	21.50
4. Permanent pastures and other grazing land	3.75	3.34	1.98	1.45	1.25	1.10
5. Cultivable waste	8.70	6.60	4.15	3.08	3.25	3.50
6. Land under miscellaneous tree crops and groves	2.49	2.64	2.15	1.82	2.25	2.60
7. Current fallow	6.60	6.12	5.31	7.03	10.93	9.5
8. Other fallow	6.60	6.12	5.31	7.03	10.93	15.00
9. Net area sown	56.38	60.26	61.35	56.22	56.32	50–52
10. Area sown more than once	10.31	11.74	13.21	10.55	10.97	8.08
11. Gross cropped area	66.69	72.00	74.56	66.77	67.29	58–60
12. Cropping intensity	118.30	119.50	121.56	118.80	119.46	114–00
13. Total area (geographical)	**129.54**	**130.13**	**130.06**	**130.06**	**130.16**	**130.27**

Source: Season and Crop reports published by Directorate of Statistics, Chennai (Annual Publication).

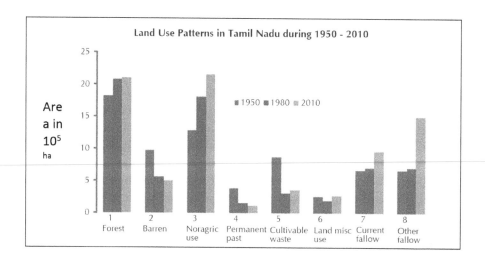

TABLE 11.9 Land Utilization in Tamil Nadu, During 2003–2010 (Area in 000' Hectares).

S. No.	Classification	2003–04	2004–05	2005–06	2006–07	2007–08	2008–09	2000–2010
I	Total geographical area (village papers) (%)	13,027 (100)	13,027 (100)	13,027 (100)	13,027 (100)	13,027 (100)	13,027 (100)	–
II	**Land utilization**							
i.	Forests	2122	2122	2111	2106	2106	2106	–
	(%)	(16.3)	(16.3)	(16.2)	(16.2)	(16.2)	(16.2)	–
ii	Barren and unculturable land	509	5090	503	502	492	492	–
	(%)	(3.9)	(3.9)	(3.9)	(3.8)	(3.8)	(3.8)	–
iii	Land put to nonagricultural use	2113	2125	2139	2160	2169	2173	–
	(%)	(16.2)	(16.3)	(16.4)	(16.6)	(16.7)	(16.7)	–
iv	Culturable waste (%)	379 (2.9)	374 (2.9)	369 (2.8)	354 (2.7)	347 (2.7)	333 (2.6)	–
v	Permanent pastures and other grazing lands (%)	113 (0.9)	114 (0.9)	110 (0.9)	110 (0.9)	110 (0.9)	110 (0.9)	–
vi	Land under miscellaneous tree crops and groves not included in the net area sown (%)	283 (2.2)	290 (2.2)	274 (2.1)	268 (2.1)	261 (2.0)	259 (2.0)	–
vii	Current fallows (%)	954 (7.3)	692 (5.3)	759 (5.8)	907 (7.0)	981 (7.5)	1013 (7.8)	–
viii	Other fallow lands (%)	1863 (11.4)	1704 (13.1)	1518 (11.7	1493 (11.5)	1499 (11.5)	1497 (11.4)	–
ix	Net area sown (%)	4689 (36.0)	5097 (39.1)	5244 (40.3)	5126 (39.3)	5062 (38.9)	5043 (38.7)	50–52 10^5 ha
III	**Area sown more than once**	627	792	789	717	753	781	–
IV	**Gross cropped area**	5316	5889	6033	5843	5815	5824	58–60 10^5 ha
i	Area under food crops	3718	4226	4398	4334	4234	4268	–
ii	Area under nonfood crops	1598	1663	1634	1509	1581	1556	–
V	**Cropping intensity (%)**	113.3	115.5	115.0	114.0	115.0	115.3	–

Figures in brackets indicate percentage to the total geographical area.

Therefore, it can be concluded that not only the cultivated area has reduced, but also the irrigation area has not increased in the last 30–40 years;

but the wasteland/fallow land area has increased from 1.7 Mha to 2.5–2.6 Mha. In fact, the average net area of rainfed and irrigated area in the present decade (2002–2009) are 2.4–2.6 Mha and 2.6–2.8 Mha, respectively. At the same time, the average fallow/wasteland is about 2.6 Mha (Table 11.10). The cropping intensity and irrigation intensity is about 113–115% in both the cases during the above period (Tables 11.9 and 11.11).

TABLE 11.10   Average Area under Waste Land (in ha) in Tamil Nadu: 1997–98 to 2001–2002.

District	Cultivable waste	Current fallow	Other fallow	Total
Chennai	0.00	0.00	0.00	0.00
Kanchipuram	10,757.00	5643.00	61,785.00	78,185.00
Thiruvallur	7884.50	7159.25	50,387.00	65,430.75
Cuddalore	7156.00	36,096.25	12,046.75	55,299.00
Villupuram	11,773.50	85,052.50	16,891.75	113,717.75
Vellore	6748.75	34,724.75	51,408.25	92,881.75
Thiruvannamalai	12,464.50	76,079.75	28,633.75	117,178.00
Salem	5700.50	20,363.25	10,100.50	36,164.25
Namakkal	6475.00	12,017.75	15,172.00	33,664.75
Dharmapurai	13,998.50	59,509.75	11,292.00	84,800.25
Coimbatore	2875.25	142,794.00	14,175.25	159,844.50
Erode	566.00	140,247.25	62,636.25	203,449.50
Trichi	12,725.75	47,372.00	63,693.50	123,791.25
Karur	68,506.50	6961.50	58,463.00	133,931.00
Perambalur	9285.00	24,071.50	18,797.50	52,154.00
Pudukottai	11,192.50	16,310.75	85,867.50	113,370.75
Thanjavur	12,593.50	6381.00	26,747.00	45,721.50
Thiruvarur	2971.25	7905.75	6003.50	16,880.50
Nagapattinam	4555.00	6558.25	13,325.00	24,438.25
Madurai	5422.25	41,847.50	43,175.50	90,445.25
Theni	4421.50	22,018.75	11,808.50	38,248.75
Dindigul	8338.25	36,761.00	71,222.50	116,321.75
Ramanathapuram	4344.00	29,570.00	62,333.75	96,247.75
Virudhunagar	8926.75	65,410.25	100,194.00	174,531.00
Sivagangai	16,707.75	29,857.25	97,684.25	144,249.25

TABLE 11.10   *(Continued)*

District	Cultivable waste	Current fallow	Other fallow	Total
Thirunelveli	69,805.75	31,149.25	162,693.50	263,648.50
Thoothukudi	29,833.50	51,713.00	59,353.50	140,900.00
The Nilgiris	2892.00	5524.75	4617.50	13,034.25
Kanniyakumari	103.50	1155.75	1292.50	2551.75
Total	**359,024.25**	**1050,255.75**	**1221,801.00**	**2631,081.00 or 2.63 Mha**

Source: Season and Crop Reports of Tamil Nadu, Chennai (annual publication).

The total cultivable area at present is 7.6–8.1 Mha (sown area + fallow and wasteland) and the net irrigated area is 2.6–2.8 Mha. It indicates irrigated area is 34% and the remaining area (66%) is under rainfed cultivation and fallows/wasteland. This has to be taken into account while planning the cropping and irrigation in TN for bringing all the cultivable land into production and bring more area under irrigation.

11.4.2   WATER

The total available surface water is 2.42 MHM of which about 0.74 MHM is sourced from neighboring states (Table 11.6) and only 1.68 MHM is from within the state. The groundwater potential is calculated as 2.24–2.37 MHM. Considering India's Water Resources, the surface water is 195 MHM compared to total rain of 400 MHM, that is, about 49% where as in TN it is only 1.68 of 12.00 = 14%. This is only 20% in usage even if the entire surface water is taken into account. Therefore, it is necessary to investigate the exact/accurate quantity of surface water available in the state. The ground water availability is estimated by the CGWB/State Govt. as 2.37–2.24 MHM, which is more than the surface water availability (1.68 MHM), in spite of more than 70–75% of the area in TN is in hard rock area. This again is to be studied to know the exact quantity of groundwater in the state.

11.4.3   IRRIGATION

Land and water are the main resources for food production. Though India has increased the food production from 50 MT in 1950–51 to 241 MT in

2008–2009, there is a long way to follow for producing 450–500 MT in 2050 for the projected population of 1650 million. To achieve this target, it is necessary to increase the present irrigation area from 100 Mha to 140–160 Mha (gross area) in 2025/2050. In TN, the food production target is fixed at 10 Mt in the current year and this has to be increased many fold in the coming years.

In TN, the gross irrigated area was about 2.7Mha in the 1950s and has increased to 3.5 Mha in the 1970s (Table 11.11), thanks to the first, second, and third 5-year plans, and during that period many irrigation projects were constructed and farmers dug thousands of open wells. The National Commission on Agriculture [1972–76[6]] mentioned that the potential area of irrigation for TN was about 4.0 Mha at that time. Unfortunately, in the last 30–35 years, in spite of spending lot of money by getting funds from World Bank and other agencies, the area of irrigation has not increased and over many years, the area has reduced. This is in spite of the introduction of many new efficient irrigation methods, like sprinkler and drip irrigation, advanced water management practices especially in paddy and sugarcane which consume more water and introducing/implementing watershed development and water harvesting and soil and water conservation measures spending millions of rupees in the last 30–35 years. Further, the intensity of irrigation has also come down from 130–132% to 115% in the above period. The reason for the failure has to be critically analyzed to improve the performance.

TABLE 11.11   Irrigated Area (10^5): Source Wise (1950–2009).

S. No.	Source	1950s (1950– 51 to 1959–60)	1960s (1960– 61 to 1969–70)	1970s (1970– 71 to 1979–80)	1980s (1980– 81 to 1989–90)	1990s (1990–91 to 1999– 2000)	2006– 07	2007– 08	2008– 09
I	Net area irrigated								
a	Canals	7.92	8.83	8.94	8.23	8.24	7.82	7.53	7.66
b	Tanks	7.76	9.12	8.49	9.16	6.21	5.31	5.06	5.40
c	Wells	4.97	6.45	9.18	10.38	13.13	15.66	15.94	16.14
d	Others	0.46	0.39	0.35	0.19	0.17	0.09	0.11	0.11
Total		**21.11**	**24.79**	**26.96**	**24.96**	**27.75**	**28.89**	**28.64**	**29.31**
II	Area irrigated more than once								
a	Canals	–	–	–	–	–	1.62	1.20	1.78
b	Tanks	–	–	–	–	–	0.38	0.40	0.40
c	Wells	–	–	–	–	–	3.26	2.27	2.43

TABLE 11.11   *(Continued)*

S. No.	Source	1950s (1950–51 to 1959–60)	1960s (1960–61 to 1969–70)	1970s (1970–71 to 1979–80)	1980s (1980–81 to 1989–90)	1990s (1990–91 to 1999–2000)	2006–07	2007–08	2008–09
d	Others	–	–	–	–	–	0.02	0	0.01
Total		**6.19**	**7.87**	**8.26**	**6.19**	**6.19**	**5.28**	**3.87**	**4.62**
III	**Gross area irrigated**								
a	Canals	–	–	–	–	–	9.44	8.73	9.44
b	Tanks	–	–	–	–	–	5.69	5.46	5.80
c	Wells	–	–	–	–	–	18.92	18.21	18.57
d	Others	–	–	–	–	–	0.11	0.11	0.12
Total		**27.3**	**32.66**	**35.22**	**31.15**	**33.94**	**34.16**	**32.51**	**33.93**
Irrigation intensity (%)		**129.32**	**131.75**	**130.64**	**124.80**	**122.29**	**118.4**	**113.50**	**115.76**
Average irrigation intensity		**2001–2009 = 116%**	–	–	–	–	–	–	–

Source: Season and Crop reports, Chennai; Economic Appraisal, Government of Tamil Nadu, Chennai.

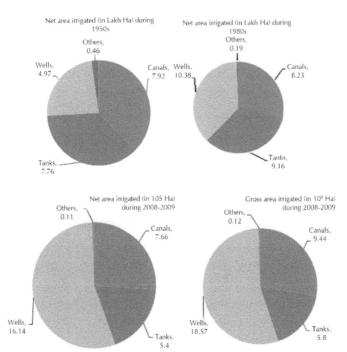

Net area irrigated (in Lakh Ha) during 1950s
Others, 0.46
Wells, 4.97
Canals, 7.92
Tanks, 7.76

Net area irrigated (in Lakh Ha) during 1980s
Others, 0.19
Wells, 10.38
Canals, 8.23
Tanks, 9.16

Net area irrigated (in 105 Ha) during 2008-2009
Others, 0.11
Canals, 7.66
Wells, 16.14
Tanks, 5.4

Gross area irrigated (in 10³ Ha) during 2008-2009
Others, 0.12
Canals, 9.44
Wells, 18.57
Tanks, 5.8

TABLE 11.12   Area Irrigated and Water Used for Different Crops in Tamil Nadu.

Crop	Area irrigated (gross)	Water used	% of water used
	Mha	MHM	%
Paddy	2.10–2.20	2.52	72–73
Sugarcane	0.30–0.31	0.48	14–14.50
Banana	0.09–0.10	0.14	4–4.50
Coconut	0.30–0.31	0.12	
Cotton	0.08–0.10	0.03	
Groundnut	0.25–0.26	0.08	} 10
Millets	0.11–0.12	0.05	
All other crops vegetables, fruits, flowers, fodders, etc.	0.17–0.20	0.05	
TOTAL	**3.40–3.60**	**3.47 or 3.50**	**100**

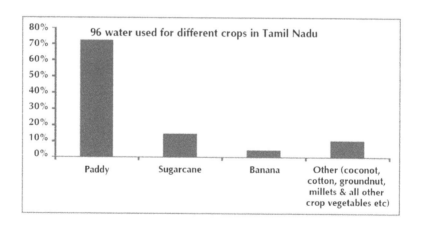

11.4.3.1   WATER UTILIZATION FOR IRRIGATION

It is estimated that about 95–97% of surface water is used and nearly 85% of ground water is extracted. Therefore, there is no scope of getting more water for the future demand. Of the available water in TN, 82–85% of the water is diverted for agriculture, that is, for irrigation at present which irrigates only (gross) an area of 3.4–3.5 Mha in good rainfall years. The remaining quantity is used for human and animal population, industries, aquaculture, power, maintaining environment, etc. The GOI, while assessing the ultimate irrigation potential in the country, has noted that the TN's potential is 5.53 Mha [Water and Related Statistics[9]].

The future demand of water other than agriculture sector is to be met with only by saving water from irrigation by better water management practices and at the same time more area is to be brought under irrigation with less water to feed the growing population in the state. It is estimated that the allotment of water for agriculture and irrigation will be reduced to 70–72% from 82–85% at present to satisfy the need of water for other sectors. In this connection, it is not out of place to mention that in advanced countries, the allocation of water for irrigation is only about 50%, and rest is for all other sectors.

In TN, paddy crop alone takes more than 72% of water diverted for agriculture (irrigation). The other crops, which use more water are sugarcane, banana, and coconut. The details of the area cultivated, water used by different crops and percent of water used by these crops are given in Table 11.12. (Data estimated by the author.)

11.4.4 PRODUCTIVITY OF MAJOR CROPS GROWN IN TAMIL NADU, INDIA, AND WORLD

Though the county has increased the agricultural production from 50 MT in 1950 to 241 MT in 2008–09, the productivity is low in all most all crops compared to world average, especially for the main food crops like rice, wheat, pulses, etc. The production of rice in TN is much above the India's average and it is less in the case of pulses. There is ample scope to increase the productivity since scientists have developed technology in both water management and in agricultural crop production areas. In fact, China has cultivable area of 129 Mha produces more than 500 Mt of food grain compared to India's production of 241 Mt from 142 Mha.[1] Table 11.13 indicates the yield of important crops in TN, India, and the world.

TABLE 11.13 Productivity (kg/ha) of Selected Crops in the World, India, and Tamil Nadu.

Crop	Crop productivity (kg/ha)				
	World	India	Tamil Nadu	Rank (world)	Highest yield in the world
Cotton (lint/ha)	600	347/419	343	57	4527
Jowar	1300	897	874	51	6182
Maize	5130	1408	3626	105	1863
Paddy	3895	2811	2817	51	10,000

TABLE 11.13   *(Continued)*

Crop	Crop productivity (kg/ha)				
	World	India	Tamil Nadu	Rank (world)	Highest yield in the world
Pulses	819	608	312	118	4769
Sugarcane	60,000	65,892	108,000	34	121,381
Wheat	2493	2698	–	32	8997

Source: Dr. M. S. Swaminathan, 2000. *For an evergreen revolution, Survey of Indian Agriculture*. The Hindu; Economic Appraisal, Govt. of Tamil Nadu, Chennai, 2008–2009.[8]

11.5   WATER POLICY: TAMIL NADU

The objective of TN water policy is within the framework provided by the national water policy by GOI. There are

a.   Plan for augmentation of utilizable water resources.
b.   Established allocation practices for water use by different section with provision of drinking water being given highest priority.
c.   Maximize multipurpose benefits from surface and ground water.
d.   Maintain water quality to established standards.
e.   Provide equity and social justice among users of water.
f.   Plan for economic and financial sustainability based upon the principle that those who benefit from projects should pay for them.
g.   Promote users participation in all aspects of water planning and management.

Action plan must be mentioned in the water policy report, such as a nodal agency to collect and store data related to water resources for the state of TN; regulation and control of ground water and surface water on sound footing including enacting laws; maximizing the benefits from the available water; augmentation of water resources including flood control and drainage component in the basin planning and participation of the beneficiaries in the water use sector.

Author emphasizes that the water policy/action plan to solve or overcome the water problem in TN should include following salient features:

1.   Concerted efforts should be made to ensure that the irrigation potential created is fully utilized and the gap is removed. For this, the

command area development program should be adopted in all irrigation projects and tank commands and implemented fully.

2. The tank-irrigated area is declining due to poor maintenance of
   tanks, supply canals, siltation, water spread area encroachment,
   etc.

3. The total capacity of all the 39,000 tanks in TN is about 180 TMC: 2
   times the capacity of Mettur reservoir. Even if it is increased by 10%,
   the quantity is substantial.

The thrust area and strategies are

1. Maintenance of the water related structures.
2. Water conservation.
3. Water management.
4. Water augmentation.
5. Trans-basin diversion.
6. Recycling and reuse of sewage and effluent water.
7. Minimizing evaporation in storage.
8. Drainage improvement works.
9. Crop/cropping pattern changes including crop calendar.
10. Appropriate methods of irrigation.

The other activities, which are to be included in the water policy, are
intensive research efforts in various areas, training of farmers and officials
on water and water use, management information system, and starting a
journal on water and water management in the local language for the use by
farmers and public.

11.6 METHODS TO INCREASE THE AREA UNDER IRRIGATION AND FOOD PRODUCTION

The available water is only 650 M^3/person/year and the quantity available
per person will be further decreased when the population increases in the
coming years. More water is required to satisfy the need of the growing
population in food, municipal water supply, water for the industries, environment, etc. The following are some of the measures that are suggested by
the author and can be implemented on war footing as the problem is very
serious in TN:

11.6.1 WATER AUGMENTATION—SUPPLY MANAGEMENT

1. Modernizing all old irrigation projects and canals in order to increase the duty and desilting the irrigation tanks to maintain the designed capacity to store the flood waters.
2. Watershed development and management works, that is, soil and water conservation, water harvesting, and storing the flood water including farm ponds on watershed basis.
3. Utilization of monsoon floods: About 177 TMC of water on an average every year is joining the sea.
4. Waste water reclamation and reuse sewage water from municipalities and effluent water from industries: about 160 and 200 TMC in 2025 and 2050, respectively.
5. Salt water and sea water utilization. Growing salt tolerance crops and desalination of sea water for drinking and industries use in coastal regions.
6. Cloud seeding—to get more rain especially in the west, south, and eastern part of the state.
7. Interlinking of state rivers in India. It is being taken up at present. *For in depth study, reader may refer to Chapter 10.*
8. Finally, interlinking of peninsular rivers including west flowing rivers in Karnataka and Kerala to east. *For in depth study, reader may refer to Chapter 10.*

11.6.2 WATER MANAGEMENT ESPECIALLY IN IRRIGATION— DEMAND MANAGEMENT

1. Water management in paddy cultivation by using only *system of rice intensification* (SRI) method. Area of paddy can be restricted to 1.60–1.70 Mha and increasing the yield from 4.25 to 7–9 t/ha. The entire paddy crop should be under SRI method, thereby 30–40% of water saving and about 60–100% more yield can be obtained.
2. Changing crop and cropping pattern especially in tank irrigation and Parambikulam Aliyar Project depends on the availability of water and adopting suitable irrigation methods, namely drip and sprinkler.
3. **Tank irrigation:** Reduce paddy area in tank fed cultivation; sprinkler and drip irrigation can be introduced. Where the paddy yield is less than 2 t/ha, paddy should not be cultivated. Convey water by pipes and not through earthen channels.

4. Bring 20–25% of the total irrigated area under drip/sprinkler: 0.7–0.8
    Mha in drip (sugarcane, banana, coconut, vegetables, fruit crops
    under drip irrigation) and 0.3–0.4 Mha in sprinkler (pulses, millets,
    oil seed/fodder under sprinkler irrigation).
5. Water management by paired row method for all row crops and paired
    row and sustainable sugarcane initiative method can be adopted.
6. Water management in municipal water supply, especially leakages in
    the pipes should be avoided.
7. Only the water consumed by the industries should be given daily and
    not any quantity as desired by the industrialists.

11.6.3   INCREASING THE CROP PRODUCTIVITY

Increasing the productivity and production in all crops and increasing the
gross area of irrigation from 3.4–3.5 Mha to 5.1–5.2 Mha and increasing the
irrigation intensity to 125–130% using only 70–71% of the available water
(4.74 MHM).

11.6.4   INCREASING THE CROPPING INTENSITY

Increase the cropping intensity from 113 to 120–125% and give protective
irrigation in rainfed agriculture. Try to take two crops in rainfed cultivation,
where the rainfall is more than 300 mm in each monsoon period. Analyze the
rainfall pattern critically, fix the crop-sowing time, and select the crop suitable for the moisture available period in the area/watershed. All the waste/
fallow lands should be brought under cultivation after implementing watershed development works by saturating the works: An amount of Rs. 25,000–
30,000/ha may be required for watershed development works to carry the
above activities including farm ponds.

11.6.5   INTENSIFIED RESEARCH EFFORTS

Intensive research on land, water, and suitable extension methods should
be taken up and the findings can be adopted by the farmers. The farmers
use presently only 5–7% of the research findings. Creation of Technology
Mission for irrigation and water management and employment of extension
officers for water management for each block to use the water optimally/
efficiently to give advice to the farmers.

11.6.6 ANTICIPATED OUTCOME

If the above suggestions and recommendations are implemented, it is possible to increase the area of irrigation to about 5.0–5.2 Mha from 3.4 to 3.5 Mha by using only 70–72% of the total available water (surface and ground water) of 4.74 MHM. The irrigated area, crops grown, water required, and percentage of water for each crop are shown in Table 11.14.

TABLE 11.14   Irrigation Area and Water Needed for Different Crops in Tamil Nadu: Proposed.

Crop	Area	Water needed	% of water required
	Mha	**MHM**	**%**
Paddy	1.60–1.70	1.40	40.00
Drip irrigation			
Sugarcane/Banana	0.35	0.29	8.30
Vegetables/Cotton	0.10		4.90
Coconut/Fruit crops	0.25	0.17	
Sprinkler irrigation			
Millets, pulses, oil seeds, and others	0.35	0.15	4.30
Surface irrigation			
Water saved from paddy crop	1.60	1.10	31.40
Water saved from sugarcane, banana, coconut crops	0.50	0.23	6.50
Cotton, groundnut, pulses millets, fodder, fruits, vegetables, etc., from the balanced water	0.35	0.15–0.17	4.60
Total	**5.10–5.20**	**3.40–3.50**	**100.00**

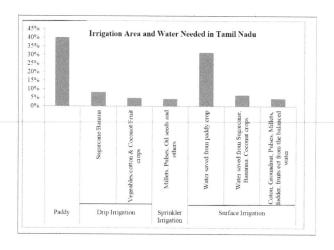

11.7 CONCLUSIONS

The water availability in India is plenty, but it is unevenly distributed and hence the water scarcity problem exists in some parts of the country especially in the southern states, particularly in TN. The unused water which is flowing into the sea in the southern States appears to be of the order of 1000 TMC, that is, 50% of the water flowing in Arabian Sea in Karnataka; 450 TMC, the excess over the demand in Kerala; and 810 TMC from Mahanadhi and Godavari Rivers as surplus water (estimated by NWDA) totaling to 2260 TMC, which can irrigate about 10 Mha. For India to become a developed country in 2020/25, the farmers should get water to irrigate about 150–160 Mha for which inter-linking of rivers is required.

Author with his experience and foresight has developed the *Tamil Nadu Water Vision* based on rainfall, water potential, tank, and storage capacities, land use, and food production. He has clearly shown what could be done and achieved. This chapter will be useful instrument to planners, administrators, farming public, and the society for the prosperity of TN. Similar model can be adopted by other developing countries.

As advocated in the chapter before this, India as a whole is aware of the need and the possible immense benefits of interlinking of the rivers as a permanent long term solution to address water scarcity, especially for the Southern states.[2–5,7] Educators and policy makers agree that the state and central government should take actions to implement the two proposals detailed by author in these chapters to solve the water, food, and energy problems of India.

ACKNOWLEDGMENTS

Author expresses his sincere thanks to: Dr. N. Mahalingam, for a source of inspiration and encouragement;[3] Dr. S. K. Mishra (Editor of Journal of Indian Water Resources Society) for his encouragement on the theme of interlinking of Indian rivers; Dr. S. R. Sreerangasamy (Former Dean at Tamil Nadu Agricultural University) to offer useful suggestions and comments; Thiru O. Arumughaswamy (Chairman, Vijalakshmi Charitable Trust, Coimbatore) for printing and publishing booklet on "Tamil Nadu water vision and interlinking of Indian rivers – need and importance." This chapter is an edited version of Part I of this booklet.

KEYWORDS

- available water
- BCM
- CGWB
- CWC
- drip irrigation
- GOI
- ground water
- ground water resources
- India
- inter-linking of rivers
- million hectare meter
- river basin
- sprinkler irrigation
- surface water
- Tamil Nadu
- tributaries
- Utilizable water
- water crisis
- water resources of India
- water scarcity
- water stress

REFERENCES

1. China Water Vision. Second World Water Forum, 20th March 2000, *Meeting the Water Challenge in Rapid Transition*. The Hague, The Netherlands, 2000.
2. IWRS. *Water Resources Day, Theme paper on Water—Vision 2050*. Indian Water Resources Society (IWRS): New Delhi, 1999.
3. Mahalingam, N. *Reporting India for Faster Economic Development through Inter-linking River Basins*. Kisan World: Chennai, 1999.
4. Mohan, S.; et al. Interlinking of Indian Rivers—Operational and Legal Issues and Options, Seminar Presentation at Indian Institute of Technology, Chennai, 2004.
5. Prabhu, S. Chairman: Task Force on Interlinking of Rivers: Press Report. The Dinamani, Coimbatore Edition, India, April 17, 2003.

6. Ministry of Agriculture, Govt. of India. *Report of the National Commission on Agriculture, Resource Development, Part V*. Ministry of Agriculture, Govt. of India: New Delhi, 1976.

7. Sivanappan, R. K. *Tamil Nadu Water Vision and Interlinking of Indian Rivers—Need and Importance*; Shree Vijalakshmi Charitable Trust: Coimbatore, Tamil Nadu, India, 2012; p 50.

8. The Hindu. *Newspapers. These Covered Numerous Articles by Many Scientists in the Last 10–20 years*, 1999, 2010, 2011, 2012.

9. Central Water Commission. *Water Related Statistics.* Central Water Commission: New Delhi, 2000.

10. National Commission for Integrated Water Resources Development Plan. *Water Resources Development Plan of India, Policy and Issues*. National Commission for Integrated water resources Development Plan, Government of India, Ministry of Water Resources: New Delhi, 1999.

APPENDIX 11.I   WATER USED AND AREA IRRIGATED IN TAMIL NADU: SAMPLE CALCULATION.

A. Water available

- Total average annual available water (surface and ground water and from other resources)   = 4.70/5.00 MHM

- Entire surface water is used and 80% of GW is extracted at present. Therefore, the ground water balance   = 0.53 or say 0.50 MHM

- The water used at present for all sectors   = 5.00 − 0.50
  = 4.50 MHM

- Water used for agriculture (i.e., 80%)   = 3.50 MHM

- Water used for drinking, industry, and other purposes   = 4.5 − 3.5
  = 1.0 MHM

- Water yet to be taped from ground water   = 0.50 MHM

                                                    Total   = 5.00 MHM

B. Area irrigated at present

- Paddy   = 2.0/2.10 Mha

- Other crops   = 1.40 Mha

                            Total, gross area   = 3.40 Mha

C. Total the area under irrigation will be as follows after implementation of the suggestions on Tamil Nadu Water Vision:

- Paddy   = 1.60/1.70 Mha

- Sprinkler and drip irrigation area, 1/2 Mha for each = 1.00 Mha

- Area irrigated from saved water   = 1.25 Mha

- Existing area irrigated other than rice, = 3.4 − 2.1   = 1.25–1.30 Mha

                            Total (Mha)   = 5.1–5.15 or 5.2

APPENDIX 11.II ILLUSTRATIONS [PHOTO COURTESY: AUTHOR]

Front wrapper (**inside**)   Back wrapper (**inside**)

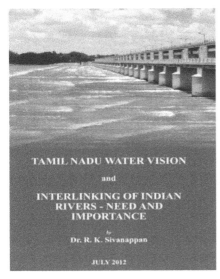

Front wrapper (front side)   Back wrapper (backside)

CHAPTER 12

WATER CONSERVATION AND INCREASED CROP PRODUCTION UNDER PRECISION FARMING

R. K. SIVANAPPAN

College of Agricultural Engineering & Technology, Tamil Nadu Agricultural University (TNAU), Coimbatore, Tamil Nadu, India. E-mail: sivanappanrk@hotmail.com

CONTENTS

Abstract 306
12.1 Introduction..306
12.2 Current Status of Crop Production...306
12.3 Precision Farming 308
12.4 Precision Farming Project in Tamil Nadu.......................................310
12.5 Results And Discussions 313
Keywords 315
References...315

ABSTRACT

Water is becoming a scarce commodity. Our productivity in almost all crops are far below in the world's productivity. It is time now to use high technology to save water and increase productivity. Precision farming (PF) now implemented in Dharmapuri and Krishnagiri project is an eye opener for all Indian farmers to following a big way in the coming years to solve the water scarcity and to get more income to the framers. It is a complete farmer's participatory project. Based on this demonstration, PF can be taken up in all districts and by all farmers.

12.1   INTRODUCTION

Land and water are the two important natural resources, which are the basic needs of agriculture and economic development. These two resources are constant, but the demand for agricultural production is increasing due to continuously escalating population. In addition, the land and water are needed for other development projects, that is, land for roads and buildings and water for numerous industries and municipal needs. During 2001, the per capita cultivable land has come down from 0.25 ha in 1950 to 0.09 ha in Tamil Nadu, India. Total utilizable water resource (both surface and ground water) of the Tamil Nadu state is estimated about 5 MHM and the availability per capita is reducing due to increased population in the last 50 years. The present per capita available water is about 650 M^3/person/year. With this land and water, it is necessary to satisfy the need of the population not only at present but also in the future. To achieve the objectives and goals, it is necessary to introduce and adopt the precision farming (PF), with integrated intensive and hi-tech farming system. Further the micro-irrigation technology (more crop per drop of water) and more crop per unit input should also be introduced systematically to feed the growing population and enhancing the living conditions of the people.

This chapter focuses on introduction of PF that is water conservation and more crop per unit input technology.

12.2   CURRENT STATUS OF CROP PRODUCTION

The land area can be used for taking 2 or 3 crops in a year, that is, 200–300% cropping intensity, but water availability is the limiting factor especially in the southern India. Crop production in India must increase to about 450–500 million tons from present production of 241 million tons to feed the expected population of 1650 million in 2050. In this connection, the agricultural productivity at present of most of the crops in the country is very less compared to world's average productivity (Table 12.1).

TABLE 12.1 Productivity of Different Crops in the World, India, and Tamil Nadu in kg/ha.

Crops	World	India	Tamil Nadu	Rank of India in the world	Highest yield in the world
Cotton lint kg/ha	600/759	347	343	57	4527
Jowar	1300	897	874	51	6182
Maize	5130	1408	3626	105	18,636
Paddy	3895	2811	2817	51	10,000
Potato	–	16,478	–	51	43,969
Pulses	819	608	312	118	4769
Sugar cane	60,000	65,892	108,000	34	121,381
Wheat	2493	2698	–	32	8997

In order to double the production from 240 to 450–500 million tons in the year 2050, India has long way to follow. To achieve this goal especially when more land and water are necessary, the Hi-tech PF with micro-irrigation and fertigation can help to increase the productivity per unit area and per unit water to solve this problem.

The world grain production in 2001 was about 1840 million tons for a population of 6 billion (i.e., 305 kg/person), compared to about 2325 million tons for a population of 7 billion in 2011–2012 (332 kg/person). Therefore, when it comes to food production, the world is doing well compared to grain production in India. The India's food production in 2010–11 was 240 million tons to feed a population of 1210 million (200 kg/person), and in 2050 for a population of 1650 million, it is estimated between 450 and 500 million tons (272–300 kg/person).

In 1974, the World Food Summit in Rome predicted that the world would not able to feed itself in 2000. This did not happen. In fact, food production almost doubled in 27 years (2000–2001). We were able to increase the food production at about 3% per annum compared with the

2.25–2.50% of population growth rate. Since 1960, average yield has gone up >150%, due to

1. Modern agricultural practices.
2. Use of more fertilizer.
3. Better water control and management.
4. Different species of crops and disease-resistant varieties.
5. Different cultivation of traditional crop.
6. Advanced irrigation methods.

Therefore, the government has initiated a number of steps to promote on farm management of land and water through the use of modern farming techniques, namely, PF and drip irrigation system which also a main component/application of PF.[3] Drip irrigation is suited for all crops except rice and helps in water saving by 40–70% and increase yield up to 100% and has less incidence of insects, pests, and diseases with improved management systems. This method can be adopted in undulating topography with limited skill power. Drip irrigation added with fertigation reduces the amount of fertilizers (by about 30–35%) and increases the yield by optimization of nutrients use at critical stage. Therefore, use of micro-irrigation/drip irrigation coupled with fertilization will double the cropped areas and tripled the production with same quantity of water.

This trend is continuing by introducing drip irrigation on a large scale throughout the world (6 Mha in the world and 2.0 Mha in India). India has 35% of the total irrigated area of the world, whereas the drip irrigated area is only about 33%, which must be increased in the coming years.

12.3   PRECISION FARMING

The cultivation practices, crops grown, and irrigation methods are as old as civilization in the country and in Tamil Nadu. The scientists have developed new crop varieties/advanced agronomical and plant protection methods/practices and improved water management technology including use of advanced methods of irrigation. It is time now to change the crop and cropping pattern, using plant protection measures and irrigation methods in the changing world to save water and get more production per unit water/unit area and/unit time. This will not only conserve the land and water but also offer more profit/income to the farmers for their better livelihood.

The **PF** implies using all the high technology available for cultivation: optimal spacing, timely availability of inputs, like fertilizers, pesticide chemicals, seeds, and seedlings. It also includes use of drip irrigation by delivering daily optimum quantity of water and fertilizer through fertigation with marketing strategy to get more income to the farmers. The intent of PF is to match agricultural inputs and practices as per crop and agroclimatic conditions to improve the accuracy of the application. Therefore, PF or precision agriculture is about doing the right thing, in the right place, in the right manner/right way at the right time. PF can

1. enhance productivity of agricultural and horticultural crops;
2. prevent soil degradation in cultivable land;
3. reduce fertilizer use in crop production;
4. conserve water resources by increasing the efficiency of use; and
5. disseminate modern farm products by improving quality, quantity, and reducing cost of production of agricultural/horticultural crops.

Several advantages of PF are agronomical perfection versus technology application; technical perfection versus time management; environmental perfection versus eco-friendly practices; and economic perfection versus profit to the farmers. Applications of PF are

1. lining of channels/pipe conveyance of water;
2. drip/micro-sprinkler/sprinkler irrigation;
3. mulching;
4. green house;
5. shade net;
6. use of plastics in nursery management; and
7. bird protection nets.

Precision agriculture has witnessed unprecedented growth in the last decade, especially in the USA, Germany, Israel, and other countries. While the rest of the world has been relatively slow in adopting PF, the change has been evident. From Australia to Zimbabwe, PF is growing across the globe. This is clearly evident by the number and diversity of technical publications in the area of PF in the International Journals and conferences/seminars. The PF was introduced the developing countries for about 2.5 decades ago. With increasing global population and limited or decreasing arable land available for crop production, the question arises: Will we be able to overcome the future challenges and seize them as opportunities? Precision agriculture management coupled with genetic improvement in crop trails and better water management will play a crucial role in meeting global demand for food, feed, fiber, shelter, and fuel in the near and distant future.

The author personally visited the project with the project directors (PDs)/ advisor in both the districts and studied the crop performance, water used, crop yields and marketing of the products. The author with his team also interacted with the farmers in detail to get the feedback information to analyze the performance of the project.

12.4 PRECISION FARMING PROJECT IN TAMIL NADU[1,2,4-7]

During 2004–2007, the Government of Tamil Nadu implemented "Precision Farming Scheme" in Dharmapuri and Krishnagiri districts of Tamil Nadu for various vegetable crops under drip irrigation with fertigation. The farmers, who have required water in the wells to grow crops, are selected and all the inputs (costing about Rs. $1.0–1.25 \times 10^5$ per ha) were provided free, including free market services, etc. Since it was a pilot project (demonstration project), the cost of the layout of drip irrigation with fertigation at a cost of Rs. 75,000/ ha and soluble fertilizers at a cost of Rs. 40,000 were given free to all the selected farmers to provide incentive to adopt the system. In the coming years, the subsidy was reduced step by step. Tamil Nadu Agricultural University (TNAU), district collector, and state Departments of Agriculture, Horticulture, Agricultural Engineering, were involved at all phases of the scheme.

The Commissioner of Horticulture and Plantation Crops, Govt. of Tamil Nadu was the PD and TNAU was the project-implementing agency (PIA). District Collectors of concerned districts were the project coordinators. Five officials of the State Departments of Horticulture and Agricultural Engineering were coordinating between the PD, project coordinator, and PIA. There was a perfect coordination among all these agencies. The technology and techniques included layout of drip irrigation with fertigation, and services to market the products were provided by TNAU.

12.4.1 PROJECT AREA

The undivided Dharmapuri district (now Dharmapuri and Krishnagiri districts) is in the north west of Tamil Nadu and the maximum elevation is about 960 m. The average rainfall is about 850 mm (NE—311 mm and SW 366 mm). Red, clay, and mixed soils are predominant in the district.

The crops grown are paddy, sugarcane, millets, groundnut, oilseeds, pulses, tapioca, all vegetables including hill vegetables, fruit trees like mango, banana, flowers, etc.

There are two river basins namely Cauvery and Ponnaiyar. There are 175 tanks, 350 canals and 11 reservoirs in the district. The total annual ground water recharge is about 978 mcm and the stage of development as on 2003 is 118%, that is, extraction is more than recharge.

Since there are no big projects and the ground water is fully utilized, it is very essential to use the water in an effective/efficient way in order to bring more area under irrigation and increase the productivity per unit quantity of water. Therefore, the PF project was implemented as a demonstration project (pilot project) in the district mainly for vegetable crops under drip (micro) irrigation with fertigation.

12.4.2 SELECTION OF FARMERS

The district collector with the help of Department of Horticulture selected the farmers based on the following criteria.

-   Enough water in the well for irrigation.
-   Pumpset, which can discharge about 11,500 L/h at 1.5 kg/cm^2 pressure to operate drip irrigation system.

12.4.3 SELECTION OF VEGETABLE CROPS

The selection of vegetable crop was based on the demand in the market in the beginning and subsequently, the farmers were to select/introduce the crop. The vegetables selected in the first year were tomato, chilli, cucumber, gerkin, cauliflower, cabbage, capsicum, baby corn, etc. The drip irrigation layout was designed to suit any crop selected by the farmers for cultivation. Therefore, the drip system could be easily adopted to each crop without making any changes.

12.4.4 SELECTION OF IRRIGATION METHOD

In India, more than 90% is under surface irrigation, though advanced irrigation methods like micro-irrigation are suitable for all row crops and for almost all soils. It was found by the scientists that the water saving in drip irrigation method was about 40–60% and the yield increase can be about 10–100%. Further, fertilizers can be applied through the irrigation water (i.e., by fertigation), by which about 30% of fertilizer use can be reduced or saved. This is the requirement for the PF.

12.4.5 PRECISION FARMS

The project was for 3 years starting from 2004 to 2007 in both Dharmapuri and Krishnagiri districts. The 50 farmers, each one to cultivate in 1 ha, were selected in Dharmapuri district. Cultivation in the precision farm was done as per advice of the scientist of the College of Horticulture, TNAU, Coimbatore. The crops selected were hybrid tomato variety—US618 in 20 ha and hybrid chilli NSI701 in 30 ha. In Krishnagiri district, also 50 farmers were selected in 2 villages. The vegetable crops introduced under this program were cauliflower—NS66/NS60N in 5.50 ha, cabbage in 28.50 ha, and Hari rani Empire hybrid tomato Abinav in 16 ha.

At the time of reporting of data in this chapter, the crops in Dharmapuri district namely tomato and chilli were yet to be completed but in Krishnagiri cabbage and cauliflower were fully harvested.

The hybrid varieties were used with hi-tech cultivation technologies like fertilizers, through fertigation, optimum spraying of pesticides, irrigating the required quantity of water by drip, and marketing the produces in different market places to get more price. The required extension staff were employed to help the farmers to raise good crop, arrest the diseases and to give water through the drip system (micro-tube) without any break. They were provided with vehicles and communication gadgets so that the extension staff can visit and advice the farmers in time. The TNAU professor/experts visited the farms every now and then to help the farmers. In addition, timely availability of inputs like fertilizers, plant protection chemicals, seeds, and seedlings were provided by the staff to the farmers.

Since it was a pilot project (demonstration), the cost of layout of drip irrigation with fertigation cost of Rs. 75,000/ha and soluble fertilizers at a cost of about Rs. 40,000/– are given free to all the farmers to have incentives to adopt this system. In the coming years, the subsidy will be reduced step by step.

12.5 RESULTS AND DISCUSSIONS

Scientists visited the fields in both the districts and studied the performance of crops, water use, crop yield, and marketing of the product. They also interacted with the farmers in detail to get the feedback information to analyze the performance. Based on their reports, following observations and feedback is noteworthy:

- The crop performance was very good.
- The farmer's perception of the drip irrigation was excellent. All persons in the project appreciated the quality of the material used, after sale service, and the cost of the system compared to earlier years.
- Farmers will adopt drip irrigation with PF, even after the expert staff are withdrawn from the project area after 3 years.
- According to the farmer's version, the water saving is about 50%, which is also the result of the research studies made by the scientists at TNAU.
- Fertilizer saving is also noticed by farmers. Recommendations are provided based on soil test analysis.
- Many farmers around the pilot/demonstration plot visited the demonstration farm to know/understand the system.
- The crop yield increased by 20–30%.
- The quality of produce was good and uniform in size. The substandard grade of fruits was about 30% before the project implementation, and it reduced to only 2–3%, after the project.
- Farmers are given training to establish community nurseries under shade net in portrays.
- To avoid the middle man in marketing: Farmers' Registered Associations are being formed, that is, "Athiaman Precision Farming Farmers' Association" at Dharmapuri and "Maruthi Precision Farming Farmers' Association" at Krishnagiri. This is a positive step.
- The harvested fruits were packed scientifically after grading without any damage and sent to the market to get premium price.
- Nurseries are maintained well. Vigorous and uniform seedlings are supplied to the farmers, with 100% establishment in the field without failures.
- Crop diversification is taking place and accepted by the farmers. It is a good sign to get more income/profit.

-   The field staff is making arrangements to have a tie up with processing industries for the crops grown by the framers, for example: onion with Jain dehydration plant at Jalgoan and Gerkin with some industries in Bangalore.

After the successful implementation of PF in Dharmapuri and Krishagiri Districts, it was then implemented in Madurai District and other districts during 2007–2008 onward through Krishi Vidya Kendra (KVK) and line departments. In Madurai district, the PF technology was disseminated in about 2500 ha. The crop yield was drastically increased by adoption of PF technology in the farmer's field. The crop yield obtained in the conventional and PF is shown in Table 12.2.

TABLE 12.2 Yield of Selected Crops under Precision Farming.

Crop	Yield (t/ha)	
	Conventional	Precision farming
Tomato, Madurai Dt	40	110
Brinjal, Madurai Dt	60	150
Banana, Madurai Dt	75	110
Sugarcane (erode district)	106	250

The adoption of PF technique is now being disseminated in all districts of Tamil Nadu. The area expansion has been slowly increasing due to awareness among the farmers. The farmers are slowly gaining knowledge on implementation of advanced technology by effective resource management through transfer of technology by KVKs in all the districts of Tamil Nadu. It has been reported that about 70,000 ha are cultivated under this technology in Tamil Nadu for various crops: mainly vegetables, fruit crops, and commercial crops like sugarcane, banana, cotton, etc.

12.5.1 RESEARCH IN INDIA

The study on PF has been initiated in many research institutions in India. For instance, space application center at Indian Space Research Organization, Ahmedabad has started research studies in the Central Potato Research Station Farm at Jalandhur, Punjab to study the role of remote sensing, GIS and GPS for mapping the variability. MSS foundation in Chennai in collaboration with NABARD has adopted a village in Dindigul district of Tamil Nadu for variable rate input application. The IARI, New Delhi has drawn up plans to conduct experiments on PF in the institute's farm. Project directorate for cropping system research—Modipuram and Meerut (UP) has initiated a project on PF in collaboration with Central Institute in Agricultural Engineering, Bhopal. These are some examples of the ongoing research work in India.

KEYWORDS

- drip irrigation
- India
- precision farming
- Tamil Nadu
- TNAU
- vegetable crop
- water scarcity

REFERENCES

1. Asokaraja, N. *High-tech Cultivation of Banana with Tissue Culture.* Department of Agronomy, Tamil Nadu Agricultural University (TNAU): Coimbatore, India, 2013.
2. Kannan, M. *Precision Farming Technologies.* Horticulture Department, TNAU: Coimbatore, India, 2014.
3. Sharward, U. K.; Patil, V. C. *Precision Farming Dreams and Realities for Indian Agriculture.* University of Agricultural Sciences: Dharward, Karnataka, India, 2004.
4. Sivanappan, R. K. High-tech Precision Farming in Agriculture and Horticulture Crops. *J. Kissan World* **2013,** *40*(1).
5. Sivanappan, R. K. Precision Farming—Global Scenario: Microirrigation to Double the Yield in Agricultural and Horticultural Crops. Key Note Address in Seminar at the *International Institute of Horticulture Management,* White field, Bangalore, India, 2012; pp 1–10.
6. Vadivel, E. *Tamil Nadu Precision Farming Project: Expertise Shared and Experience Gained.* Bulletin by Horticulture Department, TNAU: Coimbatore, India, 2006.
7. Vadivel, E. *Tamil Nadu Precision Farming Project—Turnkey Project.* Horticulture Department, TNAU: Coimbatore, India, 2008.

PART IV
Principles of Soil and Water Engineering

BASICS OF SOIL AND WATER CONSERVATION ENGINEERING

R. K. SIVANAPPAN

College of Agricultural Engineering & Technology, Tamil Nadu Agricultural University (TNAU), Coimbatore, Tamil Nadu, India. E-mail: sivanappanrk@hotmail.com

CONTENTS

Abstract 320
13.1 General Hydrology 320
13.2 Forest Hydrology 327
13.3 Watershed and Watershed Development....................................328
13.4 Soil Erosion and Sedimentation...330
13.5 Soil and Water Conservation Measures 335
13.6 Check Dams and Percolation Ponds 342
13.7 Micro-Level (In Situ) Soil and Moisture Conservation Measures 345
13.8 Soil and Water Conservation Measures for Different
      Zones of Watersheds 349
13.9 Instruments and Equipments Needed for Soil and
      Water Conservation Works 350
Keywords 352
References..352
Appendix 13.I   US Postal Stamp Advocating Water Conservation 354

ABSTRACT

In this chapter, the author (with +60 experience in soil and water conservation engineering) has discussed topics, such as general hydrology, forest hydrology, watershed and watershed development, soil erosion and sedimentation, soil and water conservation measures check dams and percolation ponds, farm ponds, micro-level (in situ) soil and moisture conservation measures, soil and water conservation measures for different zones, and instruments and equipment needed for soil and water conservation works.

13.1   GENERAL HYDROLOGY

Hydrology is the science that deals with the process governing the depletion and replenishment of the water reserves of the land area of the earth. A basic knowledge of hydrology is necessary to handle problems related to the supply and use of water and is of great use in engineering, forestry, agriculture, and other related fields.

13.1.1   HYDROLOGIC CYCLE

The circulation of water from the ocean and land surface to the air, from the air to the land and back to the ocean over the land surface or underground is called hydrologic cycle. By different natural processes, the total water supply of the earth is in constant circulation. The different processes in this cycle are shown in Figure 13.1. The essential components of the hydrologic cycle are precipitation, infiltration, runoff, evaporation, transpiration, and soil moisture content. These may be, in simple terms, expressed by the following water balance equation:

$$P{\downarrow} = E{\uparrow} + E_T{\uparrow} + I{\downarrow} + R_{\rightarrow} + \Delta S \tag{13.1}$$

where P = precipitation, E = evaporation, E_T = evapotranspiration, I = infiltration, R = runoff, and ΔS = change in soil moisture storage.

13.1.2   RAINFALL

Of all the forms of precipitation, rainfall has an important bearing on forestry and agriculture and knowledge of measuring the rainfall and analyzing the data are, therefore, necessary. The amount of rain which falls on a level surface is expressed in mm or cm. It can be measured by using either an

- **Ordinary or non-recording rain gauge**: the total rainfall during a fixed period can be measured.
- **Automatic or recording rain gauge**: the amount of rain that falls with respect to time is recorded on a graph paper.

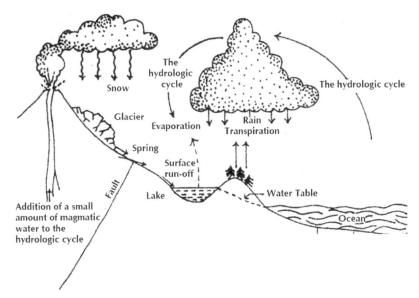

FIGURE 13.1   The hydrologic cycle.[1,2]

The **mean annual rainfall** is the arithmetic average of the annual rainfalls of several consecutive years. Larger the number of years, the more representative will be the value. A 50-year average will give a fairly accurate value.

Rainfall intensity is the rate at which rainfall occurs and is expressed in mm/h. If an automatic (recording) rain gauge is used, it is possible to calculate the intensity for any selected time interval from the rainfall chart.

13.1.2.1   RELATION BETWEEN INTENSITY AND DURATION OF RAINFALL

For calculating the runoff from a small area, the values of maximum intensity for different period are required. The intensity of rainfall is an inverse function of its duration. The longer the duration of rainfall, the less will be its intensity. Also, the intensity of the rainfall will be greater if a longer period, say 20–200 years, is considered than a short period of 1–2 years.

13.1.2.2   FREQUENCY OF RAINFALL

The frequency of occurrence of maximum rainfall or the recurrence interval is the period expressed in years during which one spell of a given duration and intensity can be expected to occur. For example, a 10-year frequency for a 1-h rainfall is the magnitude of hourly rainfall, which can be expected to be equalized or exceeded once in 10 years. It does not mean that this magnitude of rainfall will occur at a regular interval of every 10 years. It merely means that there is every chance that such a magnitude of rainfall may occur once in 10 years.

13.1.3   RUNOFF

A portion of the precipitation that reaches the land infiltrates into the soil and gets stored. The remaining portion moves down the slope as runoff. While considering the runoff from a given area, the total runoff and also the rate of runoff have to be considered. The area contributing to the runoff is variously known as the catchment, watershed, or drainage basin. The rate of runoff and its volume from a given area are influenced by two sets of factors:

A. The climatic factors are

- type of precipitation,
- rainfall intensity,
- duration of rainfall,
- distribution of rain,
- direction of storm movement,
- antecedent rain and soil moisture,
- evaporation/transpiration,
- wind velocity, and
- solar radiation.

B. The physiographic factors are

- soil type,
- land-use pattern,
- area and shape,
- elevation,
- slope,
- orientation, and
- type of drainage.

13.1.3.1 RELATIONSHIP BETWEEN RAINFALL AND RUNOFF

From the rainfall data, runoff can be estimated as a percentage of the total rainfall of any given number of years. Based on the different runoff percentage data, catchments have been classified as good, average, or bad. Table 13.1 is the Strange's table, which gives the percentage of runoff with reference to total rainfall and yield from one hectare for different catchment conditions.

13.1.3.2 ESTIMATION OF RUNOFF

Example: Calculate the yield from a degraded hill forest having a watershed of 100 ha with a monsoon rainfall of 750 mm in a year.

Watershed area	= 100 ha
Monsoon rainfall	=750 mm

Since the watershed is a degraded hill forest, it is a good catchment as good runoff is expected from it. From Table 13.1 (Strange's table), the percentage of runoff of a good catchment with 750 mm rainfall = 26.30.

Total yield from Catchment = Area in sq. meters × total monsoon rainfall in meters × % of runoff = 100 × 10,000 × 750/1000 × (26.30/100) = 1972.5 m³. In Table 13.1, yield from 1 ha = 1972.5 cu m.

Therefore, for 100 ha = 197,250 m³.

TABLE 13.1 Strange's Table: Percentage of Runoff to Rainfall and Yield of Runoff from a Catchment as a Function of Total Monsoon Rainfall.

Total monsoon rainfall	Good catchment		Average catchment		Bad catchment	
	Percentage of runoff to rainfall	Yield of runoff from catchment	Percentage of runoff to rainfall	Yield of runoff from catchment	Percentage of runoff to rainfall	Yield of runoff from catchment
mm	%	per ha in cu m	%	per ha in cu m	%	per ha in cu m
25.00	0.10	0.25	0.10	0.25	0.05	0.13
50.00	0.20	1.00	0.15	0.75	0.10	0.50
75.00	0.40	3.00	0.30	2.25	0.20	1.50
100.00	0.70	7.00	0.50	5.00	0.30	3.00
125.00	1.00	12.50	0.70	8.75	0.50	6.25
150.00	1.50	22.50	1.10	16.50	0.70	10.50
175.00	2.10	36.75	1.50	26.25	1.00	17.50
200.00	2.80	56.00	2.10	42.00	1.40	28.00
225.00	3.50	78.75	2.60	53.50	1.70	38.25
250.00	4.30	107.75	3.20	80.00	2.10	52.50
275.00	5.20	143.00	3.90	107.25	2.60	71.50
300.00	6.20	136.00	4.60	138.00	3.10	93.00
325.00	7.20	234.00	5.40	175.50	3.60	117.00
350.00	8.30	290.50	6.20	217.00	4.10	143.50
375.00	9.40	325.50	7.00	262.50	4.70	176.25
400.00	10.50	420.00	7.80	312.00	5.20	208.00
425.00	11.60	493.00	8.70	369.75	5.80	232.00
450.00	n80	576.00	9.60	432.00	6.40	288.00
475.00	13.90	660.25	10.40	494.00	6.90	327.75
500.00	16.00	800.00	11.25	562.50	7.50	377.75
525.00	16.10	845.25	12.00	630.00	8.00	420.00
550.00	17.30	951.50	12.90	709.50	8.60	473.00
575.00	18.40	1058.00	13.80	793.50	9.20	529.00
600.00	19.50	1170.00	14.60	878.00	9.70	582.00
625.00	20.60	1287.50	15.40	962.50	10.30	643.75
650.00	21.80	1417.00	16.30	1059.50	10.90	708.50
675.00	22.90	1545.75	17.10	1154.25	11.40	769.50

TABLE 13.1   *(Continued)*

Total monsoon rainfall	Good catchment		Average catchment		Bad catchment	
	Percentage of runoff to rainfall	Yield of runoff from catchment	Percentage of runoff to rainfall	Yield of runoff from catchment	Percentage of runoff to rainfall	Yield of runoff from catchment
700.00	24.00	1618.00	18.00	260.00	12.00	8110.00
725.00	25.10	1819.75	18.80	1363.00	12.50	906.25
750.00	26.30	1972.50	19.70	1477.50	13.10	982.50
775.00	27.40	2123.50	20.50	1580.50	13.70	1061.75
800.00	28.50	2218.50	21.30	1704.00	14.20	1136.00
825.00	29.60	2442.00	22.20	1831.50	14.80	1221.00
850.00	30.80	2618.00	23.10	1963.50	15.40	1309.00
875.00	31.90	2791.25	23.90	2090.25	15.90	1391.00
900.00	33.00	2917.00	24.70	2223.00	16.70	1485.00
925.00	34.10	3154.25	25.50	2358.75	17.00	1572.00
950.00	35.30	3363.50	26.40	2508.00	17.06	1672.00
975.00	36.40	3549.00	27.30	2661.75	18.20	1774.00
1000.00	37.50	3750.00	28.10	2810.00	18.70	1870.00
1125.00	43.10	4348.00	32.10	3633.75	21.50	2418.00
1250.00	48.00	6100.00	36.60	4575.00	24.40	3050.00
1375.00	54.40	7480.00	40.80	5610.00	27.70	3740.00
1500.00	60.00	9000.00	45.00	6750.00	30.00	4500.00

13.1.3.3   ESTIMATION OF PEAK RUNOFF

Peak runoff can be estimated using the rational equation:

$$Q = \frac{[CIA]}{[360]} \qquad (13.2)$$

where Q = discharge in m³/s, I = intensity of rainfall in mm/h, and A = watershed area in ha. Tables 13.2 and 13.3 give the value of C for different land uses. Effective November 25, 2013, the estimation of discharge by rational method equation is mobile-device-friendly.[3]

TABLE 13.2 Values of C for use in Rational Formula.

Soil type	Cultivation	Pasture	Forests
Above average infiltration sandy or gravelly	0.29	0.15	0.10
Average infiltration on clay pan, loam, and similar soil	0.40	0.35	0.30
Below average infiltration heavy clay, clay pan, shallow soil above impervious rock	0.50	0.45	0.40

TABLE 13.3 Values of C for Different Slopes, Land use and Soil.

Vegetation	Topography (%)	Sandy loam	Clay and silt loam	Clay
		Forest land		
Flat	0–5	0.10	0.30	0.40
Rolling	5–10	0.25	0.35	0.50
Hilly	10–30	0.30	0.50	0.60
		Grazing land		
Flat	0–5	0.10	0.30	0.40
Rolling	5–10	0.16	0.36	0.55
Hilly	10–30	0.22	0.42	0.60
		Cultivated land		
Flat	0–5	0.30	0.50	0.60
Rolling	5–10	0.40	0.60	0.70
Hilly	10–30	0.52	0.72	0.82

13.1.3.4 TIME OF CONCENTRATION

It is the time required for water to flow from the most remote point of the catchment to the outlet. It is calculated using following empirical formula:

$$TC = 0.02L^{0.77} \times S^{-0.388} \tag{13.3}$$

where TC = time of concentration, L = length of travel in min, and S = average slope in m/m. An example is given below to calculate peak runoff with the following data:

Intensity of rainfall, I = 175 mm/h
Frequency   = 10 years

Area of watershed	= 25 ha = 15 ha under cultivation + 5 ha under forests + 5 ha under grass. Watershed is clayey.
Fall	= 5 m fall in a distance of 700 m
Distance from the	= 700 m = L
	remote point to outlet

Weighted value of C for the entire watershed of 25 ha (from Table 13.1):
$$= [(15 \times 0.5) + (5 \times 0.4) + (5 \times 0.45)]/25 = 0.47.$$

$$TC = 0.22 \times (700)^{0.77}(5/100)^{-0.385} = 25 \text{ min}$$

Peak runoff = $CIA/360 = [0.47 \times 175 \times 25]/360 = 5.7$ cu m/s or = 5.7 cumec

13.2   FOREST HYDROLOGY

The total amount of rainfall over any area is more or less fixed. The moment rain water reaches land, its distribution and storage can be manipulated by man and this has been done to a large extent for ages. Water harvesting has been a part of the traditional water management system in this area "with tanks, percolation ponds, and anicuts" well distributed over the plains. However, in the hills and forest areas, the manipulations are of a negative type. Over large parts, the forests are heavily degraded and this degradation has an important consequence, namely, that the rain falling on these areas tends to runoff rapidly. Consequently, less water will infiltrate and recharge the ground water storage will be poor. Therefore, the base flow from the catchment will decrease and the period of non-flow in the rivers will become longer. Further, degradation accelerates erosion thereby leading to a reduction in the total amount of soil cover available for ground water storage in the catchments.

By biological manipulations along the slopes of the hilly watersheds, it is possible to increase water flow in the streams and also prolong the duration of flow. It is claimed that afforestation will increase the availability of water in downstream areas. This theory is however debatable since hydrological response depends on a wide range of physical conditions in an area. The most likely impact on the hydrological parameters, if an existing forest cover is removed, is shown in Figure 13.2. Where deforestation is followed by severe land degradation, some reduction in the volume and duration of dry season flow may ultimately be produced.

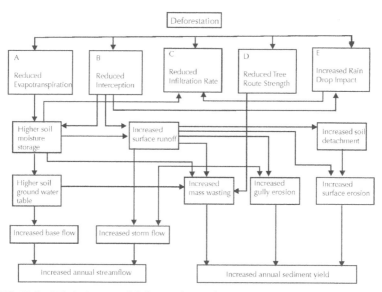

FIGURE 13.2 Likely impact of deforestation on hydrological parameters.

It is generally stated that forest cover has no influence on the total amount of rainfall, but that it can exert an influence on the intensity of rainfall and on the number of rainy days and that a decrease of forest cover will lead to a decrease in rainy days and increase the intensity. This will have its impact on erosion. In places where clearing of forests for farming was done, it resulted in a 50% reduction of light rainfall events though the total rainfall remained the same. There are some indications in Tamil Nadu state that there has been a decrease in summer rainfall during the last century. Although its relation to deforestation has not been clearly established, it is possible that summer rainfall, which is often of local origin, has indeed become lower because of the reduction in evapotranspiration due to less vegetative cover.

13.3   WATERSHED AND WATERSHED DEVELOPMENT

13.3.1   WHAT IS A WATERSHED?

Part of the rain water that falls on the ground is absorbed by the soil and the rest of it flows, as runoff, into small gullies and rivulets and joins the streams to form a river system ultimately. This forms the natural drainage system which, consists of the river basin at the macro level and the water-shed or sub-watershed at the micro level. A watershed can be defined as an area from which runoff resulting from precipitation, flows past a single point into a stream, a river, a lake or an ocean. The terms watershed, catchment or drainage basin convey the same meaning. A watershed may be only a few hectares in extent as in the case of small ponds or streams or even hundreds of square kilometers as in the case of rivers or big reservoirs. Based on the size, watersheds are classified as regions, basins, catchments, sub-catchment watershed or sub-watershed, mini and micro watersheds. As each watershed or sub-watershed is an independent hydrological unit by itself, any modification in land use in it will be reflected on the water and sediment yields from it. For more details, the reader is advised to consult Chapters 9 and 12.

13.3.2 WATERSHED DEVELOPMENT

The aim of watershed development is to combine various technologies and apply them in the watershed to achieve optimal development of land, water, and plant resources to meet at least the minimum basic needs of the people in a sustained manner. In the past, many development programs like afforestation soil and water conservation and rural water supplies were implemented in isolation to suit the needs of the individual program. As a result, the benefits of the programs did not last long after implementation. On the contrary, when various programs are integrated and implemented in unison in a watershed, the results are spectacular. Maheswaram watershed in Andhra Pradesh, *KabilNalla* watershed in Karnataka, and Ayyalur watershed in Tamil Nadu are examples for the success of integrated approach.

The concept of treating land for development whether by afforestation, soil and water conservation, or water harvesting is a recent one. In the past, many of the schemes, not executed on watershed basis, were adversely affected because the runoff and soil erosion were not controlled from the top to the bottom of the watershed. From the lessons learnt in the past, now most of the land development programs are implemented taking the watershed as the unit of treatment, irrespective of the size of the watershed.

An entire watershed must be developed in any land development program. Hence, the size of the proposed watershed for treatment must be determined taking into account the following factors:

- Resources available,
- existing infrastructure,
- availability of labor, and
- time schedule.

Past experience indicates that about 500–1500 ha in forest areas and about 1000–5000 ha in agricultural lands will be optimal.

13.3.3   INTERFACE FORESTRY AND WATERSHED DEVELOPMENT

Interface forestry, besides regeneration of degraded forests, has as its objectives, such as soil conservation, water harvesting, fodder, and cattle development, buffer-zone development, enhancing employment opportunities and overall development of the quality of life of the local people. If the treatment is based on watershed basis, there will be overall development of the treated watershed. This is the very reason why a watershed, whether it is sub or mini is taken as the unit of treatment in interface forestry program. In many other reforestation/afforestation program areas, based on target and not on watershed, the watershed is still the unit of treatment.

13.3.4   SELECTION OF WATERSHED

In Tamil Nadu, it is estimated that about 0.38 million ha of recorded forest area is degraded. The present interface forestry program, which is a component of the *SIDA assisted Social Forestry Project*, plans to treat only a fraction of these degraded forests with downstream communities which will participate in the development activities and benefit from them. The program does not restrict its activities to the reserved forest alone but extends them to the abutting buffer zone also, comprising revenue and private lands. The watershed selected may consist of upper (eco-restoration), middle (asset creation), and lower (interface) slopes with an abutting buffer zone. The criteria considered are selecting watersheds for treatments are from the satellite data/remote sensing, it is possible to prioritize the watersheds as very high, high, medium, low, and very low for development projects.

13.4   SOIL EROSION AND SEDIMENTATION

Soil, which is the disintegrated and weathered layer forming earth's crust, is exposed to atmospheric forces particularly wind and water. The loosening of the soil from its place and its transportation to another place by atmospheric forces is called soil erosion.

Soil erosion may take place due to one or more causes that are listed below:

- Destruction of the natural protective vegetal cover,
- indiscriminate felling of trees,
- over grazing leading to degradation of vegetal cover,
- forest fires, and
- improper land use like faulty agronomic practices and crop selection.

13.4.1 EROSION PROCESS

The soil erosion is of two types: geologic erosion (also known as natural or normal erosion) and accelerated erosion. Failure to distinguish between them has led to failure to comprehend the seriousness of erosion-related damages and the importance of soil conservation measures.

Geologic erosion represents the erosion when the land is in its natural and undisturbed environment under the cover of good vegetation. It includes both soil forming and eroding processes. A certain amount of erosion does take place even under a good vegetal cover but it is such a slow process and is compensated for by the formation of soil under the natural process. Hence it is not of great consequence.

Accelerated erosion is man's creation. Indiscriminate removal of the protective vegetal cover upsets the natural balance that exists between the soil, vegetation, and climate. Removal of surface soil takes place at a much faster rate than it can be built up by the natural soil-forming process. Similarly, faulty cultivation practices or land use may lead to abnormal erosion. Erosion caused by such activities is accelerated erosion and is dependent on the velocity of the agencies like wind and water.

13.4.2 TYPES OF EROSION

- Rain drop or splash erosion,
- sheet erosion,
- rill erosion,
- gully erosion,
- stream erosion,
- landslide erosion,
- erosion by waves, and
- wind erosion.

13.4.3   EROSIVITY AND ERODIBILITY

Erosion is the interaction of two entities namely, the atmospheric agencies like wind, water, and soil. On the same land, one storm may cause more erosion than another. Similarly, the same storm may cause more erosion in one field than in another. While the effect of atmospheric agencies like wind and water is called erosivity, the effect of soil is termed as erodibility. Figure 13.3 indicates the effects of soil erosion and the inter-relationship that exists among the erosion problems.

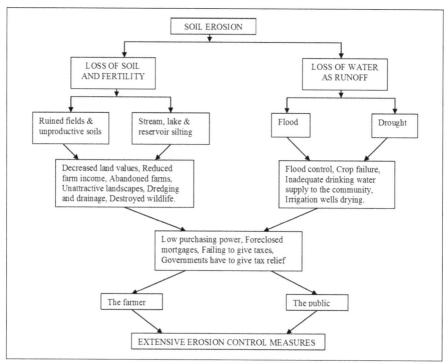

FIGURE 13.3   Effects of soil erosion.

Universal soil loss equation (USLE) by Schwab et al.[7–9,10] is given below is defined below:

$$A = RKLSCP \qquad\qquad (13.4)$$

where A = annual average soil loss (Mg/ha), R = rainfall and erosivity index for geographic location, K = soil erodibility factor, L = slope length factor,

S = slope steepness factor, C = cropping factor, and P = conservation practice factor. Values of parameters on RKS of this equation should be for local conditions.

13.4.3.1 SOIL ERODIBILITY FACTOR (K):

$$K = 2.8 \times 10^{-7}M^{1.14}(12 - a) + 4.3 \times 10^{-3}(b - 2) + 3.3 \times 10^{-3}(c - 3) \quad (13.5)$$

where, M = particle size parameter (% silt + % very fine sand) × (100 − % clay), a = % organic matter, b = soil structure code, c = profile permeability.

13.4.3.2 CONSERVATION PRACTICE FACTOR (P):

$$P = P_c \times P_s \times P_t \quad (13.6)$$

where P_c = contouring factor based on slope, P_s = strip cropping factor for crop strip widths recommended in (1.0 for contouring only or for alternating strips of corn with small grain; 0.75 for 4-year rotation with 2 years of row crop; 0.50 for 4-year rotation with 1 year of row crop), and P_t = terrace sedimentation factor (1.0 for no terraces; 0.2 for terraces with graded channel sod outlets, 0.1 for terraces with underground outlets).

13.4.3.3 SLOPE STEEPNESS FACTOR (S):

$S = 3.0(\sin\theta)^{0.8} + 0.56$, for slopes <4-m long

$S = 10.8(\sin\theta) + 0.03$, for slopes >4-m long, $s < 9\%$

$S = 16.8(\sin\theta) - 0.50$, for slopes >4-m long, $s > 9\%$ $\quad (13.7)$

Slope length is measured from the point where surface flow originates (usually the top of the ridge) to the outlet channel or a point downstream where deposition begins.

The USLE is a widely used mathematical model that describes soil erosion processes. Erosion models play critical roles in soil and water resource conservation and nonpoint source pollution assessments, including: sediment load assessment and inventory, conservation planning and design for sediment control, and for the advancement of scientific understanding.

The USLE or one of its derivatives are main models used by United States government agencies to measure water erosion [USDA-SCS now USDA-NRSC]. The USLE was developed in the United States based on soil erosion data collected beginning in the 1930s by the US Department of Agriculture (USDA) Soil Conservation Service. The model has been used for decades for purposes of conservation planning both in the United States where it originated and around the world. The *Revised Universal Soil Loss Equation (RUSLE)* and the *Modified Universal Soil Loss Equation (MUSLE)* also continue to be used for similar purposes.[4-6]

13.4.4 SEDIMENTATION

13.4.4.1 SOIL AND NUTRIENT LOSSES

An analysis of the soil erosion in India reveals that about 5333 million tons (16.33 t/ha) of top soil are detached annually and about 24% of this quantity is carried by rivers as sediments and deposited in the sea. Nearly 10% is deposited in reservoirs reducing their storage capacity by 1–2%. Annually plant nutrients to the tune of 5.37 million tons valued at about Rs. 10,000 millions are lost.

13.4.4.2 SEDIMENTATION CONTROL

There are following two basic approaches for controlling sedimentation:

Controlling soil erosion from the drainage area: Maintaining a good vegetal cover on the watershed is an effective way of preventing sedimentation. This can be achieved by preserving the existing natural forests, afforestation of degraded open lands and reforestation of degraded forests. In addition to these measures, suitable agronomic practices and engineering measures can also be taken up. These are described in detail in Chapters 3–5.

Handling of sediments: Sedimentation can be minimized by adjusting the operations in such a way that water is drawn off from the locations when the sediments are still in suspension but this is a difficult process. Once the sediments are deposited in a reservoir, it is extremely difficult to dispose them. In some special cases, they can be removed by excavation, dredging, draining, and flushing but these operations are very expensive and cumbersome.

13.5   SOIL AND WATER CONSERVATION MEASURES

Soil conservation is the prevention of soil from erosion or reduced fertility caused by overuse, acidification, salinization, or other chemical soil contamination. Slash-and-burn and other unsustainable methods of subsistence farming are practiced in some lesser developed areas.[4] A sequel to the deforestation is typically large scale erosion, loss of soil nutrients and sometimes total desertification. Techniques for improved soil conservation include crop rotation, cover crops, conservation tillage, and planted windbreaks and affect both erosion and fertility. When plants, especially trees, die, they decay and became part of the soil.

Water conservation encompasses the policies, strategies, and activities to manage fresh water as a sustainable resource, to protect the water environment, and to meet current and future human demand. Population, household size, and growth and affluence all affect how much water is used.[5] Factors such as climate change will increase pressures on natural water resources especially in manufacturing and agricultural irrigation.[11] The goals of water conservation efforts include: (1) To ensure availability for future generations, the withdrawal of fresh water from an ecosystem should not exceed its natural replacement rate; (2) energy conservation: water pumping, delivery, and waste-water treatment facilities consume a significant amount of energy. In some regions of the world over 15% of total electricity consumption is devoted to water management; (3) habitat conservation: minimizing human water use helps to preserve fresh fire habitats for local wildlife and migrating waterfowl, as well as reducing the need to build new dams and other water diversion infrastructures.

A watershed (sub/mini/micro) should always be the unit of treatment. To have a lasting effect, execution of works must commence from the upper reaches and end in the lower reaches. Soil and water conservation measures can be adopted at the macro and micro levels to conserve as much of rainfall as possible in the soil profile and reduce the quantity and velocity of surface runoff. This section deals with measures at macro level. Figure 13.4 represents typical soil and water conservation measures that can be adopted in a watershed.

13.5.1   *CONTOUR TRENCHING*

Contour trenches are suggested for hill slopes, degraded lands, and barren lands. They break the slope at intervals, trap moisture and sediments, and reduce the velocity of surface runoff. Also, they are suitable for all slopes and varying rainfall conditions and soil types. These can be

- continuous,
- interrupted,
- in series, and
- staggered.

Continuous contour trenches are suggested for low rainfall areas and require extreme care in laying. Interrupted contour trenches in series or staggered are suited for high rainfall zones. The commonly adopted dimensions are as follows:

Cross section: 1000–2500 sq. cm in area
Shape: square—30-cm wide and 30-cm deep
        Trapezoidal—30-cm wide at bottom with 1:1 side slope.
Length: 5–300 m
Horizontal interval: 15–30 m depending on slope.

Figure 13.4 shows, respectively, the cross section, layout, and location of contour trenches, respectively. Following considerations should be taken account about contour trenches:

- The horizontal or vertical interval between the contour trenches depends on rainfall, slope, and depth of soil.
- If the intensity and total rainfall are high, the large number of trenches has to be dug.
- In shallow soils, if the depth of the trenches is restricted then comparatively large number of trenches has to be dug.
- No gradient should be given in the catchment (contour) to the trenches.
- Add cross ties at every 10–15 m to further restrict the movement of water but the height of the cross ties should be 15–20 lower than the surface.
- After fixing the horizontal interval, lay the contour lines using a leveling instrument like Ghat Tracer, U frame with water tubes popularly called "Pinky-Pinky" or hand level. Hand level is the most convenient instrument of all these for use in forest areas. The work should be started from the top (ridge).

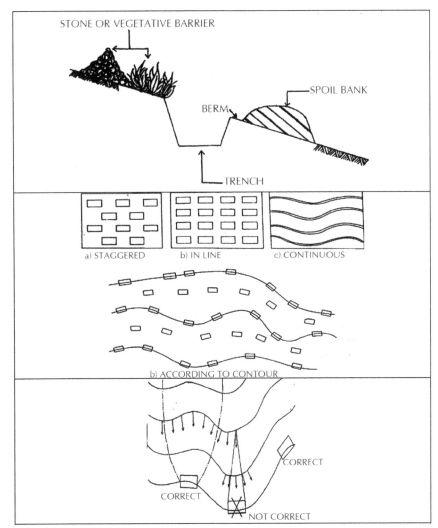

FIGURE 13.4   Contour trenches: Top—Cross section of a contour trench; **center**—layout of different contour trenches; and **bottom**—proper location of trenches.

Contour stone walls or stone barriers: Using the stones lying scattered in the area, walls to a height of 30–45 cm are constructed at 15–30-m horizontal intervals along the contour lines. A shallow foundation to a depth of 15–20 cm is dug, the dug up earth piled on the upstream side of the wall as spoil bank and stones available locally are carefully packed. Agave, vetiver, grasses, or legumes can be planted on the spoil bank (Fig. 13.5).

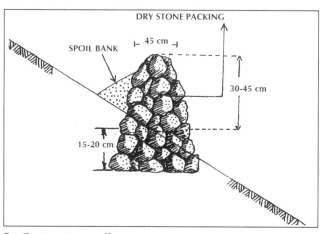

FIGURE 13.5   Contour stone wall.

Contour/Graded bunds are small embankments constructed along the contours and suited for red soils and low rainfall zones. Graded bunds are small embankments constructed with a grade (slope) of 0.1–0.3% toward the outlets. They are recommended for clayey soils and high rainfall zones.

Vegetative barriers: The efficacy and cheapness of contour vegetative barriers have been amply demonstrated in Maheswaram Watershed Development Project in Andhra Pradesh. Plants like Vetiver, Malabar lemon Grass, Lemon Grass, or Agave are planted at a close spacement along contours at suitable horizontal intervals. Vetiver is recommended for gentle slopes. Staggered planting of Agave at 25-cm spacing, in three rows is suggested for steep slopes. Besides conserving soil and moisture, the plants yield the much needed non-wood forest produce. Figure 13.6 shows the layout of contour vegetative barriers.

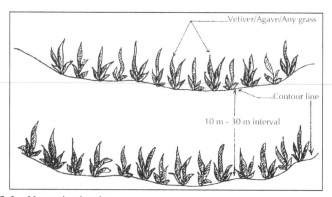

FIGURE 13.6   Vegetative barriers.

13.5.2   TEMPORARY GULLY CONTROL MEASURES

Gully erosion generally starts as a small rill, gradually develops into a deep crevice. If left unchecked ends up as a ravine, it damages land resources and contributes a large amount of sediment load to the river system. Temporary gully control structures can be described as follows:

Brushwood: Locally available vegetative cuttings can be used in their construction either as a single post row or as double post row.

Loose stone check dams: In loose stone check dams, relatively small rocks are placed across the gully or nalla. The main objective of these dams is to control channel erosion along the gully bed and to stop waterfall erosion by stabilizing the gully heads. The maximum height is 0.75 m and its foundation is about half of the height. The thickness of the dam at the spill way is about 0.75 m and the inclination of its downstream face is 20%. The upstream is vertical. Numerous dams can be constructed if stones are available (Fig. 13.7).

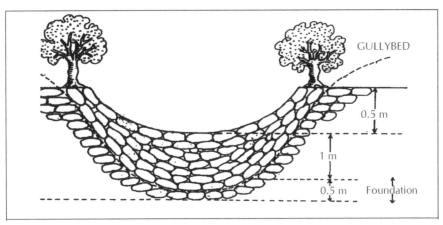

FIGURE 13.7   Front view of loose stone check dam.

Boulder check dams: Boulder check dams constructed across the gully are intended mainly to control channel erosion and to stabilize gully heads. The maximum total height of the dam is about 2 m and the foundation height at least half of the effective height. The width at the top is 0.75–1.0 m and at the bottom about 1.5–1.7 m. The upstream face is usually in vertical (Fig. 13.8).

The site selected for loose stone boulder check dam should be such that the banks of the gully are at a higher level. The slope of the dam should be as shown in Figure 13.8, so that the runoff water passes only through the middle portion of the dam and on any account, not on either side. Big stones should be placed at the center of the dam and small stones should not be, placed at the top as loose stones since they are likely to be washed away. It can be constructed at 50–200-m intervals depending on the need and the site conditions of the gully. Necessary permanent check dams can be provided after five loose stone/boulder check dams have been constructed based on the need and the site location.

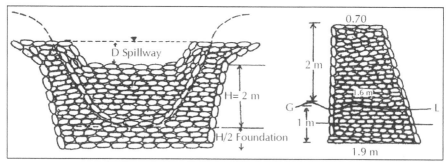

FIGURE 13.8   Boulder check dam (front view: left and section: right).

Rough stone with earth backing check dams: To reduce the cost and to increase the stability and life of the check dams, rough stone dry packing of sufficient top (60 cm), and bottom widths (1–1.5 m) can be constructed. To give support, earth backing of sufficient width may be provided in front of the dam (Fig. 13.9).

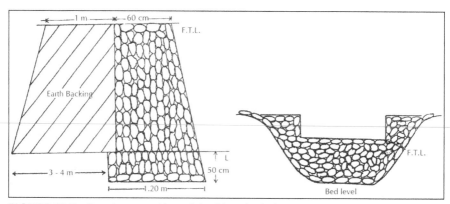

FIGURE 13.9   Rough stone with earth backing check dams.

13.5.3 PERMANENT GULLY CONTROL STRUCTURES

These are constructed in medium to large gullies. They help to protect the gullies from further development and also to store water for some time.

Gabions are wire-mesh boxes, which are mostly rectangular in shape and filled with stones. The sizes of fill stones are kept larger than the wire-mesh size, to avoid any spilling out of the stone barrier from the interstices of the wire net. Gabions were first developed in Italy and are now in common use as soil erosion control structures. They are flexible to handle, permeable, economical to construct (being only sacks full of stones), and do not create any thrust load problems. Thus, these structures are very popular throughout the world, particularly for the management of forest watersheds. Galvanized wires are generally used for constructing the mesh, because they are rust-proof and have a longer life. Being flexible, they bend without breaking at uneven settlements. Being permeable, they help relieve the hydrostatic pressure of groundwater. Gabions are most commonly used to control soil erosion caused by actions of torrents. They are also used in prevention of landslides and landslips, control of soil erosion within the gullies and at gully heads, and protection of the banks of streams from erosion. For these types of control measures, gabions are used in the following forms: (1) retaining walls, (2) spurs, (3) flexible aprons, (4) revetments, and (5) channel lining.

Gabion retaining walls are preferred in those areas where the foundation consists of unstable materials, and stones are available in sufficient quantities. Gabion walls are erected to retain to the soil mass which is likely to slide down, to protect the earth banks of torrents from getting eroded (Fig. 13.10), and to stabilize the debris (stream bank stabilization).

Gabion drop structures: Gabions are extensively used as drop structures or check dams, especially in forest and hilly watersheds, where stones are available in plenty. Gabion boxes are used for constructing drop structures on gullies and torrents. Similar to the design of conventional drop structures, aprons made of gabions are used for drops more than 1 m. Sometimes, to protect the gabions, the drop structure is plastered with a 5-cm thick cement mortar. The design procedure for gabions drop structures is similar to that for conventional drop structures.

Gabion aprons: In gullies, the bed slopes and side slopes vary with up and down levels. A flexible apron is likely to be the ideal solution for such situations. Aprons made of 1/3–1/2-m thick gabions, are flexible, and well suited to such situations. The length of the apron varies with the depth of the scour. Generally, the length is kept about twice the depth of the scour. A graded stone material is generally recommended for filling of gabions. A graded material is less permeable, and therefore less susceptible to seepage through it.

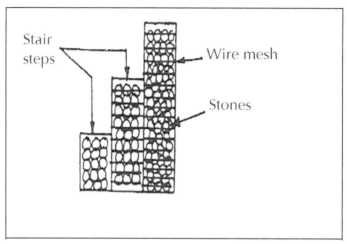

FIGURE 13.10   A gabion retaining wall with stair steps.

13.6   CHECK DAMS AND PERCOLATION PONDS

13.6.1   *SIMILARITY AND DIFFERENCE BETWEEN A CHECK DAM AND PERCOLATION POND*

Both discharge the functions of checking soil erosion and reducing the velocity of surface runoff. While a check dam is not intended to ·store large quantity of water, a percolation pond is normally designed to store about one-third of annual yield from its catchment.

13.6.2   *CHECK DAMS*

Temporary check dams are cheap and constructed with locally available materials like brush wood and stones. There are no designs for their construction. The different types are vegetative/brushwood check dam, loose-stone check dam, boulder check dam, and rough stone with earth backing check dam.

Permanent check dams are masonry ones and have to be specially designed to suit the watersheds. Based on the area of the watershed, its condition and the rainfall, the weir (length of the dam) is designed to dispose off the runoff. Utmost care should be exercised in selecting the site to reduce the cost. A check dam is designed on the same lines adopted for designing the weir of a percolation pond.

13.6.3   PERCOLATION PONDS

Following considerations should be taken into account in site selection:

- It may be, as far as possible, located in government or poromboke lands.
- Within about one kilometer radius, which is normally its zone of influence, there should not be any exiting minor tanks or other water storage systems.
- A number of wells should exist within 1-km radius of its location.
- Heavy soils and impervious strata must be avoided. The top soil must be highly porous.
- Suitable and adequate soil for forming embankment should be available within the pond area.
- Narrow portion of the steam with high ground on either side is preferable.
- The catchment area and the number of fillings available for the pond in a year decide its size (normally two fillings).

13.6.3.1   PROCEDURE TO SET UP A POND

a. Gathering of basic necessary information includes: Topographical map, field map, village map, data on the number of wells within 1-km radius, and rainfall data like total monsoon rainfall, intensity of rainfall and distribution.

b. The weir must be aligned perpendicular to the direction of stream flow.

c. The area of the catchment, its slope and density of vegetation are assessed.

d. A contour map is prepared after a detailed contour survey of the pond site using 10 m × 10 m grids.

e. The annual yield is estimated using Strange's table. The annual yield is calculated based on the monsoon rainfall and nature of catchment: good, average, or bad: See example in this chapter.

f. The capacity of the pond is designed for holding 1/3 of the annual yield of the catchment. Normally, it is assumed that the pond gets two fillings during the monsoon.

g. Using the contour map of the pond site, the full tank level (FTL) is decided based on the designed storage capacity.

h. Based on peak flood discharge and availability of adequate length for surplus arrangements, the maximum water level (MWL) is decided.

i. The height of flow over the crest of weir is taken as 30–60 cm.

j. Depending on the storage, the tank bund level (TBL) is taken up to the highest point available on either side of the stream.

$$MWL = FTL + (0.30–0.60 \text{ m})\qquad\qquad\qquad (13.8)$$

$$TBL = MWL + 1 \text{ m (minimum)}\qquad\qquad\qquad (13.9)$$

k. Using the appropriate formula, the following parameters are calculated:

- weir length,
- base width of body wall,
- depth of foundation,
- length of apron,
- thickness of apron,
- top and bottom widths of abutment and its height,
- top and bottom widths, height, and length of upstream return wall, and
- Top and bottom widths, height, and length of downstream return wall.

l. After completing the design of pond, the civil works are executed.

m. *Note:* Design must be carried by an agricultural engineer.

13.6.3.2 SILT DETENTION TANKS

These are constructed to trap and detain silt and reduce siltation of downstream reservoirs. The design and construction of permanent check dam, percolation pond, and silt detention tanks are more or less the same.

13.7 MICRO-LEVEL (IN SITU) SOIL AND MOISTURE CONSERVATION MEASURES

In addition to the various soil and water conservation structures at the watershed level described in this chapter, certain simple and cost-effective soil and moisture conservation measures can be implemented at micro level in the forest areas and in the agricultural lands of the adjoining buffer zone. They are described briefly in this section (Fig. 13.11).

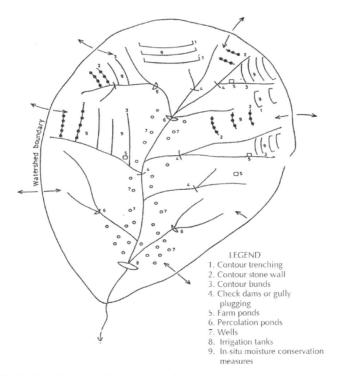

LEGEND
1. Contour trenching
2. Contour stone wall
3. Contour bunds
4. Check dams or gully plugging
5. Farm ponds
6. Percolation ponds
7. Wells
8. Irrigation tanks
9. In-situ moisture conservation measures

FIGURE 13.11   Treatments in a watershed.

13.7.1 FOREST AREAS

13.7.1.1 MICRO-CATCHMENTS

Along gentle slopes and plains, the area is divided into a number of square shaped micro-catchments with the help of simple earthen bunds. Along steep slopes, the micro-catchments are V-shaped. Each micro-catchment will have a tree in it.

13.7.1.2 SEMICIRCULAR, CRESCENT-SHAPED BUNDS AND SAUCER BASINS

These measures are less costly than micro-catchments. In sloppy lands, just below the sapling, bunds are formed in a semicircular or preferably crescent shape. Stones, if locally available, can be used for bunding. Or else, earthen bunds can be formed. In plain areas, saucers are formed around the saplings with earthen bunds on the periphery of the saucers.

13.7.1.3 SUNKEN PLANTING

While planting the seedlings/cuttings, instead of filling the pits up to the ground level, the top 10–15 cm of the pits are left unfilled. This method will help to trap more rain water and conserve moisture for a longer duration than flush planting. This can be easily practiced when the pits are 45-cm deep and polypots are 30-cm or less tall.

13.7.1.4 CATCH PITS

Along sloppy terrains and in areas with rocky outcrops, pits of 1 m × 1 m × 1 m or of smaller sizes (if the soil is shallow) may be dug and left as such. These are called catch pits. No regular spacing needs be followed. During the rain, runoff water gets collected, and silt gets deposited in these pits. After some time (6–12 months) when the pits get filled with silt, seedlings, or cuttings are planted in these catch pits.

13.7.1.5 TIE-RIDGING

Holes, 10–30 cm diameter, are dug to a depth of 10–15 cm at spacements varying from 50 to 100 cm depending on the space available. The holes can even be square or rectangular.

13.7.1.6 V-DITCH PLANTING

At 4–6-m intervals, V-shaped ditches are dug along the contour using machine or animal or human power. An earthen bund is formed below the ditch leaving 20–25-cm space (berm), that is, on the downhill side and tree seedlings are planted in these ditches at desired intervals. This method has been successfully employed in Maheswaram watershed development project in Andhra Pradesh.

13.7.2 AGRICULTURAL LANDS (BUFFER ZONE)

13.7.2.1 CONTOUR RIDGES AND FURROWS

Ridges are formed along contour lines spaced conveniently, with furrows on the uphill side of ridges. These enhance the availability of moisture to the crops.

137.2.2 BROAD BED AND FURROWS

Beds of about 1.5-m width are formed with furrows on the sides to enhance and prolong moisture availability to crops raised on the beds.

13.7.2.3 FORMING BASINS

Rectangular basins are formed across the slope to store rain water. The size depends on the slope, soil type, and rainfall.

13.7.2.4 WATER SPREADING

During rainy season, water from the stream is diverted and spread on the lands situated on either side of it to prolong and increase the moisture availability to the crops.

13.7.2.5 TIE RIDGING

This method has already been described above in this chapter.

13.7.2.6   FIELD BUNDING

The fields, if large, can be divided into compartments and bunds formed around the compartments. If the fields are small, the height of existing bunds can be increased.

13.7.2.7   FARM PONDS

This is slightly costly. Ponds of size 8 m × 8 m with a depth of 1.5 m are dug at the rate of one pond for every 1–2 ha depending on the soil type and rainfall. In porous soils, the sides, and bottom may be lined.

13.7.2.8   STRIP PLOUGHING

For barren lands or lands that have been left fallow for years, this method has been found useful, particularly for agroforestry. Strips of 2–3-m width are ploughed at convenient intervals. Tree seedlings are planted at the desired spacing along the row, in the middle of the ploughed strips. The runoff from the unploughed area gets partly stored in the ploughed area improving moisture availability.

13.7.3   VEGETATIVE BARRIERS

Vegetative barriers of Vetiver or any other suitable grass or Agave can be raised along the contours in forest areas and agricultural lands.

13.7.4   AN ADVICE BY THE AUTHOR

As far as possible, the in situ soil and moisture conservation measures should be carried out before the on-set of rains to derive the maximum benefit. This principle is equally applicable for the construction of soil conservation structures at the watershed level also.

13.8 SOIL AND WATER CONSERVATION MEASURES FOR DIFFERENT ZONES OF WATERSHEDS

In Joint Forest Management program, the watershed is developed by roughly dividing it into:

- upper zone (eco restoration),
- middle zone (asset creation),
- lower zone (utilization), and
- buffer zone (private agricultural fields and porombokes).

Soil and water conservation measures for each specific zone and common for the entire watershed are indicated below. The suggestions need not be rigidly followed and the measures that are to be executed in a zone depend on the local conditions. The local field staff is the best judge to decide the measures to be adopted.

13.8.1 UPPER ZONE (ECO-RESTORATION ZONE)

- Contour trenches (preferably interrupted and staggered) at about 15-m intervals,
- contour stone walls,
- vegetative barriers along contour, and
- catch pits (60 cm × 60 cm × 60 cm or less depending on space available and soil depth).

13.8.2 MIDDLE ZONE (ASSET CREATION ZONE)

- Contour trenches at 15–20-m intervals depending on slope,
- contour stone wall,
- vegetative barriers along contour,
- temporary check dams,
- catch pits (1 m × 1 m × 1 m or less depending on soil depth), if rocky outcrops occur,
- crescent shaped/semicircular bunds, and
- micro-catchment, square or V-shaped depending on slope.

13.8.3 LOWER ZONE (INTERFACE ZONE)

- Contour trenches at about 30-m intervals,
- vegetative barriers along contour,
- check dams—temporary and permanent,
- micro-catchments,
- saucer basin,
- crescent shaped bunds, and
- percolation ponds.

13.8.4 BUFFER ZONE

- Check dams
- percolation pond,
- micro-catchments,
- saucer basins,
- tie ridging,
- broadbed and furrows,
- field bunding,
- water spreading, and
- Farm ponds.

13.8.5 COMMON MEASURES

- Temporary check dams:
  a. brushwood,
  b. loose stone,
  c. boulder,
  d. rough stone with earth backing, and
  e. masonry check dams, for every 5–10 temporary check dams.

13.9 INSTRUMENTS AND EQUIPMENTS NEEDED FOR SOIL AND WATER CONSERVATION WORKS

Soil and water conservation measure and water harvesting need various instruments and a variety of techniques for gathering necessary data and executing the works. Such instructions and techniques are listed below:

13.9.1   ALIGNMENT OF CONTOUR LINES

- Ghat tracer,
- wooden frames with water filled tube (also called pinky–pinky)/a frame, and
- hand level.

13.9.2   METEOROLOGICAL EQUIPMENT

- Rain gauge to measure rainfall. It may be the ordinary or non-recording type or automatic or recording type,
- anemometer to measure wind force and speed, and
- maximum and minimum thermometers to measure temperatures,
- psychrometer to measure humidity,
- evaporimeter to measure evaporation, and
- automatic weather station. This is a sophisticated system which measures and records temperature, rainfall, wind speed and direction, relative humidity, and solar radiation.

13.9.3   STREAM/CHANNEL FLOW MEASUREMENTS

- V notch,
- weir,
- Parshall flume,
- current meter, and
- automatic stage level recorder.

13.9.4   GROUND WATER MEASUREMENTS

- Automatic water level indicator, and
- piezometer tube.

13.9.5   SOIL MOISTURE MEASUREMENTS

- Tensiometer (irrometer),
- gypsum block,

- neutron probe,
- pressure membrane and pressure plate apparatus,
- infiltrometer, and
- oven.

KEYWORDS

- catchment area
- check dam
- forest hydrology
- hydrology
- macro-watershed
- micro-watershed
- moisture conservation
- percolation pond
- rainfall
- runoff
- sedimentation
- soil and water conservation
- soil erosion
- soil moisture
- tensiometer
- time of concentration
- watershed
- watershed development

REFERENCES

1. Davis, S. N.; De Wiest, R. J. M. Hydrogeology. Wiley: New York, 1966.
2. http://www.ilri.org/InfoServ/Webpub/fulldocs/IntegratedWater/IWMI/Documents/related_doucments/HTML/x5524e/x5524e04.htm.
3. http://www.lmnoeng.com/Hydrology/rational.php.
4. https://en.wikipedia.org/wiki/Soil_conservation.
5. https://en.wikipedia.org/wiki/Water_conservation.
6. https://www.youtube.com/watch?v=T05djitkEFI.

7. Huffman, R. L.; Fangmeier, D. D.; Elliot, W. J.; Workman, S. R.; Schwab, G. O. *Soil and Water Conservation Engineering*, sixth ed. American Society of Agricultural & Biological Engineers (ASABE): St Joseph, MI, 2011.

8. Michael, A. M. *Irrigation, Theory and Practice*. South Asia Books, 1999.

9. Panigrahi, B. *A Handbook on Irrigation and Drainage*. New India Publishing Agency, 2013.

10. Schwab, G. O. *Manual of Soil and Water Conservation Engineering*. W. C. Brown Company, 1952.

11. USPS—USA. *United States Postal Stamp Advocating Water Conservation*. United States Postal Service and United States—USDA—Soil Conservation Service: Washington, DC, USA, no date.

APPENDIX 13.1   US POSTAL STAMP ADVOCATING WATER CONSERVATION

APPENDICES

(Modified and reprinted with permission from: Goyal, Megh R., 2012. Appendices. Pages 317–332. In: *Management of Drip/Trickle or Micro Irrigation* edited by Megh R. Goyal. New Jersey, USA: Apple Academic Press Inc.)

APPENDIX A CONVERSION SI AND NON-SI UNITS

To convert the Column 1 in the Column 2 Multiply by	Column 1 Unit SI	Column 2 Unit Non-SI	To convert the Column 2 in the Column 1 Multiply by

LINEAR

0.621 _____	kilometer, km (10^3m)	miles, mi _____	1.609
1.094 _____	meter, m	yard, yd _____	0.914
3.28 _____	meter, m	feet, ft _____	0.304
3.94×10^{-2} ___	millimeter, mm (10^{-3})	inch, in _____	25.4

SQUARES

2.47 _____	hectare, he	acre _____	0.405
2.47 _____	square kilometer, km^2	acre _____	4.05×10^{-3}
0.386 _____	square kilometer, km^2	square mile, mi^2 _____	2.590
2.47×10^{-4} ___	square meter, m^2	acre _____	4.05×10^{-3}
10.76 _____	square meter, m^2	square feet, ft^2 _____	9.29×10^{-2}
1.55×10^{-3} ___	mm^2	square inch, in^2 _____	645

CUBICS

9.73×10^{-3} ___	cubic meter, m^3	inch-acre _____	102.8
35.3 _____	cubic meter, m^3	cubic-feet, ft^3 _____	2.83×10^{-2}

6.10×10^4 ____ cubic meter, m³	cubic inch, in³ _____ 1.64×10^{-5}
2.84×10^{-2} ____ liter, L (10^{-3} m³)	bushel, bu _____ 35.24
1.057 _____ liter, L	liquid quarts, qt _____ 0.946
3.53×10^{-2} ____ liter, L	cubic feet, ft³ _____ 28.3
0.265 _____ liter, L	gallon _____ 3.78
33.78 _____ liter, L	fluid ounce, oz _____ 2.96×10^{-2}
2.11 _____ liter, L	fluid dot, dt _____ 0.473

WEIGHT

2.20×10^{-3} ____ gram, g (10^{-3} kg)	pound, _____ 454
3.52×10^{-2} ____ gram, g (10^{-3} kg)	ounce, oz _____ 28.4
2.205 _____ kilogram, kg	pound, lb _____ 0.454
10^{-2} _____ kilogram, kg	quintal (metric), q ___ 100
1.10×10^{-3} ____ kilogram, kg	ton (2000 lbs), ton ___ 907
1.10^2 _____ mega gram, mg	ton (US), ton _____ 0.907
1.10^2 _____ metric ton, t	ton (US), ton _____ 0.907

YIELD AND RATE

0.893 _____ kilogram per hectare hectare	pound per acre _____ 1.12
7.77×10^{-2} ___ kilogram per cubic meter	pound per fanega ____ 12.87
1.49×10^{-2} ___ kilogram per hectare	pound per acre, _____ 67.19 60 lb
1.59×10^{-2} ___ kilogram per hectare	pound per acre, _____ 62.71 56 lb
1.86×10^{-2} ___ kilogram per hectare	pound per acre, _____ 53.75 48 lb
0.107 _____ liter per hectare	galloon per acre _____ 9.35
893 _____ ton per hectare	pound per acre _____ 1.12×10^{-3}
893 _____ mega gram per hectare	pound per acre _____ 1.12×10^{-3}

0.446 _____ ton per hectare   ton (2000 lb) per _____ 2.24
                                          acre

2.24 _____ meter per second   mile per hour _____ 0.447

SPECIFIC SURFACE

10 _____ square meter per   square centimeter ____ 0.1
               kilogram   per gram

10^3 _____ square meter per   square millimeter ____ 10^{-3}
               kilogram   per gram

PRESSURE

9.90 _____ megapascal, MPa   atmosphere _____ 0.101

10 _____ megapascal   bar _____ 0.1

1.0 _____ megagram per   gram per cubic _____ 1.00
                cubic meter   cubic centimeter

2.09×10^{-2} ___ pascal, Pa   pound per square ____ 47.9
                                          feet

1.45×10^{-4} ___ pascal, Pa   pound per square ____ 6.90×10^3
                                          inch

To convert the column 1 in the Column 2, Multiply by	Column 1 Unit SI	Column 2 Unit Non-SI	To convert the column 2 in the column 1 Multiply by

TEMPERATURE

1.00 _____ Kelvin, K   centigrade, °C _____ 1.00
(K-273)   (C+273)

(1.8 C _____ centigrade, °C   Fahrenheit,°F _____ (F-32)/1.8
+ 32)

ENERGY

9.52×10^{-4} ___ Joule J

0.239 _____ Joule, J

0.735 _____ Joule, J

2.387×10^5 ___ Joule per square meter

10^5 _____ Newton, N

BTU _____ 1.05×103

calories, cal _____ 4.19

feet-pound _____ 1.36

calories per square ___ 4.19×10^4 centimeter

dynes _____ 10^{-5}

WATER REQUIREMENTS

9.73×10^{-3} ___ cubic meter

9.81×10^{-3} ___ cubic meter per hour

4.40 _____ cubic meter per hour

8.11 _____ hectare-meter

97.28 _____ hectare-meter

8.1×10^{-2} ____ hectare centimeter

inch acre _____ 102.8

cubic feet per _____ 101.9 second

galloon (US) per _____ 0.227 minute

acre-feet _____ 0.123

acre-inch _____ 1.03×10^{-2}

acre-feet _____ 12.33

CONCENTRATION

1 _____ centimol per kilogram

0.1 _____ gram per kilogram

1 _____ milligram per kilogram

milliequivalents _____ 1 per 100 grams

percents _____ 10

parts per million _____ 1

NUTRIENTS FOR PLANTS

2.29 _____ P

1.20 _____ K

1.39 _____ Ca

1.66 _____ Mg

P_2O_5 _____ 0.437

K_2O _____ 0.830

CaO _____ 0.715

MgO _____ 0.602

NUTRIENT EQUIVALENTS

Column A	Column B	Conversion A to B	Equivalent B to A
N	NH_3	1.216	0.822
	NO_3	4.429	0.226
	KNO_3	7.221	0.1385
	$Ca(NO_3)_2$	5.861	0.171
	$(NH_4)_2SO_4$	4.721	0.212
	NH_4NO_3	5.718	0.175
	$(NH_4)_2 \cdot HPO_4$	4.718	0.212
P	P_2O_5	2.292	0.436
	PO_4	3.066	0.326
	KH_2PO_4	4.394	0.228
	$(NH_4)_2 \cdot HPO_4$	4.255	0.235
	H_3PO_4	3.164	0.316
K	K_2O	1.205	0.83
	KNO_3	2.586	0.387
	KH_2PO_4	3.481	0.287
	KCl	1.907	0.524
	K_2SO_4	2.229	0.449
Ca	CaO	1.399	0.715
	$Ca(NO_3)_2$	4.094	0.244
	$CaCl_2 \cdot 6H_2O$	5.467	0.183
	$CaSO_4 \cdot 2H_2O$	4.296	0.233
Mg	MgO	1.658	0.603
	$MgSO_4 \cdot 7H_2O$	1.014	0.0986
S	H_2SO_4	3.059	0.327
	$(NH_4)_2SO_4$	4.124	0.2425
	K_2SO_4	5.437	0.184
	$MgSO_4 \cdot 7H_2O$	7.689	0.13
	$CaSO_4 \cdot 2H_2O$	5.371	0.186

Friction Loss (m per 100 m Length of Main Line) of Portable Aluminum Pipe with Couplings: Based on Scobey's Formula, for K_S = 10 m.

Flow		Pipe diameter (cm)						
L	GPM	7.5	10	12.5	15	17.5	20	25
2.52	40	0.658	0.157	–	–	–	–	–
3.15	50	1.006	0.239	–	–	–	–	–
3.79	60	1.423	0.339	–	–	–	–	–
4.42	70	1.906	0.449	0.150	–	–	–	–
5.05	80	2.457	0.584	0.193	–	–	–	–
5.68	90	3.073	0.731	0.242	–	–	–	–
6.31	100	3.754	0.893	0.295	0.120	–	–	–
7.57	120	5.307	1.263	0.413	0.170	–	–	–
8.83	140	7.113	1.693	0.560	0.227	–	–	–
10.10	160	9.169	2.182	0.721	0.293	–	–	–
11.36	180	11.47	2.729	0.967	0.366	–	–	–
12.62	200	14.01	3.333	1.102	0.448	0.209	–	–
13.88	220	16.79	3.996	1.321	0.537	0.251	–	–
15.14	240	19.81	4.713	1.558	0.633	0.296	–	–
16.41	260	23.06	5.448	1.814	0.737	0.344	–	–
17.67	280	26.55	6.316	2.089	0.849	0.397	–	–
18.93	300	30.27	7.203	2.381	0.967	0.452	0.235	–
20.19	320	34.22	8.142	2.692	1.094	0.511	0.265	–
21.45	340	38.39	9.137	3.020	1.227	0.573	0.298	–
22.72	360	42.80	10.18	3.366	1.368	0.639	0.332	–
23.98	380	47.43	11.29	3.731	1.516	0.708	0.368	–
25.24	400	52.28	12.44	4.113	1.671	0.781	0.399	0.136
26.50	420	–	13.95	4.513	1.833	0.857	0.445	0.149
27.76	440	–	14.57	4.930	1.988	0.936	0.486	0.163
29.03	460	–	16.23	5.364	2.179	1.019	0.529	0.177
30.29	480	–	17.59	5.815	2.363	1.104	0.573	0.192
31.55	500	–	19.01	6.284	2.554	1.193	0.620	0.208
34.70	550	–	22.79	7.532	3.060	1.430	0.742	0.249
37.86	600	–	26.88	9.886	3.611	1.687	0.876	0.294
41.01	650	–	31.30	10.35	4.204	1.965	1.020	0.342

Flow		Pipe diameter (cm)						
L	GPM	7.5	10	12.5	15	17.5	20	25
44.17	700	–	36.03	11.91	4.839	2.262	1.174	0.394
47.32	750	–	41.08	13.58	5.517	2.520	1.339	0.449
50.48	800	–	–	15.35	6.237	2.915	1.513	0.507
53.60	850	–	–	17.32	6.999	3.71	1.698	0.569
56.79	900	–	–	19.20	7.801	3.646	1.893	0.635
59.94	950	–	–	21.28	8.645	4.041	2.097	0.703
63.10	1000	–	–	23.45	9.530	4.454	2.312	0.775
69.49	1100	–	–	28.11	11.42	5.338	2.771	0.929
75.72	1200	–	–	31.75	13.58	6.298	3.269	1.096
82.03	1300	–	–	–	15.69	7.333	3.806	1.277
88.34	1400	–	–	–	18.06	8.441	4.382	1.470
94.65	1500	–	–	–	20.59	9.624	4.996	1.675
101.0	1600	–	–	–	23.28	10.88	5.648	1.894
107.3	1700	–	–	–	26.12	21.21	6.337	2.125
14.0	1800	–	–	–	–	13.61	7.064	2.369
120.0	1900	–	–	–	–	15.08	7.829	2.625
126.0	2000	–	–	–	–	16.62	8.630	2.894

Friction Loss (m per 100-m Length of Lateral Lines) of Portable Aluminum Pipe with Couplings: Based on Scobey's Formula.

Flow (L)	Pipe diameter (cm)				
	5.0	7.5	10	12.5	15
	$K_S = 0.34$	$K_S = 0.33$		$K_S = 0.32$	
1.26	–	–	–	–	–
1.89	0.32	–	–	–	–
2.52	2.53	–	–	–	–
3.15	4.40	0.565	0.130	–	–
3.79	6.85	0.858	0.198	–	–
4.42	9.67	1.21	0.280	–	–
5.05	12.9	1.63	0.376	0.122	–
5.68	16.7	2.10	0.484	0.157	–

Flow (L)	Pipe diameter (cm)				
	5.0	7.5	10	12.5	15
	$K_S = 0.34$	$K_S = 0.33$		$K_S = 0.32$	
6.31	20.8	2.63	0.605	0.196	–
7.57	25.4	3.20	0.738	0.240	0.099
8.83	–	4.54	1.04	0.339	0.140
10.10	–	6.09	1.40	0.454	0.188
11.36	–	7.85	1.80	0.590	0.242
12.62	–	9.82	2.26	0.733	0.302
13.88	–	12.0	2.76	0.896	0.370
15.14	–	14.4	3.30	1.07	0.443
16.41	–	16.9	3.90	1.26	0.522
17.67	–	19.7	4.54	1.47	0.608
18.93	–	22.8	5.22	1.70	0.700
20.19	–	25.9	5.96	1.93	0.798
21.45	–	29.3	6.74	2.18	0.904
22.72	–	32.8	7.56	2.45	1.02
23.98	–	36.6	8.40	2.74	1.13
25.24	–	40.6	9.36	3.03	1.26
26.50	–	44.7	10.3	3.34	1.38
27.76	–	–	11.3	3.66	1.521
29.03	–	–	12.3	4.00	1.66
30.29	–	–	13.4	4.35	1.80
31.55	–	–	14.6	4.72	1.95
34.70	–	–	15.8	5.10	2.12
37.86	–	–	18.9	6.12	2.52
41.01	–	–	22.2	7.22	2.98
44.17	–	–	25.9	8.40	3.46
47.32	–	–	29.8	9.68	3.99
50.48	–	–	33.8	11.0	4.54
53.63	–	–		12.5	5.15
56.79	–	–		14.0	5.78
59.94	–	–		15.6	6.44
63.10	–	–		17.3	7.14

APPENDIX D PSYCHOMETRIC CONSTANT (γ) FOR DIFFERENT ALTITUDES (Z)

$$\gamma = 10^{-3} \left[\frac{(C_p \cdot P)}{(\varepsilon \cdot \lambda)} \right] = (0.00163) \times \left[\frac{P}{\lambda} \right]$$

γ, psychrometric constant [kPa C^{-1}]
c_p, specific heat of moist air = 1.013 [kJ kg^{-10}C^{-1}]
P, atmospheric pressure [kPa]

ε, ratio molecular weight of water vapor/dry air = 0.622
λ, latent heat of vaporization [MJ kg^{-1}]
= 2.45 MJ kg^{-1} at 20°C

z (m)	γ (kPa/°C)	z (m)	γ (kPa/°C)	z (m)	γ (kPa/°C)	z (m)	γ (kPa/°C)
0	0.067	1000	0.060	2000	0.053	3000	0.047
100	0.067	1100	0.059	2100	0.052	3100	0.046
200	0.066	1200	0.058	2200	0.052	3200	0.046
300	0.065	1300	0.058	2300	0.051	3300	0.045
400	0.064	1400	0.057	2400	0.051	3400	0.045
500	0.064	1500	0.056	2500	0.050	3500	0.044
600	0.063	1600	0.056	2600	0.049	3600	0.043
700	0.062	1700	0.055	2700	0.049	3700	0.043
800	0.061	1800	0.054	2800	0.048	3800	0.042
900	0.061	1900	0.054	2900	0.047	3900	0.042
1000	0.060	2000	0.053	3000	0.047	4000	0.041

APPENDIX E SATURATION VAPOR PRESSURE [E_s] FOR DIFFERENT TEMPERATURES (T)

Vapor pressure function = e_s = [0.6108] × exp{[17.27 × T]/[T + 237.3]}							
T (°C)	e_s (kPa)	T (°C)	e_s (kPa)	T (°C)	e_s (kPa)	T (°C)	e_s (kPa)
1.0	0.657	13.0	1.498	25.0	3.168	37.0	6.275
1.5	0.681	13.5	1.547	25.5	3.263	37.5	6.448
2.0	0.706	14.0	1.599	26.0	3.361	38.0	6.625
2.5	0.731	14.5	1.651	26.5	3.462	38.5	6.806
3.0	0.758	15.0	1.705	27.0	3.565	39.0	6.991
3.5	0.785	15.5	1.761	27.5	3.671	39.5	7.181

APPENDIX C   PERCENTAGE OF DAILY SUNSHINE HOURS: FOR NORTH AND SOUTH HEMISPHERES.

Latitude	Jan.	Feb.	Mar.	Apr.	May	Jun.	Jul.	Aug.	Sep.	Oct.	Nov.	Dec.
North												
0	8.50	7.66	8.49	8.21	8.50	8.22	8.50	8.49	8.21	8.50	8.22	8.50
5	8.32	7.57	8.47	3.29	8.65	8.41	8.67	8.60	8.23	8.42	8.07	8.30
10	8.13	7.47	8.45	8.37	8.81	8.60	8.86	8.71	8.25	8.34	7.91	8.10
15	7.94	7.36	8.43	8.44	8.98	8.80	9.05	8.83	8.28	8.20	7.75	7.88
20	7.74	7.25	8.41	8.52	9.15	9.00	9.25	8.96	8.30	8.18	7.58	7.66
25	7.53	7.14	8.39	8.61	9.33	9.23	9.45	9.09	8.32	8.09	7.40	7.52
30	7.30	7.03	8.38	8.71	9.53	9.49	9.67	9.22	8.33	7.99	7.19	7.15
32	7.20	6.97	8.37	8.76	9.62	9.59	9.77	9.27	8.34	7.95	7.11	7.05
34	7.10	6.91	8.36	8.80	9.72	9.70	9.88	9.33	8.36	7.90	7.02	6.92
36	6.99	6.85	8.35	8.85	9.82	9.82	9.99	9.40	8.37	7.85	6.92	6.79
38	6.87	6.79	8.34	8.90	9.92	9.95	10.1	9.47	3.38	7.80	6.82	6.66
40	6.76	6.72	8.33	8.95	10.0	10.1	10.2	9.54	8.39	7.75	6.72	7.52
42	6.63	6.65	8.31	9.00	10.1	10.2	10.4	9.62	8.40	7.69	6.62	6.37
44	6.49	6.58	8.30	9.06	10.3	10.4	10.5	9.70	8.41	7.63	6.49	6.21
46	6.34	6.50	8.29	9.12	10.4	10.5	10.6	9.79	8.42	7.57	6.36	6.04
48	6.17	6.41	8.27	9.18	10.5	10.7	10.8	9.89	8.44	7.51	6.23	5.86
50	5.98	6.30	8.24	9.24	10.7	10.9	11.0	10.0	8.35	7.45	6.10	5.64
52	5.77	6.19	8.21	9.29	10.9	11.1	11.2	10.1	8.49	7.39	5.93	5.43

Latitude	Jan.	Feb.	Mar.	Apr.	May	Jun.	Jul.	Aug.	Sep.	Oct.	Nov.	Dec.
North												
54	5.55	6.08	8.18	9.36	11.0	11.4	11.4	10.3	8.51	7.20	5.74	5.18
56	5.30	5.95	8.15	9.45	11.2	11.7	11.6	10.4	8.53	7.21	5.54	4.89
58	5.01	5.81	8.12	9.55	11.5	12.0	12.0	10.6	8.55	7.10	4.31	4.56
60	4.67	5.65	8.08	9.65	11.7	12.4	12.3	10.7	8.57	6.98	5.04	4.22
SOUTH												
0	8.50	7.66	8.49	8.21	8.50	8.22	8.50	8.49	8.21	8.50	8.22	8.50
5	8.68	7.76	8.51	8.15	8.34	8.05	8.33	8.38	8.19	8.56	8.37	8.68
10	8.86	7.87	8.53	8.09	8.18	7.86	8.14	8.27	8.17	8.62	8.53	8.88
15	9.05	7.98	8.55	8.02	8.02	7.65	7.95	8.15	8.15	8.68	8.70	9.10
20	9.24	8.09	8.57	7.94	7.85	7.43	7.76	8.03	8.13	8.76	8.87	9.33
25	9.46	8.21	8.60	7.74	7.66	7.20	7.54	7.90	8.11	8.86	9.04	9.58
30	9.70	8.33	8.62	7.73	7.45	6.96	7.31	7.76	8.07	8.97	9.24	9.85
32	9.81	8.39	8.63	7.69	7.36	6.85	7.21	7.70	8.06	9.01	9.33	9.96
34	9.92	8.45	8.64	7.64	7.27	6.74	7.10	7.63	8.05	9.06	9.42	10.1
36	10.0	8.51	8.65	7.59	7.18	6.62	6.99	7.56	8.04	9.11	9.35	10.2
38	10.2	8.57	8.66	7.54	7.08	6.50	6.87	7.49	8.03	9.16	9.61	10.3
40	10.3	8.63	8.67	7.49	6.97	6.37	6.76	7.41	8.02	9.21	9.71	10.5
42	10.4	8.70	8.68	7.44	6.85	6.23	6.64	7.33	8.01	9.26	9.8	10.6
44	10.5	8.78	8.69	7.38	6.73	6.08	6.51	7.25	7.99	9.31	9.94	10.8
46	10.7	8.86	8.90	7.32	6.61	5.92	6.37	7.16	7.96	9.37	10.1	11.0

Mean Daily Maximum Duration of Bright Sunshine Hours (n) for Different Months and Latitudes.

North–South	Jan.–July	Feb.–Aug	Mar–Sept.	April–Oct.	May–Nov.	June–Dec.	July–Jan.	Aug.–Feb.	Sept.–Mar	Oct.–April	Nov.–May	Dec.–June
50	8.5	10.1	11.8	13.8	15.4	16.3	15.9	14.5	12.7	10.8	9.1	8.1
48	8.8	10.2	11.8	13.6	15.2	16.0	15.6	14.3	12.6	10.9	9.3	8.3
46	9.1	10.4	11.9	13.5	14.9	15.7	15.4	14.2	12.6	10.9	9.5	8.7
44	9.3	10.5	11.9	13.4	14.7	15.4	15.2	14.0	12.6	11.0	9.7	8.9
42	9.4	10.6	11.9	13.4	14.6	15.2	14.9	13.9	12.6	11.1	9.8	9.1
40	9.6	10.7	11.9	13.3	14.4	15.0	14.7	13.7	12.5	11.2	10.0	9.3
35	10.1	11.0	11.9	13.1	14.0	14.5	14.3	13.5	12.4	11.3	10.3	9.8
30	10.4	11.1	12.0	12.9	13.6	14.0	13.9	13.2	12.4	11.5	10.6	10.2
25	10.7	11.3	12.0	12.7	13.3	13.7	13.5	13.0	12.3	11.6	10.9	10.6
20	11.0	11.5	12.0	12.6	13.1	13.3	13.2	12.8	12.3	11.7	11.2	10.9
15	11.3	11.6	12.0	12.5	12.8	13.0	12.9	12.6	12.2	11.8	11.4	11.2
10	11.6	11.8	12.0	12.3	12.6	12.7	12.6	12.4	12.1	11.8	11.6	11.5
5	11.8	11.9	12.0	12.2	12.3	12.4	12.3	12.3	12.1	12.0	11.9	11.8
0	12.1	12.1	12.1	12.1	12.1	12.1	12.1	12.1	12.1	12.1	12.1	12.1

Mean Daily Percentage (P) of Annual Daytime Hours for Different Latitudes.

Latitude North–South	Jan.–July	Feb.–Aug.	March–Sept.	April–Oct.	May–Nov.	June–Dec.	July–Jan.	Aug.–Feb.	Sept.–March	Oct.–April	Nov.–May	Dec.–June
60°	0.15	0.20	0.26	0.32	0.38	0.41	0.40	0.34	0.28	0.22	0.17	0.13
58°	0.16	0.21	0.26	0.32	0.37	0.40	0.39	0.34	0.28	0.23	0.18	0.15
56°	0.17	0.21	0.26	0.32	0.36	0.39	0.38	0.33	0.28	0.23	0.18	0.16
54°	0.18	0.22	0.26	0.31	0.36	0.38	0.37	0.33	0.28	0.23	0.19	0.17
52°	0.19	0.22	0.27	0.31	0.35	0.37	0.36	0.33	0.28	0.24	0.20	0.17
50°	0.19	0.23	0.27	0.31	0.34	0.36	0.35	0.32	0.28	0.24	0.20	0.18
48°	0.20	0.23	0.27	0.31	0.34	0.36	0.35	0.32	0.28	0.24	0.21	0.19
46°	0.20	0.23	0.27	0.30	0.34	0.35	0.34	0.32	0.28	0.24	0.21	0.20
44°	0.21	0.24	0.27	0.30	0.33	0.35	0.34	0.31	0.28	0.25	0.22	0.20
42°	0.21	0.24	0.27	0.30	0.33	0.34	0.33	0.31	0.28	0.25	0.22	0.21
40°	0.22	0.24	0.27	0.30	0.32	0.34	0.33	0.31	0.28	0.25	0.22	0.21
35°	0.23	0.25	0.27	0.29	0.31	0.32	0.32	0.30	0.28	0.25	0.23	0.22
30°	0.24	0.25	0.27	0.29	0.31	0.32	0.31	0.30	0.28	0.26	0.24	0.23*
25°	0.24	0.26	0.27	0.29	0.30	0.31	0.31	0.29	0.28	0.26	0.25	0.24
20°	0.25	0.26	0.27	0.28	0.29	0.30	0.30	0.29	0.28	0.26	0.25	0.25
15°	0.26	0.26	0.27	0.28	0.29	0.29	0.29	0.28	0.28	0.27	0.26	0.25
10°	0.26	0.27	0.27	0.28	0.28	0.29	0.29	0.28	0.28	0.27	0.26	0.26
5°	0.27	0.27	0.27	0.28	0.28	0.28	0.28	0.28	0.28	0.27	0.27	0.27
0°	0.27	0.27	0.27	0.27	0.27	0.27	0.27	0.27	0.27	0.27	0.27	0.27

Vapor pressure function = e_s = [0.6108] × exp{[17.27 × T]/[T + 237.3]}							
T (°C)	e_s (kPa)	T (°C)	e_s (kPa)	T (°C)	e_s (kPa)	T (°C)	e_s (kPa)
4.0	0.813	16.0	1.818	28.0	3.780	40.0	7.376
4.5	0.842	16.5	1.877	28.5	3.891	40.5	7.574
5.0	0.872	17.0	1.938	29.0	4.006	41.0	7.778
5.5	0.903	17.5	2.000	29.5	4.123	41.5	7.986
6.0	0.935	18.0	2.064	30.0	4.243	42.0	8.199
6.5	0.968	18.5	2.130	30.5	4.366	42.5	8.417
7.0	1.002	19.0	2.197	31.0	4.493	43.0	8.640
7.5	1.037	19.5	2.267	31.5	4.622	43.5	8.867
8.0	1.073	20.0	2.338	32.0	4.755	44.0	9.101
8.5	1.110	20.5	2.412	32.5	4.891	44.5	9.339
9.0	1.148	21.0	2.487	33.0	5.030	45.0	9.582
9.5	1.187	21.5	2.564	33.5	5.173	45.5	9.832
10.0	1.228	22.0	2.644	34.0	5.319	46.0	10.086
10.5	1.270	22.5	2.726	34.5	5.469	46.5	10.347
11.0	1.313	23.0	2.809	35.0	5.623	47.0	10.613
11.5	1.357	23.5	2.896	35.5	5.780	47.5	10.885
12.0	1.403	24.0	2.984	36.0	5.941	48.0	11.163
12.5	1.449	24.5	3.075	36.5	6.106	48.5	11.447

APPENDIX F   SLOPE OF VAPOR PRESSURE CURVE (Δ) FOR DIFFERENT TEMPERATURES (T)

$$\Delta = \frac{\left[4098 \times e^0 (T)\right]}{[T + 237.3]^2}$$

$$= 2504 \frac{\left\{\exp\left[(17.27T)/(T + 237.2)\right]\right\}}{[T + 237.3]^2}$$

T (°C)	Δ (kPa/°C)	T (°C)	Δ (kPa/°C)	T (°C)	Δ (kPa/°C)	T (°C)	Δ (kPa/°C)
1.0	0.047	13.0	0.098	25.0	0.189	37.0	0.342
1.5	0.049	13.5	0.101	25.5	0.194	37.5	0.350
2.0	0.050	14.0	0.104	26.0	0.199	38.0	0.358
2.5	0.052	14.5	0.107	26.5	0.204	38.5	0.367

T (°C)	Δ (kPa/°C)	T (°C)	Δ (kPa/°C)	T (°C)	Δ (kPa/°C)	T (°C)	Δ (kPa/°C)
3.0	0.054	15.0	0.110	27.0	0.209	39.0	0.375
3.5	0.055	15.5	0.113	27.5	0.215	39.5	0.384
4.0	0.057	16.0	0.116	28.0	0.220	40.0	0.393
4.5	0.059	16.5	0.119	28.5	0.226	40.5	0.402
5.0	0.061	17.0	0.123	29.0	0.231	41.0	0.412
5.5	0.063	17.5	0.126	29.5	0.237	41.5	0.421
6.0	0.065	18.0	0.130	30.0	0.243	42.0	0.431
6.5	0.067	18.5	0.133	30.5	0.249	42.5	0.441
7.0	0.069	19.0	0.137	31.0	0.256	43.0	0.451
7.5	0.071	19.5	0.141	31.5	0.262	43.5	0.461
8.0	0.073	20.0	0.145	32.0	0.269	44.0	0.471
8.5	0.075	20.5	0.149	32.5	0.275	44.5	0.482
9.0	0.078	21.0	0.153	33.0	0.282	45.0	0.493
9.5	0.080	21.5	0.157	33.5	0.289	45.5	0.504
10.0	0.082	22.0	0.161	34.0	0.296	46.0	0.515
10.5	0.085	22.5	0.165	34.5	0.303	46.5	0.526
11.0	0.087	23.0	0.170	35.0	0.311	47.0	0.538
11.5	0.090	23.5	0.174	35.5	0.318	47.5	0.550
12.0	0.092	24.0	0.179	36.0	0.326	48.0	0.562
12.5	0.095	24.5	0.184	36.5	0.334	48.5	0.574

APPENDIX G   NUMBER OF THE DAY IN THE YEAR (JULIAN DAY).

Day	Jan.	Feb.	Mar.	Apr.	May	Jun.	Jul.	Aug.	Sep.	Oct.	Nov.	Dec.
1	1	32	60	91	121	152	182	213	244	274	305	335
2	2	33	61	92	122	153	183	214	245	275	306	336
3	3	34	62	93	123	154	184	215	246	276	307	337
4	4	35	63	94	124	155	185	216	247	277	308	338
5	5	36	64	95	125	156	186	217	248	278	309	339
6	6	37	65	96	126	157	187	218	249	279	310	340
7	7	38	66	97	127	158	188	219	250	280	311	341
8	8	39	67	98	128	159	189	220	251	281	312	342
9	9	40	68	99	129	160	190	221	252	282	313	343
10	10	41	69	100	130	161	191	222	253	283	314	344
11	11	42	70	101	131	162	192	223	254	284	315	345

Day	Jan.	Feb.	Mar.	Apr.	May	Jun.	Jul.	Aug.	Sep.	Oct.	Nov.	Dec.
12	12	43	71	102	132	163	193	224	255	285	316	346
13	13	44	72	103	133	164	194	225	256	286	317	347
14	14	45	73	104	134	165	195	226	257	287	318	348
15	15	46	74	105	135	166	196	227	258	288	319	349
16	16	47	75	106	136	167	197	228	259	289	320	350
17	17	48	76	107	137	168	198	229	260	290	321	351
18	18	49	77	108	138	169	199	230	261	291	322	352
19	19	50	78	109	139	170	200	231	262	292	323	353
20	20	51	79	110	140	171	201	232	263	293	324	354
21	21	52	80	111	141	172	202	233	264	294	325	355
22	22	53	81	112	142	173	203	234	265	295	326	356
23	23	54	82	113	143	174	204	235	266	296	327	357
24	24	55	83	114	144	175	205	236	267	297	328	358
25	25	56	84	115	145	176	206	237	268	298	329	359
26	26	57	85	116	146	177	207	238	269	299	330	360
27	27	58	86	117	147	178	208	239	270	300	331	361
28	28	59	87	118	148	179	209	240	271	301	332	362
29	29	(60)	88	119	149	180	210	241	272	302	333	363
30	30	–	89	120	150	181	211	242	273	303	334	364
31	31	–	90	–	151	–	212	243	–	304	–	365

APPENDIX H   STEFAN–BOLTZMANN LAW AT DIFFERENT TEMPERATURES (T)

$[\sigma \times (T_K)^4] = [4.903 \times 10^{-9}]$ (MJ K^{-4} m^{-2} day^{-1})

where $T_K = \{T\,[°C] + 273.16\}$

T	$\sigma \times (T_K)^4$	T	$\sigma \times (T_K)^4$	T	$\sigma \times (T_K)^4$
UNITS					
(°C)	(MJ m^{-2} d^{-1})	(°C)	(MJ m^{-2} d^{-1})	(°C)	(MJ m^{-2} d^{-1})
1.0	27.70	17.0	34.75	33.0	43.08
1.5	27.90	17.5	34.99	33.5	43.36
2.0	28.11	18.0	35.24	34.0	43.64
2.5	28.31	18.5	35.48	34.5	43.93
3.0	28.52	19.0	35.72	35.0	44.21

T	$\sigma \times (T_K)^4$	T	$\sigma \times (T_K)^4$	T	$\sigma \times (T_K)^4$
UNITS					
3.5	28.72	19.5	35.97	35.5	44.50
4.0	28.93	20.0	36.21	36.0	44.79
4.5	29.14	20.5	36.46	36.5	45.08
5.0	29.35	21.0	36.71	37.0	45.37
5.5	29.56	21.5	36.96	37.5	45.67
6.0	29.78	22.0	37.21	38.0	45.96
6.5	29.99	22.5	37.47	38.5	46.26
7.0	30.21	23.0	37.72	39.0	46.56
7.5	30.42	23.5	37.98	39.5	46.85
8.0	30.64	24.0	38.23	40.0	47.15
8.5	30.86	24.5	38.49	40.5	47.46
9.0	31.08	25.0	38.75	41.0	47.76
9.5	31.30	25.5	39.01	41.5	48.06
10.0	31.52	26.0	39.27	42.0	48.37
10.5	31.74	26.5	39.53	42.5	48.68
11.0	31.97	27.0	39.80	43.0	48.99
11.5	32.19	27.5	40.06	43.5	49.30
12.0	32.42	28.0	40.33	44.0	49.61
12.5	32.65	28.5	40.60	44.5	49.92
13.0	32.88	29.0	40.87	45.0	50.24
13.5	33.11	29.5	41.14	45.5	50.56
14.0	33.34	30.0	41.41	46.0	50.87
14.5	33.57	30.5	41.69	46.5	51.19
15.0	33.81	31.0	41.96	47.0	51.51
15.5	34.04	31.5	42.24	47.5	51.84
16.0	34.28	32.0	42.52	48.0	52.16
16.5	34,52	32.5	42.80	48.5	52.49

APPENDIX I   THERMODYNAMIC PROPERTIES OF AIR AND WATER

1. Latent heat of vaporization (λ)

$$\lambda = [2.501 - (2.361 \times 10^{-3})T]$$

where λ = latent heat of vaporization [MJ kg^{-1}]; and T = air temperature [°C].

The value of the latent heat varies only slightly over normal temperature ranges. A single value may be taken (for ambient temperature = 20°C): λ = 2.45 MJ kg^{-1}.

2. Atmospheric pressure (P)

$$P = P_0 \left[\frac{\{T_{K0} - \alpha(Z - Z_0)\}}{\{T_{K0}\}} \right]^{(g/(\alpha \cdot R))}$$

where P, atmospheric pressure at elevation z [kPa]

P_0, atmospheric pressure at sea level = 101.3 [kPa]

z, elevation [m]

z_0, elevation at reference level [m]

g, gravitational acceleration = 9.807 [m s^{-2}]

R, specific gas constant = 287 [J kg^{-1} K^{-1}]

α, constant lapse rate for moist air = 0.0065 [K m^{-1}]

T_{K0}, reference temperature [K] at elevation z_0 = 273.16 + T

T, means air temperature for the time period of calculation [°C]

When assuming P_0 = 101.3 [kPa] at z_0 = 0 and T_{Ko} = 293 [K] for T = 20 [°C], above equation reduces to

$$P = 101.3[(293 - 0.0065Z)(293)]^{5.26}$$

3. Atmospheric density (ρ)

$$\rho = \frac{[1000P]}{[T_{Kv}R]} = \frac{[3.486P]}{[T_{Kv}]}, \quad \text{and} \quad T_{Kv} = T_K \left[\frac{(1 - 0.378(e_a))}{P} \right]^{-1}$$

where ρ, atmospheric density [kg m^{-3}], R, specific gas constant = 287 [J kg^{-1} K^{-1}], T_{Kv}, virtual temperature [K], T_K, absolute temperature [K]: T_K = 273.16 + T [°C], e_a, actual vapor pressure [kPa], T, mean daily temperature for 24-h calculation time steps.

For average conditions (e_a in the range 1–5 kPa and P between 80 and 100 kPa), T_{Kv} can be substituted by: $T_{Kv} \approx 1.01(T + 273)$.

4. Saturation vapor pressure function (e$_s$)

$$e_s = [0.6108] \times \exp\{[17.27 \times T]/[T + 237.3]\}$$

where e_s, saturation vapor pressure function [kPa], T, air temperature [°C].

5. Slope vapor pressure curve (Δ)

$$\Delta = \frac{\left[4098 \cdot e^0(T)\right]}{\left[T+237.3\right]^2}$$

$$= 2504 \frac{\left\{\exp\left[(17.27T)/(T+237.2)\right]\right\}}{\left[T+237.3\right]^2}$$

where Δ, slope vapor pressure curve [kPa °C^{-1}], T, air temperature [°C], $e^0(T)$, saturation vapor pressure at temperature T [kPa]. In 24-h calculations, Δ is calculated using mean daily air temperature. In hourly calculations T refers to the hourly mean, T_h.

6. Psychrometric constant (γ)

$$\gamma = 10^{-3}\left[\frac{(C_p \cdot P)}{(\varepsilon \cdot \lambda)}\right] = (0.00163) \times \left[\frac{P}{\lambda}\right]$$

where γ, psychrometric constant [kPa C^{-1}], c_p, specific heat of moist air = 1.013 [kJ kg^{-10} °C^{-1}], P, atmospheric pressure [kPa]: Equations (2) or (4), ε, ratio molecular weight of water vapor/dry air = 0.622, λ, latent heat of vaporization [MJ kg^{-1}].

7. Dew point temperature (T_{dew})
When data is not available, T_{dew} can be computed from e_a by

$$T_{dew} = \left[\frac{\left\{116.91 + 237.3 Log_e(e_a)\right\}}{\left\{16.78 - Log_e(e_a)\right\}}\right]$$

where T_{dew}, dew point temperature [°C], e_a, actual vapor pressure [kPa], for the case of measurements with the Assmann psychrometer, T_{dew} can be calculated from

$$T_{dew} = (112 + 0.9T_{wet})\left[\frac{e_a}{(e^0 T_{wet})}\right]^{0.125} - [112 - 0.1T_{wet}]$$

8. Short-wave radiation on a clear-sky day (R_{so})

The calculation of R_{so} is required for computing net long wave radiation and for checking calibration of pyranometers and integrity of R_{so} data. A good approximation for R_{so} for daily and hourly periods is

$$R_{so} = (0.75 + 2 \times 10^{-5} z)R_a$$

where z, station elevation [m], R_a, extraterrestrial radiation [MJ m^{-2} d^{-1}], equation is valid for station elevations less than 6000 m having low air turbidity. The equation was developed by linearizing Beer's radiation extinction law as a function of station elevation and assuming that the average angle of the sun above the horizon is about 50°.

For areas of high turbidity caused by pollution or airborne dust or for regions where the sun angle is significantly less than 50° so that the path length of radiation through the atmosphere is increased, an adoption of Beer's law can be employed where P is used to represent atmospheric mass:

$$R_{so} = (R_a)\exp\left[\frac{(-0.0018P)}{(K_t \sin(\Phi))}\right]$$

where K_t, turbidity coefficient, $0 < K_t \leq 1.0$ where $K_t = 1.0$ for clean air and $K_t = 1.0$ for extremely turbid, dusty, or polluted air. P, atmospheric pressure [kPa], Φ, angle of the sun above the horizon [rad], and R_a, extraterrestrial radiation [MJ m^{-2} d^{-1}].

For hourly or shorter periods, Φ is calculated as

$$\sin\Phi = \sin\varphi\sin\delta + \cos\varphi\cos\delta\cos\omega$$

where φ, latitude [rad], δ, solar declination [rad] (Eq. (24) in Chapter 3), ω, solar time angle at midpoint of hourly or shorter period [rad].

For 24-h periods, the mean daily sun angle, weighted according to R_a, can be approximated as

$$\sin\left(\Phi_{24}\right) = \sin\left[0.85 + 0.3\varphi\sin\{(2\pi J / 365) - 1.39\} - 0.42\varphi^2\right]$$

where $\Phi_{24,}$ average Φ during the daylight period, weighted according to R_a [rad] φ, latitude [rad], J, day in the year.

The Φ_{24} variable is used to represent the average sun angle during daylight hours and has been weighted to represent integrated 24-h transmission effects on 24-h R_{so} by the atmosphere. Φ_{24} should be limited to ≥ 0. In some situations, the estimation for R_{so} can be improved by modifying to consider the effects of water vapor on short wave absorption, so that: $R_{so} = (K_B + K_D)R_a$ where:

$$K_B = 0.98\exp\left[\left\{\frac{(-0.00146P)}{(K_t \sin \Phi)}\right\} - 0.091\left\{\frac{w}{\sin \Phi}\right\}^{0.25}\right]$$

where K_B, the clearness index for direct beam radiation, K_D, the corresponding index for diffuse beam radiation

$K_D = 0.35 - 0.33 K_B$ for $K_B \geq 0.15$
$K_D = 0.18 + 0.82 K_B$ for $K_B < 0.15$
R_a, extraterrestrial radiation [MJ m^{-2} d^{-1}]
K_t, turbidity coefficient, $0 < K_t \leq 1.0$ where $K_t = 1.0$ for clean air and $K_t = 1.0$ for extremely turbid, dusty or polluted air.
P, atmospheric pressure [kPa]
Φ, angle of the sun above the horizon [rad]
W, perceptible water in the atmosphere [mm] $= 0.14e_aP + 2.1$
e_a, actual vapor pressure [kPa]
P, atmospheric pressure [kPa]

APPENDIX J

[<http://www.fao.org/docrep/T0551E/t0551e07.htm#5.5%20field%20 management%20practices%20in%20wastewater%20irrigation>]

1. **Relationship between applied water salinity and soil water salinity at different leaching fractions (FAO 1985)**

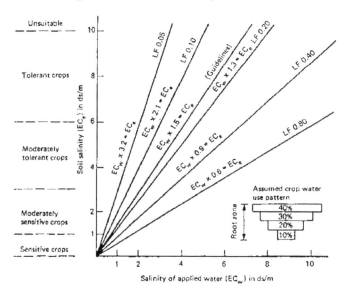

2. **Schematic representations of salt accumulation, planting positions, ridge shapes and watering patterns.**

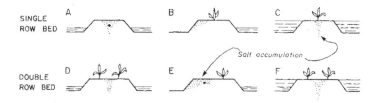

APPENDIX K   VALUES OF K_c FOR FIELD AND VEGETABLE CROPS FOR DIFFERENT CROP GROWTH STAGES AND PREVAILING CLIMATIC CONDITIONS.

Crop	Relative humidity				
		$RH_{min} > 70\%$		$RH_{min} < 20\%$	
	Crop stage	Wind speed (m/s)			
	Initial 1	0–5	5–8	0–5	5–8
	Crop development 2				
	Mid-season 3	Values of K_c			
	Late season/maturity 4	0.95	0.95	1.0	1.05
		0.9	0.9	0.95	1.0
Barley	3	1.05	1.1	1.15	1.2
	4	0.25	0.25	0.2	0.2
Beans (green)	3	0.95	0.95	1.0	1.05
	4	0.85	0.85	0.9	0.9
Beans, dry/pulses	3	1.05	1.1	1.15	1.2
	4	0.3	0.3	0.25	0.25
Beets	3	1.0	1.0	1.05	1.1
	4	0.9	0.9	0.95	1.0
Carrots	3	1.0	1.05	1.1	1.15
	4	0.7	0.75	0.8	0.85
Sweet corn (maize)	3	1.05	1.1	1.15	1.2
	4	0.95	1.0	1.05	1.1
Cotton	3	1.05	1.15	1.2	1.25
	4	0.65	0.65	0.65	0.7
Crucifers (cabbage, cauliflower, broccoli)	3	0.95	1.0	1.05	1.1
	4	0.80	0.85	0.9	0.95

Crop	Relative humidity				
	RH$_{min}$ > 70%			RH$_{min}$ < 20%	
	Crop stage	Wind speed (m/s)			
	Initial 1	0–5	5–8	0–5	5–8
	Crop development 2				
	Mid-season 3	Values of K$_c$			
	Late season/maturity 4	0.95	0.95	1.0	1.05
		0.9	0.9	0.95	1.0
Cucumber	3	0.9	0.9	0.95	1.0
	4	0.7	0.7	0.75	0.8
Lentil	3	1.05	1.1	1.15	1.2
	4	0.3	0.3	0.25	0.25
Melons	3	0.95	0.95	1.0	1.05
	4	0.65	0.65	0.75	0.75
Millet	3	1.0	1.05	1.1	1.15
	4	0.3	0.3	0.25	0.25
Oats	3	1.05	1.1	1.15	1.2
	4	0.25	0.25	0.2	0.2
Onion (dry)	3	0.95	0.95	1.05	1.1
	4	0.75	0.75	0.8	0.85
Onion (green)	3	0.95	0.95	1.0	1.05
	4	0.95	0.95	1.0	1.05
Peanuts (groundnut)	Mid-season 3	0.95	1.0	1.05	1.1
	Late season/maturity 4	0.55	0.55	0.6	0.6
Peas	3	1.05	1.1	1.15	1.2
	4	0.95	1.0	1.05	1.1
Potato	3	1.05	1.1	1.15	1.2
	4	0.7	0.7	0.75	0.75

Crop		Relative humidity				
		RH$_{min}$ > 70%		RH$_{min}$ < 20%		
	Crop stage	Wind speed (m/s)				
	Initial 1	0–5	5–8	0–5	5–8	
	Crop development 2					
	Mid-season 3	Values of K$_c$				
	Late season/maturity 4	0.95	0.95	1.0	1.05	
		0.9	0.9	0.95	1.0	
Radish	3	0.8	0.8	0.85	0.9	
	4	0.75	0.75	0.8	0.85	
Safflower	3	1.05	1.1	1.15	1.2	
	4	0.25	0.25	0.2	0.2	
Sorghum	3	1.0	1.05	1.1	1.15	
	4	0.5	0.5	0.55	0.55	
Soybeans	3	1.0	1.05	1.1	1.15	
	4	0.45	0.45	0.45	0.45	
Spinach	3	0.95	0.95	1.0	1.05	
	4	0.9	0.9	0.95	1.0	
Sugarbeet	3	1.05	1.1	1.15	1.2	
	4	0.9	0.95	1.0	1.0	
Sunflower	3	1.05	1.1	1.15	1.2	
	4	0.4	0.4	0.35	0.35	
Tomato	3	1.05	1.1	1.2	1.25	
	4	0.6	0.6	0.65	0.65	
Wheat	3	1.05	1.1	1.15	1.2	
	4	0.25	0.25	0.2	0.2	

Note: Values of K_c in this table are for field and vegetable crops; values of K_c for other crops are reported by Doorenbos and Pruitt (1977).

APPENDIX L   CROP TOLERANCE AND YIELD POTENTIAL OF CROPS AFFECTED BY IRRIGATION WATER SALINITY (EC_w) OR SOIL SALINITY (EC_E).

Field crops	100%		90%		75%		50%		0% Maximum	
	EC_e	EC_w	EC_e	EC_w	EC_e	EC_w	EC_e	EC_w	EC_e	EC_w
Barley (*Hordeum vulgare*)	8.0	5.3	10	6.7	13	8.7	18	12	28	19
Cotton (*Gossypium hirsutum*)	7.7	5.1	9.6	6.4	13	8.4	17	12	27	18
Sugarbeet (*Beta vulgaris*)	7.0	4.7	8.7	5.8	11	7.5	15	10	24	16
Sorghum (*Sorghum bicolor*)	6.8	4.5	7.4	5.0	8.4	5.6	9.9	6.7	13	8.7
Wheat (*Triticum aestivum*)	6.0	4.0	7.4	4.9	9.5	6.3	13	8.7	20	13
Wheat, durum (*Triticum turgidum*)	5.7	3.8	7.6	5.0	10	6.9	15	10	24	16
Soybean (*Glycine max*)	5.0	3.3	5.5	3.7	6.3	4.2	7.5	5.0	10	6.7
Cowpea (*Vigna unguiculata*)	4.9	3.3	5.7	3.8	7.0	4.7	9.1	6.0	13	8.8
Groundnut (Peanut) (*Arachis hypogaea*)	3.2	2.1	3.5	2.4	4.1	2.7	4.9	3.3	6.6	4.4
Rice (paddy) (*Oriza sativa*)	3.0	2.0	3.8	2.6	5.1	3.4	7.2	4.8	11	7.6
Sugarcane (*Saccharum officinarum*)	1.7	1.1	3.4	2.3	5.9	4.0	10	6.8	19	12
Corn (maize) (*Zea mays*)	1.7	1.1	2.5	1.7	3.8	2.5	5.9	3.9	10	6.7
Flax (*Linum usitatissimum*)	1.7	1.1	2.5	1.7	3.8	2.5	5.9	3.9	10	6.7
Broadbean (*Vicia faba*)	1.5	1.1	2.6	1.8	4.2	2.0	6.8	4.5	12	8.0
Bean (*Phaseolus vulgaris*)	1.0	0.7	1.5	1.0	2.3	1.5	3.6	2.4	6.3	4.2
Vegetables										
Squash, zucchini (courgette) (*Cucurbita pepo melopepo*)	4.7	3.1	5.8	3.8	7.4	4.9	10	6.7	15	10

Field crops	100%		90%		75%		50%		0% Maximum	
	EC_e	EC_w	EC_e	EC_w	EC_e	EC_w	EC_e	EC_w	EC_e	EC_w
Beet, red (*Beta vulgaris*)	4.0	2.7	5.1	3.4	6.8	4.5	9.6	6.4	15	10
Squash, scallop (*Cucurbita pepo melopepo*)	3.2	2.1	3.8	2.6	4.8	3.2	6.3	4.2	9.4	6.3
Broccoli (*Brassica oleracea botrytis*)	2.8	1.9	3.9	2.6	5.5	3.7	8.2	5.5	14	9.1
Tomato (*Lycopersicon esculentum*)	2.5	1.7	3.5	2.3	5.0	3.4	7.6	5.0	13	8.4
Cucumber (*Cucumis sativus*)	2.5	1.7	3.3	2.2	4.4	2.9	6.3	4.2	10	6.8
Spinach (*Spinacia oleracea*)	2.0	1.3	3.3	2.2	5.3	3.5	8.6	5.7	15	10
Celery (*Apium graveolens*)	1.8	1.2	3.4	2.3	5.8	3.9	9.9	6.6	18	12
Cabbage (*Brassica oleracea capitata*)	1.8	1.2	2.8	1.9	4.4	2.9	7.0	4.6	12	8.1
Potato (*Solanum tuberosum*)	1.7	1.1	2.5	1.7	3.8	2.5	5.9	3.9	10	6.7
Corn, sweet (maize) (*Zea mays*)	1.7	1.1	2.5	1.7	3.8	2.5	5.9	3.9	10	6.7
Sweet potato (*Ipomoea batatas*)	1.5	1.0	2.4	1.6	3.8	2.5	6.0	4.0	11	7.1
Pepper (*Capsicum annuum*)	1.5	1.0	2.2	1.5	3.3	2.2	5.1	3.4	8.6	5.8
Lettuce (*Lactuca sativa*)	1.3	0.9	2.1	1.4	3.2	2.1	5.1	3.4	9.0	6.0
Radish (*Raphanus sativus*)	1.2	0.8	2.0	1.3	3.1	2.1	5.0	3.4	8.9	5.9
Onion (*Allium cepa*)	1.2	0.8	1.8	1.2	2.8	1.8	4.3	2.9	7.4	5.0
Carrot (*Daucus carota*)	1.0	0.7	1.7	1.1	2.8	1.9	4.6	3.0	8.1	5.4
Bean (*Phaseolus vulgaris*)	1.0	0.7	1.5	1.0	2.3	1.5	3.6	2.4	6.3	4.2
Turnip (*Brassica rapa*)	0.9	0.6	2.0	1.3	3.7	2.5	6.5	4.3	12	8.0
Wheatgrass, tall (*Agropyron elongatum*)	7.5	5.0	9.9	6.6	13	9.0	19	13	31	21
Wheatgrass, fairway crested (*Agropyron cristatum*)	7.5	5.0	9.0	6.0	11	7.4	15	9.8	22	15

Field crops	100%		90%		75%		50%		0% Maximum	
	EC_e	EC_w	EC_e	EC_w	EC_e	EC_w	EC_e	EC_w	EC_e	EC_w
Bermuda grass (Cynodon dactylon)	6.9	4.6	8.5	5.6	11	7.2	15	9.8	23	15
Barley (forage) (Hordeum vulgare)	6.0	4.0	7.4	4.9	9.5	6.4	13	8.7	20	13
Ryegrass, perennial (Lolium perenne)	5.6	3.7	6.9	4.6	8.9	5.9	12	8.1	19	13
Trefoil, narrowleaf birdsfoot[8] (Lotus corniculatus tenuifolium)	5.0	3.3	6.0	4.0	7.5	5.0	10	6.7	15	10
Harding grass (Phalaris tuberosa)	4.6	3.1	5.9	3.9	7.9	5.3	11	7.4	18	12
Fescue, tall (Festuca elatior)	3.9	2.6	5.5	3.6	7.8	5.2	12	7.8	20	13
Wheatgrass, standard crested (Agropyron sibiricum)	3.5	2.3	6.0	4.0	9.8	6.5	16	11	28	19
Vetch, common (Vicia angustifolia)	3.0	2.0	3.9	2.6	5.3	3.5	7.6	5.0	12	8.1
Sudan grass (Sorghum sudanense)	2.8	1.9	5.1	3.4	8.6	5.7	14	9.6	26	17
Wildrye, beardless (Elymus triticoides)	2.7	1.8	4.4	2.9	6.9	4.6	11	7.4	19	13
Cowpea (forage) (Vigna unguiculata)	2.5	1.7	3.4	2.3	4.8	3.2	7.1	4.8	12	7.8
Trefoil, big (Lotus uliginosus)	2.3	1.5	2.8	1.9	3.6	2.4	4.9	3.3	7.6	5.0
Sesbania (Sesbania exaltata)	2.3	1.5	3.7	2.5	5.9	3.9	9.4	6.3	17	11
Sphaerophysa (Sphaerophysa salsula)	2.2	1.5	3.6	2.4	5.8	3.8	9.3	6.2	16	11
Alfalfa (Medicago sativa)	2.0	1.3	3.4	2.2	5.4	3.6	8.8	5.9	16	10
Lovegrass (Eragrostis sp.)	2.0	1.3	3.2	2.1	5.0	3.3	8.0	5.3	14	9.3
Corn (forage) (maize) (Zea mays)	1.8	1.2	3.2	2.1	5.2	3.5	8.6	5.7	15	10
Clover, berseem (Trifolium alexandrinum)	1.5	1.0	3.2	2.2	5.9	3.9	10	6.8	19	13
Orchard grass (Dactylis glomerata)	1.5	1.0	3.1	2.1	5.5	3.7	9.6	6.4	18	12

Field crops	100%		90%		75%		50%		0% Maximum	
	EC_e	EC_w	EC_e	EC_w	EC_e	EC_w	EC_e	EC_w	EC_e	EC_w
Foxtail, meadow (*Alopecurus pratensis*)	1.5	1.0	2.5	1.7	4.1	2.7	6.7	4.5	12	7.9
Clover, red (*Trifolium pratense*)	1.5	1.0	2.3	1.6	3.6	2.4	5.7	3.8	9.8	6.6
Clover, alsike (*Trifolium hybridum*)	1.5	1.0	2.3	1.6	3.6	2.4	5.7	3.8	9.8	6.6
Clover, ladino (*Trifolium repens*)	1.5	1.0	2.3	1.6	3.6	2.4	5.7	3.8	9.8	6.6
Clover, strawberry (*Trifolium fragiferum*)	1.5	1.0	2.3	1.6	3.6	2.4	5.7	3.8	9.8	6.6
Fruit crops										
Date palm (*Phoenix dactylifera*)	4.0	2.7	6.8	4.5	11	7.3	18	12	32	21
Grapefruit (*Citrus paradisi*)	1.8	1.2	2.4	1.6	3.4	2.2	4.9	3.3	8.0	5.4
Orange (*Citrus sinensis*)	1.7	1.1	2.3	1.6	3.3	2.2	4.8	3.2	8.0	5.3
Peach (*Prunus persica*)	1.7	1.1	2.2	1.5	2.9	1.9	4.1	2.7	6.5	4.3
Apricot (*Prunus armeniaca*)	1.6	1.1	2.0	1.3	2.6	1.8	3.7	2.5	5.8	3.8
Grape (*Vitus* sp.)	1.5	1.0	2.5	1.7	4.1	2.7	6.7	4.5	12	7.9
Almond (*Prunus dulcis*)	1.5	1.0	2.0	1.4	2.8	1.9	4.1	2.8	6.8	4.5
Plum, prune (*Prunus domestica*)	1.5	1.0	2.1	1.4	2.9	1.9	4.3	2.9	7.1	4.7
Blackberry (*Rubus* sp.)	1.5	1.0	2.0	1.3	2.6	1.8	3.8	2.5	6.0	4.0
Boysenberry (*Rubus ursinus*)	1.5	1.0	2.0	1.3	2.6	1.8	3.8	2.5	6.0	4.0
Strawberry (*Fragaria* sp.)	1.0	0.7	1.3	0.9	1.8	1.2	2.5	1.7	4	2.7

INDEX

A

Absorption, 57
Acacia senegal, 216
Accelerated erosion, 331
Acid lime, irrigation technology options, 24
Actinobacteria, 122, 123
Active compost, 110
Active solar dryers, 100
Active solar still, 88–89. *See also* Solar still
Agave, 338
Agricultural lands (buffer zone)
  broad bed and furrows, 347
  contour ridges and furrows, 347
  farm ponds, 348
  field bunding, 348
  forming basins, 347
  strip ploughing, 348
  tie ridging, 347
  water spreading, 347
Agro-biodiversity, 210
  collection and conservation
      in arid and semiarid regions, 218–221
  significance, 217
Agroecosystems in Thar desert of India "hotspots," 212
Agro-ecosystems, threats, 215–216
Agronomic parameters, 170
Air mass, 59
Aloe vera, 222–223
Anaerobic digestion (AD), 108–109
Angstrom pyrheliometer, 61–63
Animate energy sources, 38
Approach angle, 194
Archaea, 123
Arid and semiarid regions
  agro-biodiversity, collection and conservation, 218–221

diversity in wild relatives, 221
  wild species and weedy forms, 221
Artificial mechanical drying, 94
Atmospheric absorption, 57
Atmospheric pressure, 373
Autoclave, 124
Automatic or recording rain gauge, 321
Available water
  in Karnataka state of India, 258
  and utilizable water per capita per year, 252
Ayyalur watershed in Tamil Nadu, 329

B

Bacillus stearothemophilus, 125
Banana, irrigation technology options, 23
Beam radiation, 59–60
Biodiversity, 217. *See also* Agro-biodiversity
Biomass energy, 45–46
Boulder check dam, 240, 339–340
Box-type solar cooker, 79–80. *See also* Solar cooker
Brahmaputra–Ganga link, 260
Broad bed and furrows with crops, 242
Brushwood, 339
Bulk density, 145
Bureau of Indian Standards (BIS), 138

C

Cabinet dryer, 99
Cabinet-type solar dryer, 101–102
  sizing
      heat capacity of air, 102–103
Cage wheel, 177
Canal operation schedule in Hirakud command, 9
Capparis decidua, 216
C. callosus, 222–223

Central Ground Water Board (CGWB), 251
Central Water Commission (CWC), 249
C. hardwickii, 222–223
Check dams
  temporary and permanent, 342–343
Chemical energy, 38
Chemical oxygen demand (COD), 108
Citrullus vulgaris var. fistulosus, 222–223
Clostridium, 123
Cloud seeding, 296
C. melo var. agrestris, 222–223
C. melo var. momordica, 222–223
Collecting weedy form of Cyamopsis
  tetragonoloba, 217
Commercial energy, 37
Commiphora whitti tree, 225
Community dynamics of methanotrophic
  bacteria, 122
Community solar cooker for indoor
  cooking, 81. See also Solar cooker
Compost production, 108
Concentrating or focusing type solar
  collector, 68, 72. See also Solar
  collector
  advantages and disadvantage, 69
  electrical power generator, 71
  in evacuated tube, 73
Conservation practice factor, 333
Continuous contour trenches, 336–337.
  See also Contour trenches
Contour/graded bunds, 338
Contour stone wall, 240
Contour trenches, 335–337
Controlling weeds, 166
Conversion SI and non-SI units
  concentration, 358
  cubics, 355–356
  energy, 358
  linear, 355
  nutrient equivalents, 359
  nutrients for plants, 358
  pressure, 357
  specific surface, 357
  squares, 355
  temperature, 357

water requirements, 358
  weight, 356
  yield and rate, 356–357
Corchorus spp., 222–223
Correlation matrix, 193–194
Corrugated furrows for sugarcane cultiva-
  tion, 25
Cotton, irrigation technology options, 23
C. prophetarum, 222–223
Creation of Technology Mission, 297
Crop growth stages and prevailing
  climatic conditions, 380–381
Cropping intensity, 297
Crop production, current status, 306
Crop tolerance and yield potential of
  crops affected by irrigation, 381–384
Cross section of contour bund, 239
C. sativus, 222–223
Cucumis callosus, 216
Cucurbits, 222–223
Cumulative biogas productions for phase
  I experiment, 114
Cyamopsis tetrogonoloba, 216

D

Deforestation on hydrological parameters,
  328
Dew point temperature, 374
Dharmapuri and Krishnagiri project, 306
Diffuse radiation, 58–59. See also Solar
  energy
Diospyros melanoxylon, 222–223
Diospyros montana, 222–223
Direct radiation, 58–59. See also Solar
  energy
Direct solar dryers, 97
Direct solar drying, 98
  working principle, 99
Dish solar cooker, 81. See also Solar
  cooker
Drainage congestion, 4
Drip irrigation, 308
Drying, 93–94

E

Earth's atmosphere, 57
Electrical energy, 39–40

Energy cycle, 66–67
Energy resources
  classification
    based on comparative economic
      value, 37
    based on muscular energy, 38
    based on renewability, 37–38
    based on usability, 36–37
  forms
    kinetic, 39–40
    potential, 38–39
  types
    hydel, 41–44
    thermal, 41
Engine selection and specification, 171
Enhanced anaerobic digestion by inoculation with compost, 107
  discussions
    evaluation of phase II study, 123–125
    evaluation of phase I study, 120–123
  materials and methods
    anaerobic batch test, 110–111
    analytical methods, 110
    compost, 109–110
    phase I experiment, 111
    phase II experiment, 111–113
    waste-activated sludge, 109
  results
    biogas production, 113–114
    conclusion, 119–120
    confirmation of biological effects,
      117–119
    cumulative biogas production, 116
    positive effects of stable compost,
      114–115
    stimulation by stable compost,
      115–116
Eppley normal incidence pyrheliometer,
  61
Erosion process, 331
Erosivity and erodibility, 331–332. *See
  also* Soil
  conservation practice factor, 333
  slope steepness factor, 333–334
  soil erodibility factor, 333

F

Farm pests and enemies, traditional
  management, 225
Field evaluation of prototype, 180–183
Field testing, 182
Finger millet, irrigation technology
  options, 20–21
Fission process in nuclear model, 43
Flowering stage of *Capparis decidua,* 216
Fluctuation of groundwater table, 14
Forced circulation solar water heater,
  74–75
Forest
  hydrology, 327–328
  watershed showing four zones, 238
Fossil fuels, 42
Fresh water availability in TN, 277
Front view of loose stone check dam, 339
Fruit of wild *Momordica balsamina,* 224
Fruit variability in different species
  collected from Indian Thar Desert, 215
Fuel consumption of tractor, 141–142

G

Gabions, 341–342
Ganges, Brahmaputra, and Meghna rivers
  maps, 262
Geobacillus sp., 125
Geologic erosion, 331
Geothermal energy
  advantages and disadvantages, 50
  defined, 49
  power plant, 49
Ghat Tracer, 336
Gingelly, irrigation technology options
  f, 22
Government of India (GOI), 249
Gram-negative organisms, 125
Gravitational energy, 39
Greece engineer, performance of tillage
  implements, 143
Groundnut, irrigation technology options,
  21
Gulf Intracoastal Waterway, 267
Gully control structures, 241

H

Hand khurpi, 197
Hibiscus cannabinus, 222–223
High-yielding varieties (HYVs), 216
Himalayan Rivers Development and
  Peninsular Rivers Development, 249
Himalayan Rivers Development Project,
  260–262. *See also* Indian rivers,
  interlinking
Hirakud command of Odisha, 4
  waterlogging in, 6
    basic details of canal, 7
    canal operation schedule in, 9
    causes, 9–10
    climate, 8
    cropping intensity, 6
    kharif and *rabi* crop, 6
    opportunities to reduce, 14–15
    physical and chemical properties of
      surface soil, 8
    quality of irrigation water, 8
    spatial variation of groundwater
      table, 11–14
    surface and subsurface, 15–17
    topography, 6–7
    views, 6
Hi-tech PF, 307
Hydel energy, 41–44
Hydroelectric power station, 42
Hydrologic cycle, 320
Hydrology
  cycle, 320
  defined, 320
  rainfall, 320
    frequency, 322
    relation between intensity and dura-
      tion, 321–322
  runoff
    climatic factors, 322
    estimation, 324–325
    peak runoff, 325–326
    physiographic factors, 323
    relationship between rainfall, 323
    time of concentration, 326–327

I

Illinois Waterway system, 266

Inanimate energy sources, 38
Indian desert, families and species diversity in, 213
Indian rivers, interlinking
  chronological developments, 271
  Himalayan Rivers Development
    Project, 260–262
  international examples
    Gulf Intracoastal Waterway, 267
    Illinois Waterway system, 266
    Rhine–Main–Danube Canal, 266
    Tennessee–Tombigbee Waterway,
      267
  investment needs and potential
    economic impact, 273
  obstacles, 265–266
  peninsular river development (PRD),
    254–260
  proposal by Captain Dastur, 272
  recommendations, 267–268
  salient features, 262–265
Indirect solar dryers, 97
Indirect solar drying, 99
  principle, 100
Inoculants, 110
In-situ moisture conservation measures,
  241
Instruments to measure solar radiation
  and sunshine hours
  types
    Angstrom pyrheliometer, 61–63
    Eppley normal incidence pyrheliom-
      eter, 61
    pyranometer, 63–64
    sunshine recorder, 64–65
Integrated watershed development and
  management, 232
Inter-basin transfer (IBT), 248
Inter-basin water transfer links proposed
  by NWDA, 257
Intercultivator of shank design, 175–177
Irradiance, 54
Irradiation, 54
Irrigation and cropping system, 14
Irrigation area and water needed for
  different crops in TN, 298

Irrigation technology options for selected
    agricultural crops
  farmer's practice
    acid lime, 24
    banana, 23
    cotton, 23
    finger millet, 20–21
    gingelly, 22
    groundnut, 21
    maize, 21
    pearl millet, 20
    pulses, 21
    sorghum, 20
    sugarcane, 24–25
    sunflower, 22
    tomato, 24

K

KabilNalla watershed in Karnataka, 329
Karnataka state
  annual yield of west flowing rivers in,
    252
  water resources, 259
Kinara from *Clerodendron phloidis* for
    traditional seed storage, 223
Krishi Vidya Kendra (KVK), 314
Krishna and Cauvery basins, 259

L

Land
  lay in Hirakud ayacut, 10
  locked topo-system, 10
  size holdings in India and Karnataka,
    168
  use and cover, 211
Land development program, 329
Latent heat of vaporization, 90, 363,
    371-373
Lemon grass, 338
Local cultivars, 222–223
Local landraces collected in different crop
    groups, 219–220
Loose stone check dams, 240, 339
Loss of agro-biodiversity
  due to over grazing, 224

    at Jaisalmer district of western Rajas-
        than, 225
Luffa hermaphrodita, 222–223
Lug design, 173–174

M

Maheswaram Watershed Development
    Project in Andhra Pradesh, 338
Mahindra tractor models, 133
Maize, irrigation technology options, 21
Malabar lemon Grass, 338
Mean annual rainfall, 321
Mean daily maximum duration of bright
    sunshine hours, 366
Mean daily percentage, 367
Mechanical energy, 39–40
Methano bacterium thermos auto
    trophicum, 122
Methano thermobacter sp., 122
Methods of solar energy utilization, 56
Micro-irrigation technology, 306–307
Micro-level *(in situ)* soil and moisture
    conservation measures
  forest areas
    catch pits, 346
    micro-catchments, 345
    semicircular, crescent-shaped bunds
        and saucer basins, 346
    sunken planting, 346
    tie-ridging, 346
    V-shaped ditches, 346–347
Modified Universal Soil Loss Equation
    (MUSLE), 334
Moisture content, 144
Moldboard (MB), 139
Mulching
  benefits, 26–27
  labor input for, 28
  materials for, 27–28
  methods, 28
  potential drawbacks, 27
  vertical, 28–30

N

National Center for Agricultural Mecha-
    nization (NCAM), 143

National Center for Water Resources, 268
National Institute or Center for Water Resources, Research, Development & Management, 268
National Water Development Agency (NWDA), 248, 249
Natural Drainage System, 233
NBPGR/IPGRI Collaboration Program, 218
Noncommercial energy, 37
Non-concentrating/flat plate type solar collector, 68. *See also* Solar collector advantages, disadvantages, and applications, 68–69
components of, 70–71
Nonrenewable energy, 38
Nonsystem tank irrigation system, 25–26
Nuclear energy, 39, 43
Number of day in year (Julian day), 370–371

O

Ocean energy
advantages and disadvantages, 51
defined, 50
OTEC, 50
power plant, 51
Ocean thermal eddy currents (OTEC), 50
Open sun drying (OSD), 94. *See also* Solar dryer
principle of, 97–98
Open-surface drain, 15
Orans, 214
Ordinary or non-recording rain gauge, 321
OTEC. *See* Ocean thermal eddy currents (OTEC)

P

Parallel field surface drains, 16
Parambikulam Aliyar Project, 296
Passive solar still, 87–88. *See also* Solar still
P. cineraria, 216
Pearl millet, irrigation technology options, 20

Penetrometer resistance (PR), 142
Peninsular River Development (PRD), 254–260. *See also* Indian rivers, interlinking
component, 249
Percentage of daily sunshine hours, 364–365
Percentage-wise contribution of financing organizations, 133
Percentage-wise tractor model available in Jalgaon Jamod, 134
Percolation ponds
procedure to set up pond, 343–344
silt detention tanks, 344
site selection, 343
Permanent check dams, 342–343
Permanent gully control structures, 341–342
"Pinky-Pinky" or hand level, 336
friction loss, 360–362
Planting methods in banana cultivation, 23
Plowing using animal power and tractor power, 140
Power transmission system, 172
Precision farming (PF), 306, 308
advantages, 309
application, 309
farms, 312
International Journals and conferences/ seminars, 309
projects area, 310–311
projects in TN, 310
results and discussions
observations and feedback, 313–314
research in India, 314–315
yield of selected crops under, 314
selection of
farmers, 311
irrigation method, 311–312
vegetable crop, 311
Primary and secondary sources of energy, 36–37
Productivity
of different crops in World, India, and Tamil Nadu, 307

of major crops grown in India, 293–294
of major crops grown in world,
    293–294
Project-implementing agency (PIA), 310
Proposed inter-basin water transfer links,
    261
Proposed Peninsular links, 256
Prosopis cineraria L., 212–213
Proteobacteria, 122
Prototype self-propelled intercultivator,
    179
Psychometric constant for different alti-
    tudes, 368
Psychrometric chart at sea level, 376
Psychrometric constant, 374
Pulses, irrigation technology options, 21
Pyranometer, 63–64

R

Radiant energy, 54
Radiation energy, 39–40
Rainfall in Tamil Nadu, 281
Reflected radiation, 58–59. *See also* Solar
    energy
Reflection, 57
Renewable energy
    advantages of, 40
    comparison with nonrenewable energy,
        41
    disadvantages of, 40–41
    sources, 37
Research at Akola, 144
Research Development & Management,
    268
Reversible mold board plow performance,
    137–138
    cost operation, 159–162
    data sheet and evaluation data, 163
    data sheet for evaluation, 159
    field evaluation, 152–155
        depth of cut, 149
        effective field capacity, 148
        field efficiency, 148
        width of cut, 148
    fuel consumption, 149–150

general considerations
    field operational pattern, 146
    field parameters, 147
    selection of plot, 146
    selection of test samples, 146
    speed of operation, 146
    test duration, 147
    wheel slip, 147
literature review, 140–144
methods and material
    mechanical analysis of soil, 144–145
results and dissussion, 150
    bulk density, 151
    moisture content, 151
    soil texture, 151
Reversible plow, 139
*Revised Universal Soil Loss Equation
    (RUSLE),* 334
Rhine–Main–Danube Canal, 266
Rough stone with earth backing check
    dams, 340

S

Salt accumulation, planting positions,
    ridge shapes and watering patterns, 377
Sand dunes at Western Rajasthan, 213
Saturation and seepage zone in contour
    bund area, 239
Saturation vapor pressure
    for different temperatures, 363
Scattering, 57
Sedimentation
    control, 334
    soil and nutrient losses, 334
Seed treatment with ash for pest free
    storage of leguminous seeds, 224
Self-propelled intercultivator
    bulk density, 183–184
    cost of operation
        assumptions, 203
        for cotton crop, 204
        fixed cost, 203
        for red gram crop, 204–205
    development, 177
    economics of weeding operation, 186,
        197

effective field capacity, 185
farm operations, 167
field
    efficiency, 185
    evaluation, 195–196
fuel consumption, 185
literature review, 168–169
materials and methods
    agronomic and soil parameters,
        169–171
    design of self-propelled weeder,
        171–179
optimization of operational parameters,
    179–180
performance, 165
    of index, 186
plant damage, 185–186
power requirement, 184
results and discussion
    optimization of operational param-
        eters, 186–195
soil moisture content, 183
specifications, 178
theoretical field capacity, 184–185
traveling speed, 184
weeding efficiency, 184
Self-propelled weeder
    optimum design and operational param-
        eters, 195
Short-wave radiation on clear-sky day,
    374–376
SIDA assisted Social Forestry Project,
    330
Slope of vapor pressure curve for
    different temperatures, 369–370
Slope steepness factor, 333–334
Slope vapor pressure curve, 374
Soil
    compaction, 143
    erodibility factor, 333
    erosion, 330
        causes, 331
        process, 331
        types, 331
    erosivity and erodibility, 331–332
        conservation practice factor, 333

    slope steepness factor, 333–334
    soil erodibility factor, 333
    resistance, 170
    sedimentation
        control, 334
        soil and nutrient losses, 334
    texture, 145
    water salinity at different leaching frac-
        tions, 376
Soil and water conservation measures,
    335
    check dams and percolation ponds,
        342–344
    contour trenching, 335–338
    for different zones of watersheds
        buffer zone, 350
        common measures, 350
        lower zone (interface zone), 350
        middle zone (asset creation zone),
            349
        upper zone (eco-restoration zone),
            349
    instruments and equipments needed,
        350
        alignment of contour lines, 351
        ground water measurements, 351
        meteorological equipment, 351
        soil moisture measurements, 351–352
        stream/channel flow measurements,
            351
    micro-level (in situ) soil and moisture
        conservation measures, 345–348
    permanent gully control structures,
        341–342
    temporary gully control measures,
        339–340
Solanum spp., 222–223
Solar collector. See also Solar energy
    thermal energy, 66
    types
        components of, 69–71
        concentrating or focusing type, 68
        differences between, 70
        non-concentrating or flat plate type,
            68
        in various ranges and applications, 69

Solar constant, 54
Solar cooker, 78. *See also* Solar energy
  advantages, 79
  disadvantages, 79
  energy produced, 86
  energy required for cooking, 83–84
  estimating time required to cook food
    assumptions, 84–85
  materials used, 82
  parts, 81–82
  principle of operation, 83
  simple payback period
    energy content, 85
  types
    box-type solar cooker, 79–80
    community solar cooker for indoor
    cooking, 81
    dish solar cooker, 81
Solar dryer, 93
  parameters considered for, 95
  principle of operation, 97–100
  requirements of
    postharvest losses, 95
    problems in open sun drying, 94
  steps in drying, 96
Solar energy, 34, 43
  advantages, 43
  applications, 56–57, 65
    simple payback period, 85–86
    solar collector, 66–73
    solar cooker, 78–85
    solar dryer, 93–103
    solar still, 86–93
    solar water heater, 73–78
  biomass, 45–46
  disadvantages, 43
  energy resources (*See* Energy
  resources)
  geothermal, 49–50
  instruments to measure solar radiation
  and sunshine hours
    advantages, 61
    disadvantages, 61
    types, 61–65
    working principle, 60
  ocean, 50–51

  principles, 51
    cost, 52
    methods of solar energy utilization,
    56
    potential, 52
    radiant energy, 54
    radiation of heat energy, 53
    solar system, 53
    spectral distribution of solar radia-
    tion, 55–56
    use, 52–53
    Wien's displacement law, 54
  propagation of solar radiation in
  atmosphere
    direct radiation, 58–59
    mechanisms, 57
    terrestrial and extraterrestrial regions,
    58
  technical potential, 35
  tidal, 48–49
  wind, 46–47
Solar radiation, 54
Solar still, 86
  classification, 86–89
  components of, 90
  cost, 92
  energy requirement and efficiency,
    90–91
  estimation of payback period for, 93
  general maintenance, 93
  parameters affecting, 92
  principle of operation, 89
  sizing of, 91
Solar water heater. *See also* Solar energy
  advantages, 73
  classification of, 74
  components of, 73–74
  design and costing
    cost, 77–78
    heat capacity of water, 76–77
    simple payback period, 78
  objectives, 73
  principle of operation, 74–75
  types, 75
Soluble COD, 110

Sorghum, irrigation technology options, 20
Specific draft, analysis of variance, 189
Spectral distribution of solar radiation, 55–56
Speed of operation, 194
Stefan–Boltzmann law at different temperatures, 371–372
Stored mechanical energy, 39
Sugarcane, irrigation technology options, 24–25
Sun at zenith, 59–60
Sunflower, irrigation technology options, 22
Sweep blade design, 175
Sweep, operational parameters, 194–195
System of rice intensification (SRI) method, 296

T

Tamil Nadu Agricultural University (TNAU), 310
Tamil Nadu (TN)
    food production, 295
        anticipated outcome, 298
        cropping intensity, 297
        crop productivity, 297
        demand management, 296–297
        research efforts, 297
        water augmentation, 296
    fresh water availability, 277
    geographical area of, 285–289
    irrigation, 289–291
        water utilization, 292–293
    per capita availability of water, 254
    productivity of major crops grown, 293–294
    rainfall, 277–281
    water, 289
    water policy, 294–295
    water resources, 276, 282–285
    water used and area irrigated in
        area irrigated at present, 302
        illustrations, 303
        total area under irrigation, 302
        water available, 302

Tamil Nadu Water Vision, 269, 299
Tank irrigation, 296
Taramira *(Eruca sativa),* 222–223
T. dicoccum, 222–223
Temporary check dams, 342
Temporary gully control measures, 339–340. *See also* Soil and water conservation measures
Tennessee–Tombigbee Waterway, 267
Thar desert of India, 212
Thermal energy, 39–40, 41–42
Thermoactinomyces, 123
Thermodynamic properties of air and water
    atmospheric density, 373
    atmospheric pressure, 373
    dew point temperature, 374
    latent heat of vaporization, 372–373
    psychrometric constant, 374
    saturation vapor pressure function, 373–374
    short-wave radiation on a clear-sky day, 374–376
    slope vapor pressure curve, 374
Thermosyphon systems, 74
Tidal energy
    advantages, 48
    disadvantages, 49
Tie ridging with crops, 242
Tillage
    implements in India, 138
    systems, 139
    testing and evaluation, 139
Tomato, irrigation technology options, 24
Total chemical oxygen demand (TCOD), 110
Total global radiation, 57–58
Total monsoon rainfall, 324–325
Total suspended solids (TSS), 110
T. polonicum, 222–223
Tractor operated reversible moldboard plow, 140
Tractor utilization in vidarbha region, India, 129
    conclusions, 134–135
    growth of agriculture, 130

information in the questionnaire, 131
methodology, 131
results and discussion
credit facility, 132–134
opinion of farmers, 134
Trapezoidal tool frame, 178
Triticum durum, 222–223
Triticum spp., 222–223
T. sphaeroccum, 222–223

U

U frame with water tubes, 336
Union Ministry of Water Resources, 255
Universal soil loss equation (USLE),
332–333
US Postal stamp advocating water conser-
vation, 354

V

Vajkand (*Dioscorea* spp.), 225
"V" ditch technology, 242
Vegetative barriers, 338
on contour, 238
Vegetative propagules, 218
Vertical mulching, 28–30
Vetiver, 338
Vigna radiata var. *sublobata,* 222–223
Volatile suspended solids (VSS), 110
V. trilobata, 222–223

W

Waste-activated sludge (WAS), 108
Wastewater treatment plant (WWTP),
108–109
Water
availability in India and in Tamil Nadu,
277
crisis, 253–254
policy in TN
action plan, 294
features, 294–295
objectives, 294
research efforts, 295
thrust area and strategies, 295
resources of India, 248–249, 274
assessment, 251

CWC estimates, 251
ground water resources, 250–251
mean flow and utilizable surface,
250–251
spreading, 243
Waterlogging in Hirakud command, 6
average groundwater table, 9
basic details of canal, 7
canal operation schedule in, 9
causes, 9–10
climate, 8
cropping intensity, 6
kharif and *rabi* crop, 6
opportunities to reduce, 14–15
physical and chemical properties of
surface soil, 8
quality of irrigation water, 8
spatial variation of groundwater table,
11–14
surface and subsurface, 15–17
topography, 6–7
views, 6
Watershed development and management, 231–232
components, 235–237
concept, 234
constraints
institutional cum political, 245
sociocultural and economic, 244
defined, 329–330
experiences and challenges, 243–244
interface forestry, 330
objectives, 234–235
participation of people, 244
selection of watershed, 330
watershed, defined, 328–329
Weeding
efficiency, 193–194
operation, 166
Wien's displacement law, 54
Wind energy, 46
advantages, 47
disadvantages, 47
geothermal, 49–50
tidal, 47–49

wind mills types, 47
World Food Summit in Rome, 307–308
World grain production, 307

Y

Yield losses in various crops due to
   weeds, 168

Z

Zones of watersheds, soil and water
   conservation measures
   buffer zone, 350
   common measures, 350
   lower zone (interface zone), 350
   middle zone (asset creation zone), 349
   upper zone (eco-restoration zone), 349

Milton Keynes UK
Ingram Content Group UK Ltd.
UKHW031139141024
449569UK00024B/1202